Management and Ecology
of River Fisheries

Management and Ecology of River Fisheries

EDITED BY

I. G. COWX

Hull International Fisheries Institute
University of Hull, UK

Fishing News Books
An imprint of Blackwell Science

b

Blackwell
Science

Copyright © 2000 by
Fishing News Books
A division of Blackwell Science Ltd
Editorial Offices:
Osney Mead, Oxford OX2 0EL
25 John Street, London WC1N 2BL
23 Ainslie Place, Edinburgh EH3 6AJ
350 Main Street, Malden, MA 02148 5018,
   USA
54 University Street, Carlton, Victoria 3053,
   Australia
10, rue Casimir Delavigne, 75006 Paris, France

Other Editorial Offices:

Blackwell Wissenschafts-Verlag GmbH
Kunfürstendamm 57
10707 Berlin, Germany

Blackwell Science KK
MG Kodenmacho Building
7–10 Kodenmacho Nihombashi
Chuo-ku, Tokyo 104, Japan

The right of the Author to be identified as the
Author of this Work has been asserted in
accordance with the Copyright, Designs and
Patents Act 1988.

First published 2000

Produced by and typeset in Times
by Gray Publishing, Tunbridge Wells, Kent
Printed and bound in Great Britain by
MPG Books Ltd., Bodmin, Cornwall

The Blackwell Science logo is a trade mark of
Blackwell Science Ltd, registered at the United
Kingdom Trade Marks Registry

DISTRIBUTORS

Marston Book Services Ltd
PO Box 269
Abingdon
Oxon OX14 4YN
(*Orders:* Tel: 01865 206206
          Fax: 01865 721205
          Telex: 83355 MEDBOK G)

USA
   Blackwell Science, Inc.
   Commerce Place
   350 Main Street
   Malden, MA 02148 5018
   (*Orders:* Tel: 800 759 6102
                    781 388 8250
             Fax: 781 388 8255)

Canada
   Login Brothers Book Company
   324 Saulteaux Crescent
   Winnipeg, Manitoba R3J 3T2
   (*Orders:* Tel: 204 837–2987
             Fax: 204 837–3116)

Australia
   Blackwell Science Pty Ltd
   54 University Street
   Carlton, Victoria 3053
   (*Orders:* Tel: 03 9347 0300
             Fax: 03 93475 001)

A catalogue record for this title is available
from the British Library

ISBN 0-85238-250-2

Library of Congress Cataloging-in-Publication
Data
Management and ecology of river fisheries/
   edited by I.G. Cowx.
      p. cm
   Includes bibliographical references.
   ISBN 0-85238-250-2
   1. Fisheries Congresses. 2. Stream ecology
Congresses. 3. Fishery management
Congresses. I. Cowx, I.G. (Ian G.)
SH3.M36   1999
333.95′6152—dc21   99–38346
                                    CIP

For further information on Fishing News
Books, visit our website:
http://www.blacksci.co.uk/fnb/

Contents

Preface   ix

I   Assessment of fish community structure and dynamics

1   Selecting gear for monitoring fish assemblages   3
L.E. Miranda and H.L. Schramm, Jr

2   Longitudinal hydroacoustic survey of fish in the Elbe River
supplemented by direct capture   14
J. Kubecka, J. Frouzová, A. Vilcinskas, C. Wolter and O. Slavík

3   Summary of the use of hydroacoustics for quantifying the escapement of
adult salmonids (*Oncorhynchus* and *Salmo* species) in rivers   26
B.H. Ransom, S.V. Johnston and T.W. Steig

4   Relative selectivity of hoopnetting and electric fishing in the Lower
Mississippi River   40
H.L. Schramm, Jr and L.L. Pugh

II   Large river fisheries

5   The fish community of the River Thames: status, pressures and
management   55
S.N. Hughes and D.J. Willis

6   Cyclic behaviour of potamodromous fish in large rivers   71
R. Quiros and J.C. Vidal

7   Seasonal movements of coarse fish in lowland rivers and their relevance
to fisheries management   87
M.C. Lucas, T. Mercer, G. Peirson and P.A. Frear

8   Seasonal and diel changes of young-of-the-year fish in the channelised
stretch of the Vltava River (Bohemia, Czech Republic)   101
O. Slavík and L. Bartos

III   Habitat requirements

9   Trout, summer flows and irrigation canals: a study of habitat condition
and trout populations within a complex system   115
D.J. Walks, H.W. Li and G.H. Reeves

10   Dynamics of a population of brown trout (*Salmo trutta*) and fluctuations in physical habitat conditions – experiments on a stream in the Pyrenees; first results   126
     V. Gouraud, C. Sabaton, P. Baran and P. Lim

11   Fish habitat associations, community structure, density and biomass in natural and channelised lowland streams in the catchment of the River Wensum, UK   143
     N.T. Punchard, M.R. Perrow and A.J.D. Jowitt

12   A methodology to evaluate physical habitat for reproduction of brown trout (*Salmo trutta* L.) and relation with fry recruitment   158
     M. Delacoste, P. Baran and J.-M. Lascaux

IV   Anthropogenic impacts

13   Wimbleball Pumped Storage Scheme: integration of water resource management, engineering design and operational control to compliment the needs of the salmonid fisheries in the River Exe   177
     H.T. Sambrook and I.G. Cowx

14   Impacts of hydraulic engineering on the dynamics and production potential of floodplain fish populations in Bangladesh: implications for management   201
     A.S. Halls, D.D. Hoggarth and K. Debnath

15   Management of fisheries in rivers used for potable abstractions in drought conditions: public health versus ecology   218
     S. Axford

16   Impacts of the Ok Tedi copper mine on fish populations in the Fly River system, Papua New Guinea   232
     S. Swales, B.S. Figa, K.A. Bakowa and C.D. Tenakanai

17   Riverine fish stock and regional agronomic responses to hydrological and climatic regimes in the upper Yazoo River basin   242
     D.C. Jackson and Q. Ye

V   Rehabilitation of river fisheries

18   An assessment of anthropogenic activities on and rehabilitation of river fisheries: current state and future direction   261
     M.C. Lucas and G. Marmulla

19   Defining and achieving fish habitat rehabilitation in large, low-gradient rivers   279
     P.B. Bayley, K. O'Hara and R. Steel

20   Watershed analysis and restoration in the Siuslaw River, Oregon, USA   290
     N.B. Armantrout

21 Planning implications of a habitat improvement project conducted on a
    Newfoundland stream   306
    M.C. van Zyll de Jong, I.G. Cowx and D.A. Scruton

22 Provision for the juvenile stages of coarse fish in river rehabilitation
    projects   318
    B.P. Hodgson and J.W. Eaton

VI Management

23 Principles and approaches for river fisheries management   331
    R.L. Welcomme

24 Fisheries science and the managerial imperative   346
    P. Hickley and M. Aprahamian

25 The use of spawning targets for salmon fishery management in England
    and Wales   361
    N.J. Milner, I.C. Davidson, R.J. Wyatt and M. Aprahamian

26 The effectiveness of rod and net fishery bye-laws in reducing exploitation
    of spring salmon on the Welsh Dee   373
    I.C. Davidson, R.J. Cove and N.J. Milner

27 From sector to system: towards a multidimensional management in the
    Lower Amazon floodplain   388
    F. Castro and D. McGrath

28 The role of Nkhotakota Wildlife Reserve, Malawi, Africa, in the
    conservation of the mpasa, *Opsaridium microlepis* (Günther)   400
    D. Tweddle

29 Conservation status, threats and future prospects for the survival of
    freshwater fishes of the Western Cape Province, South Africa   418
    N.D. Impson and K.C.D. Hamman

30 Conservation of endangered fish species in the face of water resource
    development schemes in the Guadiana River, Portugal: harmony of the
    incompatible   428
    I.G. Cowx and M.J. Collares-Pereira

Subject index   439
Species index   443

Preface

During a meeting of the European Inland Fisheries Advisory Commission (EIFAC) in Dublin, Ireland in June 1996, the need to understand fully the key issues regarding the management and ecology of river fisheries was recognised because of their importance to society in general. In view of this importance, it was felt that current knowledge of the status and management of river fisheries warranted further discussion and dissemination.

To this end, the International Fisheries Institute at the University of Hull, in cooperation with the European Inland Fisheries Advisory Commission, organised a symposium and workshop on *Management and Ecology of River Fisheries* which took place in Hull, UK in late March and early April 1998. The objectives of the symposium were:

- to examine aspects of the management and ecology of river fisheries, reviewing the status of river fisheries, assessment methodology, constraints on the development, issues and options regarding management, and the benefits and problems associated with these actions;
- to identify constraints and gaps in our knowledge that affect the application of fisheries management policy in temperate and tropical river fisheries;
- to recommend and promote action to improve the management of river fisheries.

The main conclusions of the symposium and workshop are outlined below and in the selected papers which make up these proceedings. It is hoped they will stimulate fisheries scientists, managers and academics to collaborate in further research to improve our understanding of river fisheries, and promote further research and collaboration to maintain and enhance these important resources worldwide.

Stock assessment for management purposes

It was recognised that river fisheries are formed in diverse types of habitat from small streams to large rivers and extensive floodplains. Consequently, there is a need for an array of methods to provide adequate assessment of stocks in those habitats. There is also a need to define the objectives of the assessment activity to ensure that the appropriate precision in information is gained. In many instances it was questioned whether stock assessment was necessary as trends in species composition, catches etc., can provide adequate information for management at minimal cost.

The major constraints on stock assessment were inadequate development in gear technology, poor data analysis methods, weak understanding of the limitation of stock assessment procedures, and financial and human resources being redirected to other activities.

Anthropogenic activities/rehabilitation and mitigation

It was recognised that we are able to:

- identify and prioritise problems impacting upon river fisheries and aquatic communities;
- identify and carry out appropriate technical solutions to these problems (acknowledging, within this context, the influences of continued innovation and the effects of the scale of some problems).

Conversely, social and economic aspects of rehabilitation are currently more problematic, and present the greatest challenges to the maintenance and development of aquatic ecosystems and fisheries.

It was recognised that in many cases, when potentially damaging schemes are being considered, or when alterations to existing schemes might allow rehabiliation work, those operating in the water resource planning sector do not actively solicit the input of fisheries at an early stage, if at all. Fisheries specialists need to interact with other disciplines in projects at the earliest possible stage and must be more vocal in their support of fisheries interests.

Most of the factors causing problems for fish communities are outside the control of the fisheries sector. Those involved in fisheries must therefore broaden and strengthen their cause, by interacting and making alliances with other interested parties, in seeking to limit damage to aquatic ecosystems, and in promoting rehabilitation and enhancement activities.

Management issues

Several key management issues were raised which link to other outputs of the Symposium.

- While stock assessment may not be required *per se*, there is a need to improve fisheries statistical monitoring procedures to provide baseline information on exploitation levels.
- There is a need to improve communication links between fisheries managers, scientists and those utilising the resource. This can be the first step in management of the resources, which appears to be the most desirable way to manage large-scale river fisheries.
- The profile of river fisheries needs raising in general, and particularly where there is a multiple array of resource users who are often in conflict, or potential conflict. To support this action there is a growing awareness of the importance

of the economic and social evaluation of fisheries to ensure that they are well represented in all development activities.

● If water and aquatic resources are to be exploited on a sustainable basis in the future, concerted effort is needed to resolve the conflicts between user groups. Where possible, this must be based on sound scientific evidence, close liaison between user groups, full cost–benefit analysis and transparency in the decision-making process. Where scientific information is not available this should not prevent decisions being made, but the precautionary approach should be adopted. If resolution of conflict is to be successful it must involve cross-education of all user groups, recognition of stockholder participation and needs, and must probably be implemented at the local community level.

● There is also a need for robust methods for prioritising demands for the water and aquatic resources of rivers that balance human requirements against protection of the environment and biodiversity.

The production of these proceedings has involved considerable effort by a number of people. In particular, thanks must go to the following for their contribution to workshops held prior to the main symposium and support in reviewing the papers for the proceedings: M. Aprahamian, N. Armantrout, S. Axford, P.B. Bayley, D. Coates, M.J. Collares-Perriera, C.A. Dollof, V. Gourand, A.S. Halls, J. Harvey, D. Hefti, P. Hickley, S. Hughes, N.D. Impson, D.C. Jackson, R. Knösche, J. Kubecka, M.C. Lucas, G. Marmulla, L.E. Miranda, J. Munyandorero, G. Peirson, A. Pires, R. Quiros, H. Sambrook, H.L. Schramm Jr, S. Swales, D. Tweddle, R.L. Welcomme and C. Wolter. I would like to thank Jon Harvey and Emma Burnett for their considerable assistance in the running of the symposium and Julia Cowx for the production of these proceedings. Finally, I would like to thank the many international funding agencies and organisations for their financial support, thus ensuring truly international coverage of the issues and the success of the symposium.

Ian G. Cowx
University of Hull International Fisheries Institute

Section I
Assessment of fish community structure and dynamics

Chapter 1
Selecting gear for monitoring fish assemblages

L.E. MIRANDA and H.L. SCHRAMM, JR

Mississippi Cooperative Fish and Wildlife Research Unit, Post Office Box 9691, Mississippi State, MS 39762, USA (e-mail: smiranda@cfr.msstate.edu)

Abstract

Various gears have been used to sample fish in rivers, but none seem to have been demonstrated to characterise accurately fish assemblages. Often the objective of sampling is to detect changes caused by disturbances such as pollution, flow regulation or fishing. In such cases, the most suitable gear may be the one that is most sensitive to spatial or temporal differences. Dissimilarity indices were used to evaluate the sensitivity of gears to differences in fish assemblages among four secondary channels spread over 875 river kilometres of the Mississippi River, and six mainstream reaches spread over 21 km of Luxapallila Creek, USA. The analysis indicated that the gears sampled disparate but overlapping segments of the fish assemblages, yet some identified more dissimilarities than others and, thus, were more sensitive to spatial differences in assemblages. The process used to identify which gear distinguished the greatest dissimilarities is described. Although the gears considered and conclusions reached about their effectiveness are specific to the study streams, the process used is general and applicable to any community and set of gear.

Keywords: Gear selectivity, fish sampling, riverine fish assemblages.

1.1 Introduction

Fish communities are difficult to evaluate in large rivers. Various gears are used to sample fish, but all provide biased and unequal representation of the true species assemblage and size composition (Casselman *et al.* 1990; MacLennan 1992). These biases are often associated with changing gear efficiencies affected by characteristics peculiar to the gear (e.g. mesh size, passive versus active capture, ratio of fish size to sample unit size), the fish (e.g. size, behaviour), and the habitat (e.g. substrate, current, depth, turbidity).

Because all fishing gears provide a biased snapshot of the fish community, it is difficult to decide what gear or gear combination should be used to evaluate stream fish assemblages. Often the objective of sampling riverine fish is to detect differences or changes influenced by anthropogenic disturbances, such as pollution, regulation of flow or fishing. Then, the investigator may be most interested in detecting dissimilarities among sites or time periods. In such instances, the most suitable gear may be that which is most sensitive to differences between sites or time periods. Two key assumptions in this rationalisation are that: (1) gear efficiency remains constant over sites or time periods, so that differences can be attributed solely to fish assemblage dissimilarities; and (2) differences identified by the gear are not merely a function of erratic catches.

The process used to select the gear that identified the greatest dissimilarities in fish assemblages among sites within a large and a small river is described. Although the conclusions about the gears that most effectively detect differences in fish assemblages are specific to the suite of gears used and to the study rivers, the process used is applicable to any set of gear and community.

1.2   Methods

Fish were sampled in secondary channels distributed over 875 km of the Mississippi River (USA), and in main channel reaches distributed over 21 km of the Luxapallila Creek (Mississippi, USA). These two rivers were examined because they differ greatly in scale. The secondary channels sampled in the Mississippi River are large expanses of water that extend over a wide geographical area, whereas reaches in the Luxapallila Creek were smaller and located within a confined geographical area. All fish were identified to species and enumerated. Data were analysed with dissimilarity indices to identify the gears that provided the greatest dissimilarities.

1.2.1   *Mississippi River*

In the Mississippi River, sampling was restricted to secondary channels. Secondary channels are former main channels, abandoned by the river's direct flow as the river meanders and changes courses within the flood plain. These channels are separated from the main channel by an island or sand bar, and are narrower and carry less flow than the main channel. Secondary channels ranged from straight reaches shorter than the adjacent main channel, to bends considerably longer than the adjacent main channel.

Four secondary channels were sampled at corresponding habitats. Secondary channel one was located at river kilometre 630–634, secondary channel two at river kilometre 655–662, secondary channel three at river kilometre 1467–1475, and secondary channel four at river kilometre 1497–1505 (Fig. 1.1). Thus, the first pair of secondary channels (located near Vicksburg, Mississippi) and the second pair (near Hickman, Kentucky) were located within 25 km of each other, but the pairs were separated by over 800 km. The range of channel width was 200–300 m. Within each secondary channel the sampling effort was stratified into shallow sand-bar (<3 m deep) and steep-bank (<6 m deep) habitats. Current velocity in these habitats was generally slow or zero near shore, but increased to moderate velocity (1–1.5 m s^{-1}) near the deep edge of the habitat.

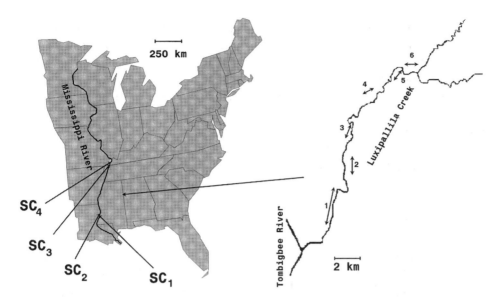

Figure 1.1 Location of the four secondary channels (SC) along the Mississippi River, and reaches (inset) in the Luxapallila Creek

Fish were sampled in October 1995 with two electric fishing configurations. Sampling was conducted with a boat-mounted Smith-Root GPP 7.5 electric fishing gear, powered with a 7.5 kW generator. The unit was configured to output 1000 V and 15 Hz (E_{15}) or 500 V and 60 Hz (E_{60}); 60 Hz was selected because it is a commonly used frequency output, and 15 Hz because in preliminary trials it was identified to yield rather different catches. These outputs were preset in the electric fishing gear but the actual output into the water was not measured because it was expected to vary depending on bottom contour, substrate etc. Five replicate 5-min samples were collected in each habitat type, with each electric fishing configuration, in each secondary channel, while systematically alternating E_{15} and E_{60} samples.

1.2.2 *Luxapallila Creek*

Sampling was conducted in six reaches of the main channel. Reach 1 was located at river kilometre 3–6, reach 2 at river kilometre 8–9, reach 3 at river kilometre 10–12, reach 4 at river kilometre 15–17, reach 5 at river kilometre 18–20, and reach 6 at river kilometre 23–24 (Fig. 1.1). Reach 1 was a deep, channellised section of stream; reach 2 was also channellised but generally shallower and contained several rock rip-rap ledges installed across the stream to forestall channel degradation; and reaches 3–6 represented a natural stream modified only by low-intensity disturbances associated with mild urbanisation. The range of channel width was 20–30 m. Sampling within each reach was conducted at three stations located in the upper, middle, and lower portion of the reach.

Fish were sampled with three gear types at corresponding habitats within each reach during July 1997. A 4.6 × 1.5 m seine with 3.2-mm square mesh was pulled 5–10 times at each station in gravel and sand bars. Electric fishing was conducted with a backpack Smith-Root 15-D electric fishing gear configured to output 600 V and 110 Hz. This output was preset in the gear, but the actual output into the water was not measured. Sampling was conducted in pools, riffles, gravel bars, and sand bars. Sampling lasted 10 min per station, or less if all available habitat was sampled. Four hoop nets were fished in each station for two consecutive 24-h periods. Two of the hoop nets measured 61 cm diameter with 2.5-cm square mesh, and two measured 122 cm diameter with 3.8-cm square mesh. Hoop nets were submerged and kept open by the current.

1.2.3   *Data analysis*

A factorial analysis of variance (ANOVA) with subsamples (i.e. replicates taken within a secondary channel or reach) was used to identify whether log-transformed catch rates differed significantly among gears, species and locations in each river. A statistically significant effect of location (secondary channels or reaches), species–location interaction, species–gear interaction, or species–gear–location interaction would signal that differences in species assemblages among locations are not simply a function of erratic catches. It would then be appropriate to use resemblance functions to compare fish assemblages. Comparison of resemblance functions in the absence of significant spatial differences (or temporal, if the analysis involves time series) would lead either to values that differ little, or to values that differ erratically.

The data were pooled by locations (i.e. each reach or secondary channel) and the Bray-Curtis proportion dissimilarity index (*PD*, Ludwig & Reynolds 1988) was used to assess dissimilarities between paired locations as:

$$PD_{jk} = 1 - [2W/(\Sigma x_{ij} - \Sigma x_{ik})],$$

where PD_{jk} is the dissimilarity between location j and location k; $W = \Sigma[\min(x_{ij}, x_{ik})]$; x_{ij} is the abundance of the ith species in the jth location and x_{ik} is the abundance of the ith species in the kth location.

Values of PD_{jk} were computed for each j and k combination. Values range from 0 to 1, where 0 indicates that two locations have completely similar species abundances, whereas 1 indicates complete dissimilarity. This index was recommended by Huhta (1979) and Beals (1984), based on successful application over a wide range of ecological studies. Preliminary analyses indicated various other similarity indices described by Ludwig and Reynolds (1988), logarithmic transformation of the data and exclusion of species accounting for less than 1% of the total catch produced analogous results. Thus, only the values for *PD* computed with untransformed data, including all species collected, are reported.

Differences among *PD*s provided by different collection methods were tested with randomised complete block ANOVA; mean separation was conducted with Student-

Newman-Keuls multiple range test. The treatment variable was gear and the blocking variable was a specific site comparison. Blocking was necessary because *PD* values were expected to vary among site comparisons (i.e. some site comparisons were expected to yield larger *PD* values than others). By blocking, the variability in *PD* values among site comparisons was removed, allowing the test to focus largely on treatment (i.e. gear) differences.

1.3   Results

1.3.1   *Mississippi River*

A total of 472 fish of 21 species were collected with E_{15} and 697 fish of 25 species with E_{60} (Fig. 1.2). Total catch according to secondary channels was 203, 199, 335, and 432 individuals for secondary channels 1–4, respectively. Factorial analysis of variance indicated significant differences in catch rates among secondary channels ($P < 0.0029$), gears ($P = 0.0287$) and species ($P = 0.0001$). However, this analysis also revealed a gear-species interaction ($P = 0.0001$), suggesting that the two gears sampled different proportions of the same species assemblage, a species-secondary channel interaction ($P = 0.0001$), suggesting that species proportions differed among secondary channels, but no gear-secondary channel interaction ($P = 0.11$), suggesting that the gears provided similar trends of abundance among secondary channels. No significant ($P = 0.26$) three-way interaction was detected.

1.3.2   *Luxapallila Creek*

A total of 179 fish of 18 species were collected with hoopnetting, 424 fish of 26 species with seining, and 1069 fish of 43 species with electric fishing (Fig. 1.3). Total catch according to reaches was 188, 639, 382, 154, 158, and 151 individuals for reaches 1–6, respectively. Factorial analysis of variance indicated no significant differences in catch rates among gear ($P = 0.19$), but differences among reaches ($P = 0.0252$) and species ($P = 0.0133$). However, this analysis also revealed significant gear–species interaction ($P = 0.0162$), suggesting that the gears sampled different proportions of the species assemblage, a reach–species interaction ($P = 0.0371$), suggesting that species proportions differed among reaches, and no gear–reach interaction ($P = 0.46$), suggesting that the gears provided similar trends of abundance among reaches. No significant ($P = 0.95$) three-way interaction was detected.

1.3.3   *Dissimilarity indices*

In as much as some significant interactions were detected among gears, location and species, it was concluded that it was appropriate to use resemblance functions to

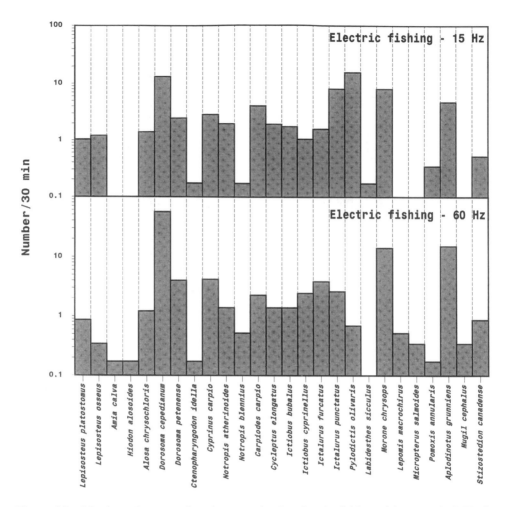

Figure 1.2 Mean catch rates of various species by electric fishing with 15 and 60 Hz in secondary channels of the Mississippi River. Each electric fishing configuration sampled 50 min in each secondary channel, but catch rates are standardised to mean number of fish per 30-min sample

compare fish assemblages. In the secondary channels of the Mississippi River the Bray-Curtis proportion dissimilarity index varied from 0.36 to 0.85 (Table 1.1). With both electric fishing configurations dissimilarity values were generally lower for the pairs of secondary channels closely located geographically than for pairs located far apart. The ANOVA indicated E_{60} ($\bar{x} = 0.71$) produced significantly higher ($P = 0.0243$) dissimilarity values than E_{15} ($\bar{x} = 0.60$).

In reaches of the Luxapallila Creek the dissimilarity index varied from 0.30 to 0.96 (Table 1.2). Hoopnetting and seining dissimilarity values tended to be lower for comparisons among reaches 3–6, but higher when reaches 3–6 were compared with reaches 1 or 2. For electric fishing, dissimilarity indices tended to be consistently high. The ANOVA indicated that dissimilarity indices differed significantly ($P = 0.0051$)

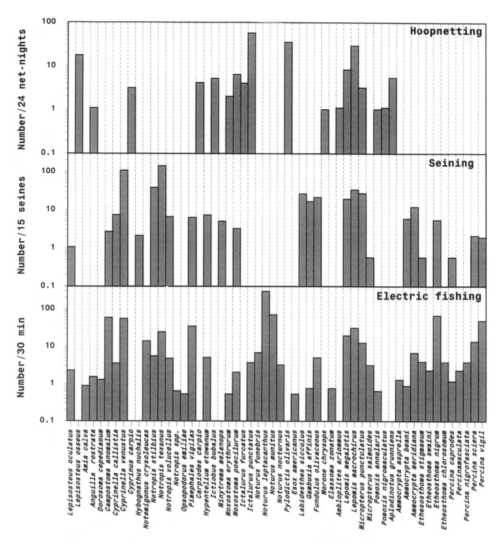

Figure 1.3 Mean catch rates of various species in Luxapallila Creek by hoopnetting (24 net-nights reach^{-1}), seining (15–30 samples reach^{-1}, standardised to 15), and electric fishing (about 30 min reach^{-1}, standardised to 30 min)

among gears. Mean separation indicated dissimilarity values generated by electric fishing ($\bar{x} = 0.72$) and seining ($\bar{x} = 0.73$) were not statistically different ($P > 0.05$), but were larger than those generated by hoopnetting ($\bar{x} = 0.53$).

1.4 Discussion

Two assumptions must be met when comparing catches to identify the gear that is most sensitive to spatial or temporal differences. The first assumption is that gear

Table 1.1 Bray-Curtis proportion dissimilarity index (*PD*) for every pair combination of secondary channels and each electric fishing configuration used in the Mississippi River sampling. Index values can range from 0 to 1, where 0 indicates that two secondary channels have equal species and abundance, whereas 1 indicates complete dissimilarity. Comparisons between adjacent secondary channels are shown in dark-shaded cells and were expected to have lower *PD*s, whereas comparisons between distant secondary channels are shown in light-shaded cells and were expected to have higher *PD*s

Secondary channel	Secondary channel		
	1	2	3
Electric fishing – 15 Hz			
2	0.58		
3	0.60	0.70	
4	0.63	0.76	0.36
Electric fishing – 60 Hz			
2	0.57		
3	0.74	0.81	
4	0.85	0.80	0.49

efficiency remains constant over space or time. Otherwise, observed differences may not be attributed to differences in fish assemblages, but instead to a combined effect of variations in fish assemblage and gear efficiency. Temporally, this assumption may be best addressed by comparing permanent stations, selected randomly or otherwise. This assumption may be more difficult to address spatially; in this study attempts were made to reduce the effects of variable gear efficiency over space by consistently sampling habitats perceived to have similar characteristics. The second assumption is that spatial and temporal differences should not be caused by erratic catches. This assumption was addressed by taking replicate samples in each secondary channel and reach, and including the within-location variability identified by the replicate samples into the factorial ANOVA. *PD* was computed only if differences among locations were statistically significant (i.e. variability between locations was larger than variability within locations).

Estimates of *PD* were generally consistent with expectations. For the Mississippi River, it was predicted that secondary channels in geographic proximity would have lower *PD* values. This prediction was upheld by both E_{15} and E_{60}. For the Luxapallila Creek, it was predicted that *PD* values would be higher between the disturbed reaches 1–2 and the undisturbed reaches 3–6. Although some comparisons of disturbed versus undisturbed reaches had low values, *PD*s generally were lower among undisturbed reaches for species assemblages measured with hoopnetting or seining. With few exceptions this prediction was true for seining and hoopnetting; electric fishing produced mainly high *PD* values regardless of which reaches were compared. These results suggest that the dissimilarity index generally behaved as

Table 1.2 Bray-Curtis proportion dissimilarity index (*PD*) for every pair combination of reaches and gears used in the Luxapallila Creek sampling. Index values can range from 0 to 1, where 0 indicates that two secondary channels have equal species and abundance, whereas 1 indicates complete dissimilarity. Comparisons between undisturbed reaches 3–6 are shown in light-shaded cells and were expected to have lower *PD*s, whereas comparisons between disturbed (reaches 1 and 2) and undisturbed reaches are shown in dark-shaded cells and were expected to have higher *PD*s

	Reach				
Reach	1	2	3	4	5
Hoopnetting					
2	0.54				
3	0.59	0.49			
4	0.72	0.63	0.54		
5	0.56	0.70	0.45	0.49	
6	0.60	0.56	0.37	0.31	0.55
Seining					
2	0.92				
3	0.79	0.92			
4	0.67	0.95	0.69		
5	0.68	0.96	0.46	0.59	
6	0.84	0.94	0.37	0.59	0.58
Electric fishing					
2	0.51				
3	0.60	0.70			
4	0.79	0.84	0.76		
5	0.78	0.84	0.78	0.30	
6	0.66	0.77	0.81	0.86	0.87

expected, supporting the validity of the proposed usage for monitoring changes or differences in fish communities.

In the analysis species relative abundance was used as the variable that described fish assemblages, and the Bray-Curtis percentage dissimilarity index was used to express differences among fish assemblages. Alternative approaches may also be appropriate. For example, fish species may be further divided into two or more size groups; conceivably, two sites may have identical species composition and abundance, but one may sustain primarily juvenile fishes whereas the other may sustain primarily adults. Species or species and size categories could be alternatively analysed in terms of presence–absence rather than relative abundance. Furthermore, functional groups (e.g. trophic levels) rather than structural groups (i.e. species and size categories) may be used to describe fish assemblages. Ecologists and statisticians have developed various resemblance functions, and their appropriateness differs depending on the type of data available. Thus, in practice optimum application may require identification of the most appropriate descriptor of fish assemblage and the most applicable resemblance function. Several authors, including Huhta (1979), Wolda (1981), Beals (1984), Pielou (1984) and Ludwig and Reynolds (1988), have provided evaluations of various resemblance functions.

Riverine fish assemblages are difficult to sample because no single gear can catch all species and sizes of fish (Casselman *et al.* 1990), and no suite of gears can adequately reflect proportions of the various fishes in the assemblages (e.g. Beamesderfer & Rieman 1988). Therefore, fish sampling often provides only an incomplete account of the fish community. To cope with gear selectivities, authors have combined catches from various gears to minimise bias introduced by each gear (e.g. Hinch *et al.* 1991), estimated selectivity of one gear relative to the catch of a suite of gears to identify the least selective gear (e.g. Yeh 1977; Dauble & Gray 1980; Layher & Maughan 1984; Schramm & Pugh, Chapter 4), and developed adjustment factors so that corrected estimates may reflect existing assemblages (e.g. Hamley 1975; Willis *et al.* 1985; Buttiker 1992; Spangler & Collins 1992). In spatial and temporal monitoring programmes, focusing on relative differences and trends or change rather than on unbiased estimates (Willis & Murphy 1996) often circumvents selectivity. The approach illustrated offers yet another option for coping with gear selectivities – it circumvents biases associated with selectivities by exploiting a gear's ability to discern differences. Although the approach is not applicable to every stream monitoring goal, it is applicable when the objective is to detect temporal changes or spatial inequalities.

References

Beals E.W. (1984) Bray-Curtis ordination: an effective strategy for analysis of multivariate ecological data. *Advances in Ecological Research* **14**, 1–55.

Beamesderfer R.C. & Rieman B.E. (1988) Size selectivity and bias in estimates of population statistics of smallmouth bass, walleye, and northern squawfish in a Columbia River reservoir. *North American Journal of Fisheries Management* **8**, 505–510.

Buttiker B. (1992) Electric fishing results corrected by selectivity functions in stock size estimates of brown trout (*Salmo trutta* L.) in brooks. *Journal of Fish Biology* **41**, 673–684.

Casselman J.M., Penczak T., Carl L., Mann R.H.K., Holcik J. & Woitowich W.A. (1990) An evaluation of fish sampling methodologies for large river systems. *Polish Archives of Hydrobiology* **37**, 521–551.

Dauble D.D. & Gray R.H. (1980) Comparison of a small seine and backpack electroshocker to evaluate nearshore fish populations in rivers. *The Progressive Fish Culturist* **42**, 93–95.

Hamley J.M. (1975) Review of gill net selectivity. *Journal of the Fisheries Research Board of Canada* **32**, 1943–1969.

Hinch S.G., Collins N.C. & Harvey H.H. (1991) Relative abundance of littoral zone fishes: biotic interactions, abiotic factors, and postglacial colonization. *Ecology* **72**, 1314–1324.

Huhta V. (1979) Evaluation of different similarity indices as measures of succession in arthropod communities of the forest floor after clear-cutting. *Oecologia* **43**, 371–376.

Layher W.G. & Maughan O.E. (1984). Comparison efficiencies of three sampling techniques for estimating fish populations in small streams. *The Progressive Fish Culturist* **46**, 180–184.

Ludwig J.A. & Reynolds J.F. (1988) *Statistical Ecology: A Primer on Methods and Computing.* New York: Wiley, 337 pp.

MacLennan D.N. (1992) Fishing gear selectivity: an overview. *Fisheries Research* **13**, 201–204.

Pielou E.C. (1984) *The Interpretation of Ecological Data.* New York: Wiley, 263 pp.

Spangler G.R. & Collins J.J. (1992) Lake Huron fish community structure based on gill-net catches corrected for selectivity and encounter probabilities. *North American Journal of Fisheries Management* **12**, 585–597.

Willis D.W. & Murphy B.R. (1996) Planning for sampling. In B.R. Murphy & D.W. Willis (eds) *Fisheries Techniques*, second edition. Bethesda, MA: American Fisheries Society, pp. 1–15.

Willis D.W., McCloskey K.D. & Gabelhouse D.W. Jr (1985) Calculation of stock density indices based on adjustments for efficiency of gill net mesh size. *North American Journal of Fisheries Management* **5**, 126–137.

Wolda H. (1981) Similarity indices, sample size, and diversity. *Oecologia* **50**, 296–302.

Yeh C.F. (1977) Relative selectivity of fishing gear used in a large reservoir in Texas. *Transactions of the American Fisheries Society* **106**, 309–313.

Chapter 2
Longitudinal hydroacoustic survey of fish in the Elbe River supplemented by direct capture

J. KUBECKA, J. FROUZOVÁ

Hydrobiological Institute, Academy of Sciences of the Czech Republic, Na Sádkách 7, 37005 Ceské Budjovice, Czech Republic (e-mail: Kubecka@hbu.cas.cz)

A. VILCINSKAS

Zoologisches Institut der Freien Universitat Berlin, Konigin-Luise-Str. 1–3, D-14195 Berlin, Germany

C. WOLTER

Institute of Freshwater Ecology and Inland Fisheries, Muggelseedamm 310, D-12587 Berlin, Germany

O. SLAVÍK

Water Research Institute, Podbabská 30, 16062 Praha 6, Czech Republic

Abstract

The fish stock in nearly 200 km of the Elbe River in the Czech Republic and 15 km in Germany was monitored by an elliptical horizontal sonar beam (Simrad EY 500, 120 kHz). The biomass of fish in the Elbe was low, averaging 40 kg ha^{-1} for all Czech sections surveyed and 10–20 kg ha^{-1} for the free-flowing German Elbe. In the Czech Elbe, it was possible to distinguish the upper parts with relatively high fish biomass (average 73 kg ha^{-1}) and the impoverished lower reaches where human impact was heavy (average biomass 11.2 kg ha^{-1}). The difference in biomass was caused mainly by the presence or absence of larger fish. Much higher fish biomass (average nearly 1000 kg ha^{-1}) was found in static side waters (harbours, sand pits, old arms), which, together with tributaries, seem to support the fish stocks of the main stream. Confluences with major tributaries were important areas of fish aggregations (average biomass 337 kg ha^{-1}; average individual weight of fish was four times higher than in surrounding reaches). Navigation lock cuts usually harbour more fish than the mainstream (average biomass 160 kg ha^{-1}), but the amount of fish here does not correlate with the fish biomass of the adjacent river. Abundance and biomass of fish estimated acoustically correlated with the catch-per-unit-of-effort by electric fishing.

Keywords: River Elbe, hydroacoustics, sonar, pollution, fish biomass, fish abundance.

2.1 Introduction

Obtaining quantitative information on the fish stock in large rivers is notoriously difficult (Cowx 1996). Most of the conventional assessment techniques are usually impossible or confined to inshore regions or smaller streams (Penáz *et al.* 1986; Hickley 1996; Harvey & Cowx 1996). Application of scientific echosounders in vertical mode (Guillard *et al.* 1994) and horizontal mode (Kubecka 1996) appear to be a promising way of obtaining quantitative information about fish stocks in large rivers. The main limitations of hydroacoustic methods in these applications are low signal-to-noise ratio (limits the detection of small fish), inability to detect fish close to the bottom and in the littoral zone, and little means of species identification. Signal-to-noise ratio has been improved dramatically with hardware development (stable electronics, low side lobe transducers, high sampling rate of processors, short pulses) and in most cases it is possible to record the fish down to the size of fry (Kubecka 1996). It appears possible to detect large fish even on the background of the bottom and surface echoes (Trevorow 1997), but for acceptable detection of all sizes of fish it is advisable to use a maximum useable range (Kubecka 1996), thus avoiding any phase boundary. Identification of freshwater species by sonar remains impossible.

The River Elbe is one of Europe's largest rivers, with a total length of 1092 km. The first 365 km are in Czech territory, the rest in Germany (Fig. 2.1). Human society has used the Elbe for sewage disposal for many decades. After the collapse of the Eastern Bloc, the multidisciplinary Elbe project was started and some positive trends in water quality have been achieved (IKSE 1996). The fish stock is best documented in the estuary (Thiel *et al.* 1995), while only qualitative data are available for the upper reaches (Vostradovský 1994; IKSE 1996). This chapter summarises preliminary findings of a quantitative hydroacoustic survey in Czech and German territories supported by electric fishing.

2.2   Materials and methods

The longitudinal hydroacoustic survey of the Elbe River (Fig. 2.1) was carried out during nights 31 May–1 June 1997 (German Elbe) and 9–16 June 1997 (Czech Elbe). A Simrad EY500 scientific sounder operating with the frequency of 120 kHz was used for horizontal beaming of the usable range. The transducer was mounted on a submersed rotator in front of the Dory 13 boat and had an elliptical beam pattern of 4.3 × 9.1 degrees. The two-way side lobe sensitivity of the transducer beam pattern was 35 dB down compared with the beam axis. Echo integration and echo counting and sizing of single targets were performed in real time and stored data were post-processed with Simrad EP500 software. Thresholds safe from highest noise signals by at least 5 dB were set in 40 log R time-varied-gain data (usually target strength – TS threshold of −62 or −59 dB) and range-dependent 20 log R thresholds were set with the same restrictions (Kubecka 1994). The elementary unit of information used in this chapter is one transect defined by one megabyte of information collected.

The Elbe in Czech territory is divided by a number of weirs (21 weirs were passed during this survey; Fig. 2.1). The section between two weirs is defined as a reach, and for every reach a number of transects was surveyed. Typical habitats recognised within the reach include lock-cut, tributary and backwaters. The last category represents all connected water areas such as old arms, harbours and sand pits.

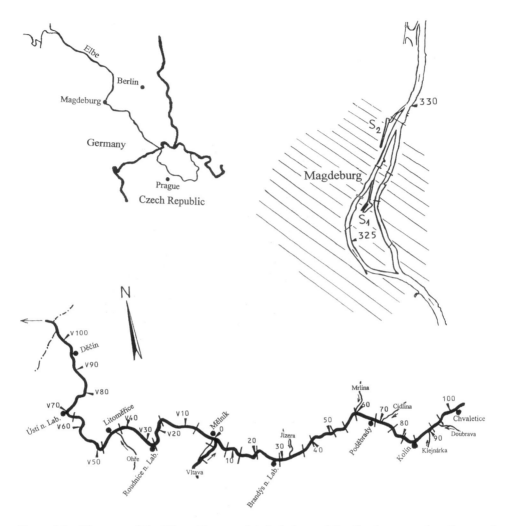

Figure 2.1 The map of the Elbe with more detailed views of the Czech surveyed sections and the section in Magdeburg. S1 and S2 are two surveyed side waters in the Magdeburg area. River kilometres below the confluence with the major Czech tributary, River Vltava, are labelled V. The confluence point is 0 for both downstream and upstream counts

Unfortunately, the entrances of most old arms were full of sediment and thus too shallow for surveying. Most side waters surveyed were harbours and sand pits. The total sampling volume was about 30×10^6 m^3 and the number of sized fish single targets was nearly 5 million.

The deconvolution technique (Kubecka *et al.* 1994) was employed for sizing the fish outside the mainstream of the Elbe. On the mainstream, the hydroacoustic vessel was steered downstream along the river with the sonar beam oriented to the right across the river (Fig. 2.2). The sonar beam was not oriented exactly perpendicular to the river flow, but it was panned 10° forward. This setup is advantageous for fish

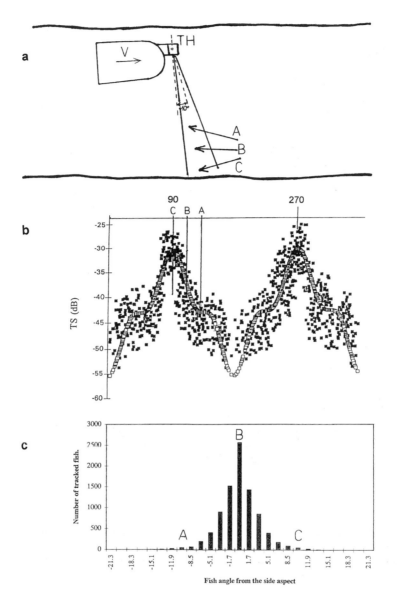

Figure 2.2 Proposed new method for sizing of acoustically detected fish in rivers: (a) The top view of the hydroacoustic vessel (V) with the transducer holder (TH) and the sonar beam oriented horizontally across the Elbe. The beam is panned by 10° from perpendicular to the river flow. The fish in the water can cross the beam in three extreme ways (the arrows show the change of the position of the fish relative to the boat; the movement of the boat – approximately 2 m s^{-1} is more important that the movement of fish). (b) An example of the relationship between fish target strength (TS) and aspect (raw data and cos^3 2α model of Kubecka 1994) and the position of hits of the three fish A, B and C shown in Fig. 2.2c. The signal from fish A will be weakest, fish C will be the strongest target and fish B will have mean TS. (c) The fish angle frequency distribution of 8801 fish tracks recorded during fixed locations along the Elbe. The potential position of three groups of fish, A, B, C, introduced in Figs 2.2a and 2.2b is shown. Very few fish were recorded outside the interval A–C

sizing. Most fish in rivers are swimming along the longitudinal axis of the river upstream or downstream, but the swimming direction of the variable proportion of fish is usually slightly angled (Kubecka 1996; Kubecka & Duncan 1998a, b). The same was true in the Elbe (Fig. 2.2b gives the fish angle frequency distribution observed during the survey at five fixed-location stations separated by approximately 30 km). The fish with zero angle from the side aspect were swimming exactly with or against the current, because the sonar beam was positioned exactly across the river in fixed-location stations. Positive and negative angles indicate angled fish trajectories. Very few fish (less than 0.5%) were panned (or sloped) by more than 10° (Fig. 2.2c) (the same was observed in the River Thames; Kubecka 1996). For the angle of 10° from the side aspect, it is possible to estimate the TS/length or weight relationship from the regressions given by Kubecka and Duncan (1998b). The size of fish swimming 10° towards the transducer (fish A) will be underestimated, as TS equivalent to the aspect 20° from the side is recorded. The size of fish swimming away from the transducer (fish C) will be overestimated as we record its true side aspect TS. Due to the symmetrical distribution of fish (Fig. 2.2c) it was concluded that panning the sonar beam slightly forward, the majority of fish will be recorded 10° from the side aspect and the under- and over-estimate of the size of angled fish will compensate each other. Panning by 10° slightly decreases the signal-to-noise ratio compared with the beaming of the side aspect, but the gain of elimination of aspect-related uncertainty fully justifies the panning.

The inshore fish community was sampled by generator-powered electric fishing equipment (5–11 kW). At each locality sampled (four localities in the Czech Republic and four localities in Germany), 1–2 km of the shoreline (width about 5 m) was fished semi-quantitatively and the catch per unit of effort was recorded.

2.3   Results

The average biomass of fish in the different sections surveyed showed considerable variation (Fig. 2.3; Table 2.1). The most dramatic changes occurred with average size and biomass (Table 2.1). The Czech backwaters contain much bigger fish than any other part (see also Fig. 2.4). The average biomass was nearly 1000 kg ha^{-1}. Side arms in Magdeburg also contained higher densities of fish, but the biomass was lower, at about 240 kg ha^{-1}.

The length frequency distribution of fish observed acoustically shows several distinct features (Fig. 2.4). All Czech localities have a distinct peak corresponding to the fish length of 15–30 mm. The proportion of larger fish decreases exponentially, presumably due to mortality and emigration. The fastest decline was found in the mainstream, whilst for the tributary zones there was a higher representation of bigger fish. Side waters exhibited a second distinct peak of fish of about 200 mm. The proportion of fish of over 500 mm was low and similar to the tributary zones. This is probably due to lower signal-to-noise ratio in free-flowing sections and consequently higher thresholds which discriminate against small targets. The data

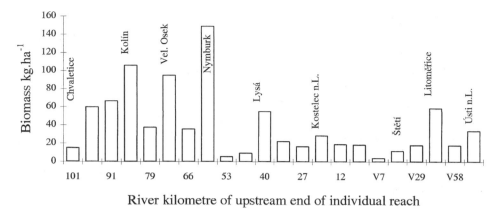

Figure 2.3 Average fish biomass in individual reaches of the Czech Elbe

for German side waters suggest three peaks in the zone of small fish equivalent to 20, 40 and 70 mm (the biggest fry were probably 1 +). The proportion of fish bigger than 25 cm was very low in both German sections.

The importance of side waters and lock cuts to the fish populations in the main river were tested through correlation of fish biomass in their habitats and that estimated in the mainstream river (Figs 2.5 and 2.6). A positive relationship was found between fish biomass in side waters and the adjacent river reach ($r^2 = 0.86$) but no such relationship was found for lock cuts ($r^2 = 0.007$).

Confluences with various tributaries were also found to be areas of higher fish biomass (Table 2.2). The differences of fish density were usually not great (less than double on average), and were not significant (paired *t*-test, $P > 0.05$). The differences in average weights and biomass were significant ($P < 0.001$ and < 0.05 respectively). Similar to side waters, tributary areas were characterised by more larger fish (see also Fig. 2.4).

Fish community composition in the various sections of the Elbe surveyed was similar. In the Czech sections, 8–14 species of fish were captured, in Germany 12–14

Table 2.1 Fish density, mean weight and biomass in certain reaches of the River Elbe and its side waters

Country	Czech	Czech	Czech	Czech	Czech	Germany	Germany
Type of habitat	Elbe all	Elbe natural	Elbe impoverished	Side waters	Lock cuts	Elbe all	Side waters
Fish density (no. m^3)	0.23	0.21	0.14	0.51	0.55	0.75	1.16
Mean weight (g)	6.7	11.8	2.8	77.0	11.9	0.9	7.8
Biomass (kg ha^{-1})	40.0	72.8	11.2	976.0	159.6	13.0	239.6

Table 2.2   Fish density, mean weight and biomass at the confluences of several tributaries of the River Elbe

River km	94	89	69	59	32	0	V45	Mean
Tributary	Doubrava	Klejnárka	Cidlina	Mrlina	Jizera	Vltava	Ohre	
Density (no m³)	0.16	0.20	0.20	1.02	0.14	0.21	0.13	0.29
Mean weight (g)	22.3	48.1	48.3	39.9	23.4	16.5	56.7	36.5
Biomass (kg ha⁻¹)	108.8	291.0	283.0	1225.0	100.6	133.2	213.7	336.5
Adjacent river reach mean								
Density (no m³)	0.08	0.15	0.18	0.48	0.06	0.16	0.13	0.18
Mean weight (g)	5.4	14.3	17.2	7.2	6.3	4.7	17.8	10.4
Biomass (kg ha⁻¹)	32.9	66.5	96.0	141.8	27.3	19.0	58.2	63.1

species were caught in the mainstream. Roach, *Rutilus rutilus* (L.), dominated in most stations, with common bream, *Abramis brama* (L.), silver bream, *Blica bjoerkna* (L.) and bleak, *Alburnus alburnus* (L.) also common. Chub, *Leuciscus cephalus* (L.) was important in the Czech Elbe, but much less so around Magdeburg. Although the sonar beam samples the open water while electric fishing is effective inshore, significant correlation (Fig. 2.7) ($P < 0.01$) was found between both abundance and

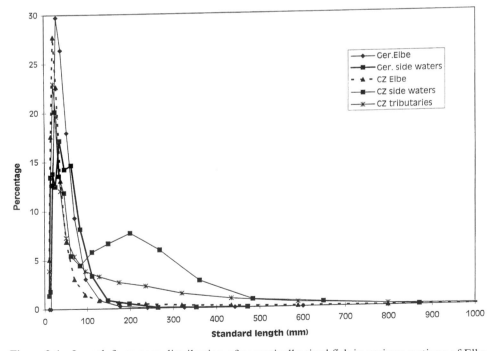

Figure 2.4   Length frequency distribution of acoustically-sized fish in various sections of Elbe

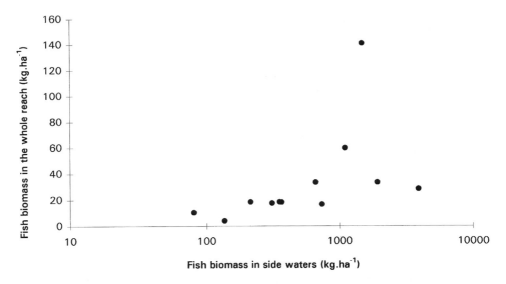

Figure 2.5 The relationship between fish biomass in side waters and in adjacent river reaches

biomass, estimated by both methods (similar observations were reported by S. Hughes, J. Kubecka and A. Duncan unpublished data). In the case of electric fishing, only one fishing was carried out in the inshore area. As electric fishing in such areas is notoriously inefficient (probability of capture <20%; Harvey & Cowx 1996) the use of calibrated gear methods (Harvey & Cowx 1996) is likely to improve the correlation and provide similar abundance estimates.

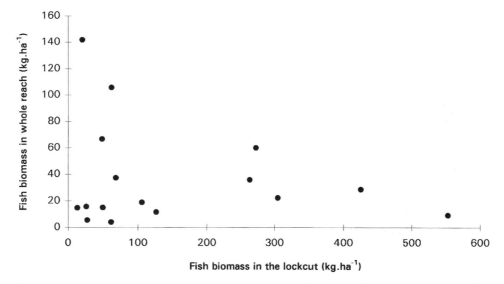

Figure 2.6 The relationship between fish biomass in lock cuts and in adjacent river reaches

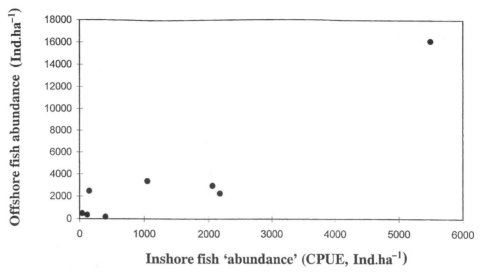

Figure 2.7   The relationship between fish abundance detected acoustically in the open water of the Elbe and abundance estimated by electric fishing in inshore regions

2.4   Discussion

Acoustic detection of fish in shallow waters is limited by unwanted signals (reverberation) in the water environment (Kubecka 1996). The amplitude of these signals is usually small, thus interfering with the detection of small targets. Large targets, which represent the greatest share on fish biomass, are less susceptible to these interferences. Consequently, fish biomass is theoretically the most robust parameter for the description of the changes in fish community.

The biomass of fish along the Elbe was lower than previously determined (Vostradovský *et al.* 1994) or found in other large rivers, e.g. Thames (Butterworth *et al.* 1996) and Vltava (Kubecka & Vostradovský 1995) This probably reflects water quality problems in the Elbe. At the start of the surveyed section the river is recovering from the pollution caused by the Pardubice industrial agglomeration (30 km upstream). The fish biomass ranged between 35 and 105 kg ha^{-1} in most reaches between kilometre 95 and 60. Further downstream, adverse human impacts build-up again as the Elbe passes through several industrial areas. Some recovery is apparent below Lysa and near Litomerice. The reaches below kilometre 53 are considered impoverished, as the fish stock is significantly reduced here.

The average size and biomass of fish in the free-flowing Elbe at Magdeburg was low. Current velocity is possibly a limiting factor for fish in the open water of free-flowing sections of the German section surveyed. Weirs are absent in free-flowing sections and the water velocity is high (50–100 cm s^{-1} compared with 10–40 cm s^{-1} in Czech sections with weirs). Under such conditions, it is likely that most stay close to the bottom where they cannot be detected with sonar. Also, very few large fish were present (Fig. 2.4). There was a similar situation in the lowermost reach of the Czech Elbe (below km V69, Fig. 2.1).

The difference in fish biomass in impoverished (below km 55) and less impoverished (above km 55) parts of the Czech Elbe was caused by different sizes of fish (Table 2.1). The average weight of fish was more than four times greater in less impoverished sections. Also fish biomass in lock cuts was four times higher than in the main stream, due both to higher fish density and average weight. (Average fish biomass in lock cuts, main stream and side waters differed significantly; paired *t*-test; $P < 0.02$.)

The distribution of fish in many rivers is generally patchy (Duncan & Kubecka 1996). In the Elbe the average coefficient of variation of the fish biomass between individual transects in all reaches surveyed was 123.2%. In the reaches with less impoverished fish stock (km 95–60), the variation in biomass along the river was significantly lower ($P < 0.001$) (average CV = 76.7%), than the impoverished sections of the middle and lower parts of the Czech Elbe (average CV = 148.1%). This difference suggests that fish can inhabit less impacted reaches more evenly, while in impoverished stretches most of fish are restricted to marginal habitats such as mouths of tributaries, side arms and weir pools (these are not always suitable for echosounding). The first reach where such distribution was noticed was around Nymburk (km 58–53), where most of the fish were in a weir pool (up to 1000 kg ha^{-1}), which is also the mouth of a tributary. In the rest of the reach, the fish biomass rarely exceeded 10 kg ha^{-1}. This type of distribution was repeated in many reaches.

Two types of relationship were found when the fish biomass in side waters or lock cuts was related to the biomass in the adjacent river reach. A highly positive correlation ($r^2 = 0.86$) was found for side waters (Fig. 2.5). These side waters may serve as nursery areas (Roux & Copp 1996; Copp 1997) and juveniles disperse into the main stream proportionally to their density in the side waters. No such relationship was found for the lock cuts (Fig. 2.6, $r^2 = 0.007$). If the conditions were bad, the fish probably used the lock cuts as refugia. The number of side waters and tributaries per kilometre of Elbe in individual reaches was weakly correlated with the average fish biomass ($r^2 = 0.22$, $P < 0.05$, 22 df). Navigation dykes (small dams coming from the shore pushing more discharge to the middle of the river) do not seem to have any positive effect on fish biomass; if added to side waters and tributaries the coefficient of determination decreases to 0.003 (dykes appeared frequently in most impoverished reaches).

This study shows the impact of pollution on the fish stock of a major industrialised river. It also shows the importance of backwaters and tributaries for the recruitment of fish populations in the mainstream river. These observations provide some insight into the diversity of habitat required to maintain fish stocks in lowland industrial rivers, albeit at a low level, and the type of habitat that will be required to promote the recovery of the river.

Acknowledgments

The authors are grateful to Mr Z. Prachar for the field help and preparation of figures and to Dr J. Matena for valuable comments. The project was supported by the International Elbe Project.

References

Butterworth A., Kubecka J. & Duncan A. (1996) Hydroacoustic techniques to assess fish populations in lowland rivers. In *Proceedings of Institute of Fisheries Management 24th Annual Study Course, Cardiff*. Nottingham: The Institute of Fisheries Management, pp. 69–79.

Copp G.H. (1997) Importance of marinas and off-channel water bodies as refuges for young fishes in a regulated lowland river. *Regulated Rivers: Research and Management* **13**, 303–307.

Cowx I.G. (ed.) (1996) *Stock Assessment in Inland Fisheries*. Oxford: Fishing News Books, Blackwell Science, 506 pp.

Duncan A. & Kubecka J. (1996) Patchiness in longitudinal fish distributions of a river as revealed by continuous hydro-acoustic survey. *ICES Journal of Marine Research* **53**, 161–165.

Guillard J.H., Boet P, Gerdeaux D. & Roux P. (1994) Application of mobile acoustic techniques fish surveys in shallow water: the river Seine. *Regulated Rivers: Research and Management* **9**, 121–126.

Harvey J. & Cowx I.G. (1996) Electric fishing for the assessment of fish stocks in large rivers. In I.G. Cowx (ed.) *Stock Assessment in Inland Fisheries*. Oxford: Fishing News Books, Blackwell Science, pp. 11–26.

Hickley P. (1996) Fish population survey methods: a synthesis. In I.G. Cowx (ed.) *Stock Assessment in Inland Fisheries*. Oxford: Fishing News Books, Blackwell Science, pp. 3–10.

IKSE (1996) Fishes in Elbe. *Internationale Komission zum Schutz der Elbe*, Magdeburg, 44 pp (in German and Czech).

Kubecka J. (1994) Simple model on the relationship between fish acoustical target strength and aspect for high-frequency sonar in shallow waters. *Journal of Applied Ichthyology* **10**, 75–81.

Kubecka J. (1996) Use of horizontal dual-beam sonar for fish surveys in shallow waters. In I.G. Cowx (ed.) *Stock Assessment in Inland Fisheries*. Oxford: Fishing News Books, Blackwell Science, pp. 165–178.

Kubecka J. & Duncan A. (1998a) Diurnal changes of fish behaviour in a lowland river monitored by dual-beam echosounder. *Fisheries Research* **35**, 55–63.

Kubecka J. & Duncan A. (1998b) Acoustic size versus real size relationships for common species of riverine fish in different aspect. *Fisheries Research* **35**, 115–125.

Kubecka J. & Vostradovský J. (1995) Effects of dams, regulation and pollution on the fish stock of the Vltava River in Prague. *Regulated Rivers: Research and Management* **10**, 93–98.

Kubecka J., Duncan A., Duncan W., Sinclair D. & Butterworth A.J. (1994) Brown trout populations of three Scottish lochs estimated by horizontal sonar and multi-mesh gillnets. *Fisheries Research* **20**, 29–48.

Penáz M., Sterba O. & Prokes M. (1986) The fish stock of the middle part of the Morava River, Czechoslovakia. *Folia Zoologica* **35**, 371–384.

Roux A.L. & Copp G.H. (1996) Fish populations in rivers. In G.E. Petts & C. Amoros (eds) *Fluvial Hydrosystems*. London: Chapman & Hall, pp. 167–305.

Thiel R., Sepulveda A., Kafemann R. & Nellen W. (1995) Environmental factors as forces structuring the fish community of the Elbe Estuary. *Journal of Fish Biology* **46**, 47–69.

Trevorow M.V. (1997) Detection of migrating salmon in the Fraser River using 100 kHz sidescan sonars. *Canadian Journal of Fisheries and Aquatic Sciences* **54**, 1619–1629.

Vostradovský J., Pivnicka K., Cihar M. & Poupe J. (1994) Species diversity, abundance, biomass and yield of fishes in the Elbe River and its tributaries. *Bohemia Centralis* **23**, 121–127.

Chapter 3
Summary of the use of hydroacoustics for quantifying the escapement of adult salmonids (*Oncorhynchus* and *Salmo* species) in rivers

B.H. RANSOM, S.V. JOHNSTON and T.W. STEIG
Hydroacoustic Technology Inc., 715 NE Northlake Way, Seattle, WA 98105, USA
(e-mail: support@htisonar.com)

Abstract

Many anadromous salmonid populations (*Oncorhynchus* and *Salmo* spp.) are declining in North America and Europe as pressure from over harvesting, habitat degradation and other sources increases. To aid the management of these stocks, hydroacoustic techniques have been used since the 1960s to estimate adult salmonid escapement in nearly 50 rivers. Initial evaluations used single-beam hydroacoustic techniques, with dual-beam techniques being introduced in the mid-1980s. Since 1992, digital split-beam hydroacoustic techniques have been used in over 50 studies in 17 rivers. Due in large part to its improved spatial resolution and three-dimensional fish tracking capabilities, the split-beam technique has proved more useful than single-beam or dual-beam acoustic techniques for monitoring escapement and behaviour at most sites. Monitoring in rivers is one of the more challenging applications for fisheries acoustics. Unlike typical marine mobile survey applications, riverine applications use stationary transducers with beams aimed in a relatively small water volume, surrounded by the acoustically reflective boundaries of the river surface and bottom. Rivers typically have uneven bottom bathymetry and non-laminar hydraulics, requiring relatively sophisticated equipment and careful deployment, calibration, and testing. The major issues to be addressed to obtain reliable estimates of escapement include hydroacoustic equipment and techniques, site selection, transducer deployment and fish behaviour. Narrow-beam transducers are typically mounted near the shore and aimed horizontally into the river, perpendicular to flow, monitoring migrating fish in side-aspect. A bottom substrate of low acoustic reflectivity (e.g. sand, small rocks) enables the acoustic beam to be aimed close to the bottom. In many cases, migrating salmonids are strongly shore- and bottom-oriented, where water velocities are slowest. Sites are sought where fish are actively migrating, not holding or milling. In addition to escapement counts, results include estimated fish sizes, diel distributions, spatial distributions and velocities.

3.1 Introduction

Many anadromous salmonid populations (*Oncorhynchus* and *Salmo* spp.) are declining in North America and Europe as pressure from habitat degradation, over harvesting, and other sources increases. To aid the management of these stocks, hydroacoustic techniques have been used since the 1960s to estimate adult salmonid escapement in nearly 50 rivers in North America and Europe (Johnston & Steig 1995).

Counting migrating fish in rivers is one of the more challenging applications for fisheries hydroacoustics. Unlike typical marine mobile survey applications (MacLennan & Simmonds 1992), migrating fish in rivers pass fixed transducers through a relatively small water volume, surrounded by the acoustically reflective boundaries of the river surface and bottom. Rivers typically have high reverberation levels, uneven bottom bathymetry and non-laminar hydraulics. Flow conditions can change rapidly, altering the area available for fish migration and increasing background noise. In addition, fish swimming characteristics, orientation and position in the river may be variable.

3.2   Methods

The four major issues that must be addressed to obtain reliable counts of salmonids in rivers include hydroacoustic equipment and techniques, site selection, transducer deployment, and fish behaviour.

3.2.1   *Hydroacoustic equipment and techniques*

Mobile survey evaluations are rarely used to count adult salmon escapement in rivers (Cheng *et al.* 1991). Fixed-location hydroacoustic techniques based on the deployment of stationary transducers are usually employed. Initial evaluations used single-beam hydroacoustic techniques, with dual-beam techniques being introduced in the mid-1980s. A number of authors have reviewed single-beam and dual-beam fixed-location techniques applied to rivers (Mesiar *et al.* 1990; Johnston & Steig 1995), and for some riverine applications these may be adequate. The first application of split-beam acoustic techniques to adult salmonid escapement estimation was in 1992 on the Yukon River (Johnston *et al.* 1993). Since then, digital split-beam hydroacoustic techniques have been used in over 50 studies in 17 rivers. For riverine monitoring, split-beam techniques offer several advantages over single-beam and dual-beam techniques (Ransom *et al.* 1995; Ehrenberg & Torkelson 1996).

Originally, the application of the split-beam technique for fisheries assessment was developed for providing *in situ* target strength (TS) estimates to scale echo integrator output from marine mobile surveys. In the early 1990s the split-beam acoustic technique was applied to riverine monitoring (Johnston *et al.* 1993; Ehrenberg & Torkelson 1996). By tracking the three-dimensional location of each fish in the beam at every ping (e.g. typically 10–20 times s^{-1}), selected echoes are grouped for individual fish, and mean TS calculated for each tracked fish. The lengths of individual fish can be estimated from the mean TS (Love 1977; Goddard & Welsby 1986). This improved split-beam spatial resolution results in improved TS estimates, as well as providing absolute direction of fish movement, permitting discrimination of upstream migrating fish from downstream fish. Ambiguous directional data are common with single-beam or dual-beam techniques (Johnston & Hopelain 1990; Harte 1993a, b). In addition, split-beam techniques provide estimates of fish velocity, trajectory and other behavioural parameters.

For any acoustic technique, excessive background noise can limit the accuracy of the fish TS measurements. The split-beam technique has better performance in the presence of noise, producing TS estimates that are more accurate and less variable than dual-beam estimates (Traynor & Ehrenberg 1990; Burwen *et al.* 1995).

Basic split-beam systems used in the studies described below included an HTI Model 240/243 Split-Beam Digital Echo Sounder, Model 340/343 Digital Echo Processor, Model 540 Split-Beam Transducers with cables, a Model 402 Digital Chart Recorder, digital audio tape recorder, oscilloscope and remote rotators for aiming transducers. All systems operated at 200 kHz. Low side-lobe transducers with elliptical beams (e.g. $3° × 10°$) were typically used. The elliptical-beams allowed more time for each target in the acoustic beam, and hence more data. Also, to maximise sample data, the systems employed ping rates up to 40 pings s^{-1}.

At the fish densities typically observed in rivers, the computer-based processing system was capable of tracking and counting individual migrating salmon in real time. Some projects employed manual tracking, using software that displayed echoes in echogram format and allowed viewing and selection of fish traces in the upstream/ downstream plane. Automatic fish tracking programs recognised a fish by examining individual echoes to see if their amplitudes were above a predetermined threshold, if they had proper pulse width and shape (i.e. matching the transmitted pulse), and then tracked echoes in three dimensions. If there were enough sequential detections (typically 4–6), a series of echoes were grouped as a fish detection. For each fish detected, the data collected included the fish's distance from the transducer, time of entry and exit from the beam, direction of travel, trajectory angle, and velocity.

With limited attention, systems operated for months at a time. In some cases, hydroacoustic systems were controlled and data and results transferred from the river to an office via modem, reducing the requirement for on-site labour.

3.2.2   Sample site selection

Virtually all of the salmon escapement evaluations to date took place at carefully selected sampling sites with smooth bottom profiles and relatively laminar flow (Johnston & Steig 1995; Ransom *et al.* 1995). Sites have typically been 1–20 m deep and 12–500 m wide. Water velocities averaged 1–2 m s^{-1}, with slower velocities observed near shore.

The ideal site has an acoustically 'soft' (silt to small cobble), gently sloping bottom, with adequate velocity but minimum turbulence and entrained air. The site should have a triangular cross-section, such that the smallest angle of the triangle (and the transducer mount location) is at shore. The bottom should have an even and gradual gradient, with no protrusions.

A bottom substrate of low acoustic reflectivity enables the acoustic beam to be aimed close to the bottom. Naturally occurring substrates of silt or mud frequently approach acoustic invisibility, relative to the higher reflectivity of the fish being monitored. Lacking this, gravel or small cobble is better than larger cobble, and smooth cobble is better than angular cobble. Large angular boulders or bedrock are the least desirable substrate.

Laminar flow with minimal entrained air is required for successful hydroacoustic monitoring. Turbulent water (e.g. immediately downstream from waterfalls or rapids) is difficult to monitor. The water velocity should be strong enough to discourage fish milling.

Occasionally, river beds have been artificially modified to facilitate aiming along a flat bottom. On the Fraser River, a sand bag substrate was laid evenly along the bottom where the acoustic beam was to be located.

3.2.3   *Transducer deployment*

Elliptical-beam transducers were deployed with the long axis of the ellipse in the horizontal plane, and the transducer aimed across the river perpendicular to flow (Fig. 3.1). The narrow axis of the beam was aimed close to the bottom, minimising interference problems associated with the water surface and bottom structure. It is typical in medium-to-large rivers for opposing transducers to be deployed on each shore.

3.2.4   *Fish behaviour*

Adult salmon must be actively migrating past the sample site, not holding or milling. Sites that experience significant spawning should be avoided. Since behaviour can be variable in areas of tidal influence, these areas should also be avoided.

For sites where the fish are highly shore orientated, small portable weirs are often placed just downstream of each transducer extending from shore out into the river a short distance (Fig. 3.1). These ensure that fish do not pass behind the transducer, in the near-field of the transducer, or where the sample volume would be too small for adequate detectability.

3.3   Results

The following examples were taken from studies conducted in the Chandalar River in 1994 monitoring chum salmon, *Oncorhynchus keta* (Walbaum) (Daum & Osborne

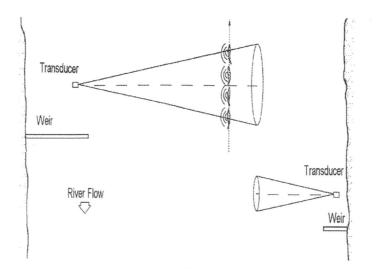

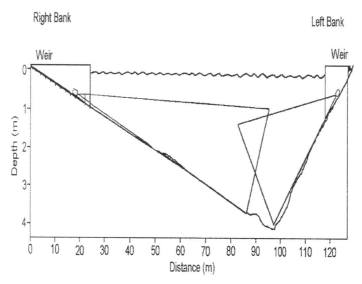

Figure 3.1   Typical riverine deployment in plan view, showing four acoustic ensonifications on an upstream migrating fish passing through the acoustic beam, and elevation view

1995); the Fraser River in 1993 monitoring chinook salmon, *Oncorhynchus tshawytscha* (Walbaum), sockeye salmon *Oncorhynchus nerka* (Walbaum), and pink salmon *Oncorhynchus gorbuscha* (Walbaum) (Johnston *et al.* 1994; Burwen *et al.* 1995); the Kenai River in 1994 and 1995 monitoring chinook salmon (Burwen & Bosch 1996); and the Yukon River in 1992 monitoring chinook and chum salmon (Johnston *et al.* 1993).

3.3.1 *Direction of movement*

At typical sites, fish traces on echograms can be easily distinguished from background noise. The slopes of the traces were typically not reliable indicators of direction of movement. Typically 10–25% of the fish monitored were moving downstream, identifying a potential source of bias in escapement estimates for acoustic techniques not able to identify direction of movement.

In the Yukon River, 85% of the chinook and chum salmon monitored were travelling upstream. In the Chandalar River, 93% of chum salmon were travelling upstream.

3.3.2 *Run timing*

Early in the season, daily fish passage estimates in the Kenai River were less than 500 fish day^{-1}, peaked at nearly 5000 fish day^{-1}, and then dropped off to less than 500 fish day^{-1} (Fig. 3.2).

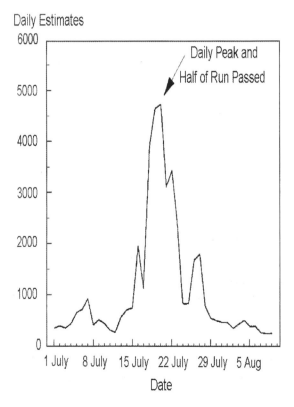

Figure 3.2 Daily fish passage rates for upstream migrating chinook salmon in the Kenai River during 1995 (Burwen & Bosch 1996)

Seasonal hourly counts of chum salmon passing in the Chandalar River exceeded 75 fish h^{-1} (Fig. 3.3). Passage rates were relatively stable in 1994 until severe flooding caused an abrupt end to data collection on 27 August.

During peak passage on the Fraser River, sockeye salmon passage rates exceeded 45 fish min^{-1} on one side of the river.

3.3.3 *Spatial distributions*

To date, medium to large fast flowing rivers typically observed shore- and bottom-oriented horizontal and vertical distributions for upstream migrating salmonids.

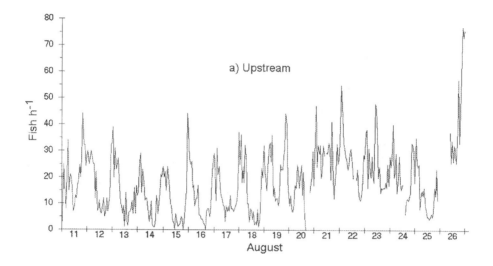

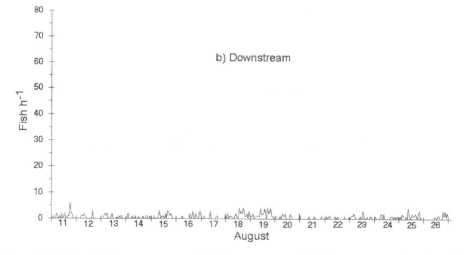

Figure 3.3 Hourly fish passage rates for (a) upstream and (b) downstream migrating chum salmon on the Chandalar River (left bank) during 1994 (Daum & Osborne 1995)

Presumably this was the result of fish attempting to conserve energy by swimming upstream in areas of slower water velocity. In smaller rivers, distributions tended to be more dispersed (Iverson 1995; Steig *et al.* 1995).

Yukon River horizontal and vertical distributions for upstream-travelling chinook and chum salmon were shore and bottom orientated (Fig. 3.4).

Upstream travelling chum salmon in the Chandalar River were shore orientated, while downstream targets were more evenly distributed across the river (Fig. 3.5).

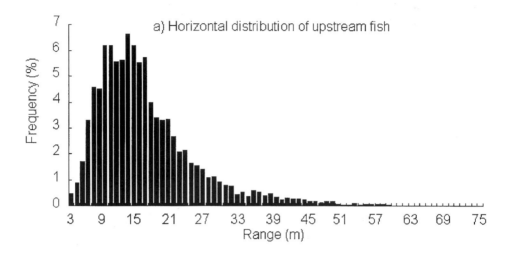

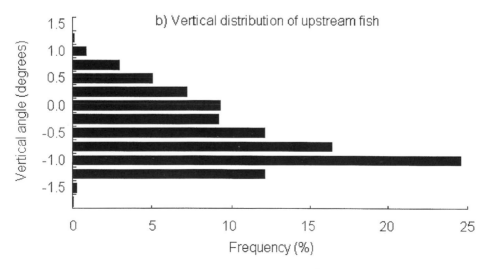

Figure 3.4   Yukon River (a) horizontal distribution (across river), and (b) vertical distribution of 5751 upstream travelling fish on the right bank, during 1992 (Johnston *et al.* 1993)

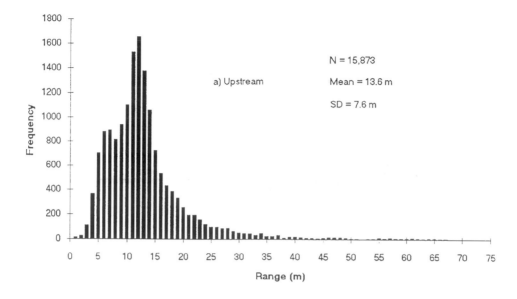

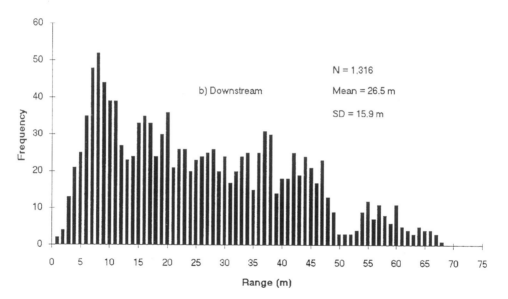

Figure 3.5 Horizontal distribution of (a) upstream and (b) downstream travelling chum salmon on the Chandalar River at the right bank (Daum & Osborne 1995)

Upstream fish were bottom orientated (Fig. 3.6). During night hours, fish tended to be located higher in the water column and nearer to shore than in daylight hours.

On the Fraser River, sockeye and pink salmon were strongly shore and bottom orientated, with most fish passing between the end of the diversion weir (extending out 4 m from the transducer) and a range of 10 m.

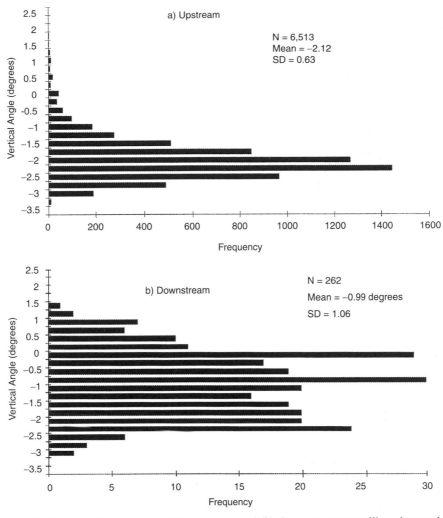

Figure 3.6 Vertical distribution of (a) upstream and (b) downstream travelling chum salmon on the Chandalar River left bank (Daum & Osborne 1995)

Spatial distributions can vary by species. On the Kenai River most chinook were located offshore, while smaller, more abundant sockeye were located predominantly near shore.

3.3.4 *Diel passage rates*

Differences in diel migration rates were observed. Chandalar River and Yukon River diel distributions were weighted toward night (Fig. 3.7). Diel distributions in the smaller Deep Creek were similar (Iverson 1995).

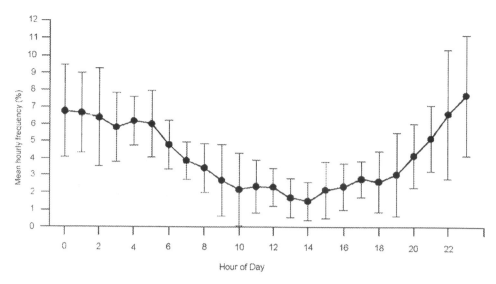

Figure 3.7   Diel distribution of chum salmon on the Chandalar River (left bank) during 1994 (Daum & Osborne 1995)

3.3.5   *Target strength*

On the Chandalar River, the overall mean TS for upstream fish from the right bank was −23 dB (Fig. 3.8). On the Fraser River, the daily average TS steadily decreased as the species composition shifted from larger chinook salmon to smaller sockeye salmon, then to even smaller pink salmon (Fig. 3.9).

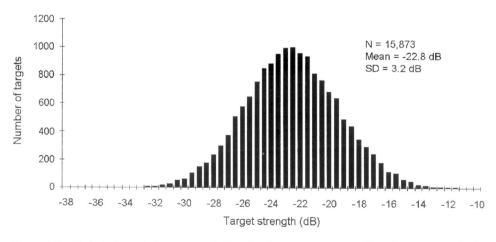

Figure 3.8   Target strength frequency distribution for upstream travelling chum salmon in the Chandalar River (right bank) during 1994 (Daum & Osborne 1995)

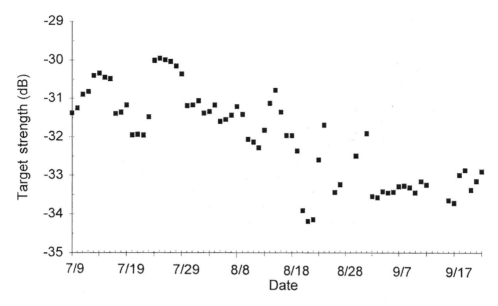

Figure 3.9 Daily mean target strength of upstream travelling fish on the Fraser River, British Columbia, during 1993 (Johnston *et al.* 1994)

3.3.6 *Fish velocity*

On the Yukon River, fish velocities (i.e. speed over ground) calculated using three dimensional target tracking techniques were within expected values (Bell 1990), and were slightly higher for downstream travelling fish (mean = 0.9 m s^{-1}) than for upstream travelling fish (mean = 0.8 m s^{-1}). Fish with higher TS had significantly higher velocities.

Chandalar River fish velocities averaged 1.0 m s^{-1} and 1.5 m s^{-1} for upstream and downstream moving fish, respectively. Taking into account the water velocity at the site, these values fall within the upper range for cruising speeds and the lower range of sustained speeds reported by Bell (1990). A decrease in mean fish velocities corresponded to an increase in river flows.

3.4 Discussion

Hydroacoustics has been effectively used to monitor adult salmonid escapement in carefully selected rivers. Fish passage rates, direction of migration, diel distributions, spatial distributions and velocities have been monitored.

While remaining a challenge, many of the problems associated with monitoring adult salmon escapement have been overcome by the availability of increasingly sophisticated hydroacoustic electronics and signal processing techniques, skilled operators and careful selection of sample sites. Due in large part to their improved spatial resolution and three-dimensional target tracking capabilities, split-beam techniques have been more useful than single-beam or dual-beam acoustic techniques for monitoring escapement and behaviour at most sites.

The proximity of the beam to the bottom is critical to enumerating salmon, since upstream travelling fish frequently swim very near the bottom. Like all acoustic techniques, the split-beam technique can be susceptible to excessive reverberation. Unless sample sites are carefully selected, excessive turbulence and entrained air can limit the usefulness of hydroacoustics. Other limitations include the lack of direct species identification. However, one can frequently use behavioural and distributional evidence, coupled with periodic net sampling, to estimate species composition. Nevertheless, the unobtrusive nature of hydroacoustics and its high sample power makes it attractive for many riverine applications.

Potential improvements in riverine monitoring capabilities include quadrature demodulation for improved spatial resolution, and FM slide/chirp signals to increase the signal-to-noise ratios, both minimising bias and variability around estimates.

Acknowledgements

The results presented above were graciously made available by the following researchers and organisations: Dave Daum, US Fish and Wildlife Service, Alaska; Debbie Burwen of Alaska Dept. of Fish and Game; and Dr Tim Mulligan of the Pacific Biological Station (Dept. of Fisheries and Oceans Canada); as well as various personnel from Hydroacoustic Technology Inc.

References

Bell M. (1990) *Fisheries Handbook of Engineering Requirements and Biological Requirements.* Portland, OR: US Army Corps of Engineers, North Pacific Division, Fish Passage Development and Evaluation Program, 307 pp.

Burwen D. & Bosch D. (1996) Estimates of Chinook salmon abundance in the Kenai River using split-beam sonar, 1995, *Fishery Data Series No. 96–9.* Anchorage: Alaska Dept. of Fish and Game, 27 pp.

Burwen D.L., Bosch D. & Fleischman S. (1995) Evaluation of hydroacoustic assessment techniques for Chinook salmon on the Kenai River using split-beam sonar, *Fishery Data Series No. 95–45.* Anchorage: Alaska Dept. of Fish and Game, 57 pp.

Cheng P., Levy D.A. & Nealson P.A. (1991) Hydroacoustic estimation of Fraser River pink salmon abundance and distribution at Mission, BC, in 1987, *Technical Report No. 3.* Vancouver, BC: Pacific Salmon Commission, 35 pp.

Daum D.W. & Osborne B.M. (1995) Enumeration of Chandalar River Fall Chum salmon using split-beam sonar, 1994, *Fisheries Progress Report No. 95–4.* Fairbanks, Alaska: US Fish and Wildlife Service, 50 pp.

Ehrenberg J.E. & Torkelson T.C. (1996) The application of dual-beam and split-beam target tracking in fisheries acoustics. *ICES Journal of Marine Sciences* **53**, 329–334.

Goodard G.C. & Welsby V.G. (1986) The acoustic target strength of live fish. *Journal du Conseil International pour l'Exploration de la Mer* **42**, 197–211.

Harte M.K. (1993a) 1992 Hydroacoustic study of the Atlantic salmon spawning run on the Moisie River, Quebec, Canada. Report to Groupe Environnement Shooner. Quebec City, Quebec, 41 pp.

Harte M.K. (1993b) 1993 Hydroacoustic study of the Atlantic salmon spawning run on the Moisie River, Quebec, Canada. Report to Groupe Environnement Shooner. Quebec City, Quebec, 41 pp.

Iverson T.K. (1995) Feasibility of using split-beam hydroacoustics to monitor adult Chinook salmon escapement in Deep Creek, Alaska. Report to Alaska Dept. of Fish and Game, Anchorage, Alaska. Seattle, WA: Hydroacoustic Technology Inc, 99 pp.

Johnston S.V. & Hopelain J.S. (1990) The application of dual-beam target tracking and Doppler-shifted echo processing to assess upstream salmonid migration in the Klamath River, California. *Rapports et Proces-verbaux des Reunions Conseil International pour l'Exploration de la Mer* **189**, 210–222.

Johnston S.V. & Steig T.W. (1995) Evolution of fixed-location hydroacoustic techniques for monitoring upstream migrating adult salmonids (*Oncorhynchus* spp.) in riverine environments. *ICES International Symposium on Fisheries and Plankton Acoustics, June 12–16, Aberdeen, Scotland*, 15 pp.

Johnston S.V., Ransom B.H. & Kumagai K.K. (1993) Hydroacoustic evaluation of adult Chinook and Chum salmon migrations in the Yukon River during 1992. Report to US Fish and Wildlife Service, Fairbanks, Alaska. Seattle, WA: Hydroacoustic Technology Inc., 113 pp.

Johnston S.V., McFadden B.D. & Kumagai K.K. (1994) Hydroacoustic evaluation of adult migrations in the Fraser River during 1993. Report to Institute of Ocean Sciences, Dept. Fish. Oceans Canada, Sydney, BC. Seattle, WA: Hydroacoustic Technology Inc., 76 pp.

Love R.H. (1977) Target strength of an individual fish at any aspect. *Journal of the Acoustic Society of America* **62**, 1397–1403.

MacLennon D.N. & Simmonds E.J. (1992) *Fisheries Acoustics*. London: Chapman & Hall, 325 pp.

Mesiar D.C., Eggers D.M. & Gaudet D.M. (1990) Development of techniques for the application of hydroacoustics to counting migratory fish in large rivers. *Rapports et Proces-verbaux des Reunions Conseil International pour l'Exploration de la Mer* **189,** 223–232.

Ransom B.H., Kumagai K.K., Steig T.W. & Johnston S.V. (1995) Installation and testing of a split-beam hydroacoustic system for counting adult Atlantic salmon in the River Wye, Wales. Report to National Rivers Authority, Welsh Region, Cardiff, Wales. Seattle, WA: Hydroacoustic Technology Inc., 56 pp.

Steig T.W., Ransom B.H. & Johnston S.V. (1995) Data report from hydroacoustic monitoring of adult Atlantic salmon in the River Spey. Report to Spey District Fishery Board, Scotland. Seattle, WA: Hydroacoustic Technology Inc., 17 pp.

Traynor J.J. & Ehrenberg J.E. (1990) Fish and standard-sphere target-strength measurements obtained with a dual-beam and split-beam echo-sounding system. *Rapports et Proces-verbaux des Reunions Conseil International pour l'Exploration de la Mer* **189**, 325–335.

Chapter 4
Relative selectivity of hoopnetting and electric fishing in the Lower Mississippi River

H.L. SCHRAMM, Jr
US Geological Survey, Biological Resources Division, Mississippi Cooperative Fish and Wildlife Research Unit, Mississippi State, MS 39762, USA (e-mail: hschramm@cfr.msstate.edu)

L.L. PUGH
Department of Wildlife and Fisheries, Mississippi State, MS 39762, USA

Abstract

Ecology and management of fisheries in large, free-flowing rivers have been constrained by safe and effective sampling methods. All fishing gears are to some extent selective and methods to quantify species selectivity in large rivers are not available. Analytical methods were developed and catch data used from two sizes of hoop nets and two electric fishing outputs to measure the relative selectivity of each of these gears in flowing-water habitats in the Lower Mississippi River (LMR). Electric fishing was found to have lower selectivity relative to hoop nets in the flowing-water, nearshore habitats sampled in the LMR and further use in other large rivers was recommended. The analytical methodology used allows calculation of the relative selectivity of other gears in other river systems.

Keywords: Mississippi River, fish sampling, selectivity, hoop nets, electric fishing.

4.1   Introduction

Effective, quantitative fish sampling methods are essential to manage and understand the ecology of riverine fish communities. Quantitative fish sampling methods for large riverine systems are less well developed than those used in other freshwater systems. From a survey of fisheries biologists, Casselman, *et al.* (1990) developed a lengthy list of gears that have been used in large rivers. Some of these sampling gears are effective in non-flowing extra-channel habitats or impounded rivers (reservoir conditions); only a few of the many gears listed can be safely and efficiently used to obtain quantitative samples in high current velocity, mainstem habitats of a free-flowing (unimpounded) river such as the Lower Mississippi River (LMR). Starrett and Barnickol (1955) compared the relative efficiency of different gears in the Mississippi River. Many of the gears were fished in lentic or slow flowing extra-channel habitats, and the comparisons provide limited guidance for sampling riverine habitats of large rivers.

An ideal sampling method for any habitat is one that obtains a representative sample of a population or community; i.e. a gear that collects 'types' (e.g. species, sizes, genders) of fish in proportion to their actual abundance in the habitat. All fish sampling methods are, to varying degrees, size and species selective. Although indirect methods are available for estimating size selectivity (e.g. Hamley 1975; Pierce *et al.* 1994; Hovgard 1996), comparison of catch with the known composition of the population or community is the only way to accurately estimate species selectivity of fish sampling gears. Methods for obtaining relatively accurate estimates of fish communities in lentic systems (draining, poisoning) or small lotic systems (blocknetting and poisoning) are not effective in flowing-water habitats in large rivers, and the open systems preclude population estimation by conventional mark–recapture, depletion, or area–density methods. Catches in controlled environments (e.g. research ponds) with known populations or communities can provide measures of species selectivity for standing-water gears, but such environments of meaningful scale to measure selectivity do not exist for lotic systems. Furthermore, comparisons with gears with known selectivity are not possible because selectivities of the few usable gears in large river flowing-water habitats are not known. Thus, an absolute measure of species selectivity of a gear for sampling riverine habitats cannot be obtained. However, the selectivity of one sampling gear relative to the catches with other usable gears (i.e. 'relative selectivity'; Yeh 1977) can be measured. Starrett and Barnickol (1955) evaluated relative selectivity of different gears in the Mississippi River by comparing catches with each gear to the catch of 25-mm mesh wing nets. Yeh (1977) measured relative selectivity of several gears by comparing the catch of each gear with the combined catch of all gears.

Hoop nets are a standard sampling gear for lotic systems (Hubert 1996) and have been used in flowing-water habitats in the LMR (Starrett and Barnickol 1955; Beckett and Pennington 1986; Baker *et al.* 1987). Pugh and Schramm (1998) demonstrated that high- and low-frequency pulsed DC electric fishing collected more species, had higher and less variable catch rates, and caught wider length ranges of fishes than hoop nets in mainstem LMR habitats. Although other gears may be useful in some large river habitats (Starrett and Barnickol 1955; Casselman *et al.* 1990), hoop nets and electric fishing are the only methods presently available for safely and efficiently collecting quantitative fish samples in riverine habitats with high current velocity, relatively deep water and diverse bottom substrates, including large rock rip rap and woody debris (large snags). For example, seines can be used in high current velocities, but their use is limited to shallow water with relatively smooth bottom substrate. Trawls can be effective in flowing water, but large rock rip-rap on revetted banks and snags hinder trawling and make it dangerous in high current velocities. Snags also interfere with drifting gill nets or trammel nets. Hydroacoustics can be used in the diverse habitats present in the LMR; however, present technology does not allow identification of species.

In the absence of measures of absolute species selectivity, relative species selectivity of different gears is useful for determining the best gear or gears for describing the diversity and relative abundance of fish in a community. Hence, relative selectivity is also useful for developing efficient sampling procedures. For example, the gear in a suite of gears with the lowest relative selectivity would provide the best description of the fish community and, hence, would be the best gear to use if sampling was constrained by time or cost. In this chapter, a procedure for measuring relative species selectivity is developed and evaluated and this procedure is used to compare the relative species selectivity of two sizes of hoop nets and two electric fishing outputs for sampling LMR fish communities.

4.2   Methods

4.2.1   *Study sites*

Fish were sampled at two main channel and four secondary channel locations in the LMR from river kilometre (RK) 635 to 1505. The main channel is the portion of the mainstem LMR that is maintained by the US Army Corps of Engineers for commercial navigation and carries the greatest portion of the discharge. As such, the main channel is generally wider, deeper and has higher current velocities than other LMR habitats. The main channel was sampled at RK 956–960 (Victoria Bend) and RK 935–940 (Rosedale Bend). Secondary channels are former main channels that no longer carry the main flows of the LMR (either as a result of natural channel meandering or intentional relocation of the main channel). Secondary channels were sampled at RK 1497–1505 (Wolf Island), RK 1467–1475 (Island 8), RK 655–662 (Middle Ground Island) and RK 630–634 (Bondurant Towhead). These secondary channel locations had flowing water at all river stages sampled. Steep bank habitat and sandbar habitat were sampled at each site. Habitat designations follow Cobb and Clark (1981).

Steep bank habitat includes both natural banks in secondary channel sites and revetted banks in main channel sites. Natural steep banks are shoreline habitats with a slope greater than one vertical unit: seven horizontal units extending from the shore–water interface to the deep channel. Substrate is heterogeneous and includes areas of consolidated silt and clay, sand, and occasionally gravel; large, woody debris is prevalent. Current velocities range from near zero close to shore to 1.5 m s^{-1} near the deep channel. Due to woody debris and irregular shoreline and bottom topography, current irregularities such as whirlpools and upstream flow are common.

Revetted steep banks have shorelines with slopes greater than 1:7 extending from the top of the revetment, or the revetment–water interface, to the deep channel. Revetted steep banks are covered by articulated concrete mattress, asphalt paving, or large limestone rock rip-rap to minimise bank erosion. Current velocities range from less than 0.3 m s^{-1} near shore to 2.0 m s^{-1} in deeper areas at the interface with the deep channel.

Sandbar habitat is a gradually sloping area (slope less than 1:7) that extends from the shore–water interface to the deep channel. Current velocity is near zero close to the shore and increases with perpendicular distance from shore (towards the deep channel) to a maximum of about 1 m s^{-1}. Bottom substrate typically is sand and, rarely, gravel or large woody debris.

4.2.2 Fish sampling methods

Fish were sampled with hoop nets and electric fishing gear in sandbar and steep bank habitats at each location. Hoop nets were 61 cm diameter, 2.5 cm square mesh (H_{61}) and 122 cm diameter, 3.8 cm square mesh (H_{122}), each with six hoops. Six hoop nets of each size were randomly set in each habitat at each location. Hoop nets were set in the sandbar habitat in depths of 1.5–3.0 m and current velocities of 0.2–1.0 m s^{-1}. In steep bank habitats, hoop nets were set within 6 m from the shore in depths of 1.5–8.0 m and current velocities of 0.2–1.5 m s^{-1}. Nets were baited with soybean meal, set at 15.00–18.00 h, and retrieved at 08.30–11.00 h the next day.

Fish were sampled by electric fishing in the areas fished with hoop nets on the day the hoop nets were retrieved. All electric fishing was with boat-mounted electric fishing gear (Smith-Root Model GPP7.5) powered by a 7.5 kW generator. In each habitat at each location five, 5-min (pedal time) samples were collected with an electric fishing output of 15 Hz and 1000 V pulsed DC (E_{15}) and five, 5-min samples were collected at 60 Hz and 500 V pulsed DC output (E_{60}). Power output was near the maximum output of the electric fishing unit: 3.8–4.5 A for E_{15} and 10.0–12.0 A for E_{60}. The outputs were alternated between E_{15} and E_{60} for successive 5-min samples at each habitat in each location, and the starting output was randomly selected. Sandbar habitats were sampled by operating the boat in a downstream direction in an S-shaped course in water depths of 0.6–3.0 m. Steep bank habitats were sampled by operating the electric fishing boat in a downstream direction parallel to the bank in depths of 1.5–8.0 m. In each habitat at each location, successive electric fishing samples began a short distance from where the previous sample ended; i.e. the same site was not electric fished in successive samples.

Fishes were sampled at three river stages: 2–4 m above Low Water Reference Plane (LWRP, a river stage corresponding to a discharge exceeded 97% of the time), falling water in August (falling stage); 1–2 m above LWRP, low water in October (low stage); and 4–7 m above LWRP, rising water in December (rising stage). Due to equipment problems, sampling during the falling river stage was conducted with all four gears only at three locations. Samples were collected at all six locations with all four gears during the low and rising river stages, except that the sandbar habitat at Wolf Island was inundated and not sampled during the rising river stage. All fish caught were identified to species and enumerated.

4.2.3 Measurement of relative species selectivity

Species selectivity of the four gears could not be measured because the actual abundance of the fishes in the fish communities at each site (a habitat at a location) was not known. In the absence of the actual abundance, the combined catch with all usable gears was used to estimate the composition and relative abundance of fishes at each site. In the flowing-water steep bank and sandbar habitats in the mainstem LMR, hoop nets and electric fishing are the only gears that safely and repeatedly can be used to enumerate the fishes. Therefore, the relative species selectivity of each of the four gears (E_{15}, E_{60}, H_{61}, and H_{122}) was estimated by comparison with the combined catch of all four gears; the gear that provided the greatest resemblance in terms of species composition and abundance with the combined catch of all gears would have the lowest relative species selectivity.

Catch rates varied considerably among the four gears. Catch rates, averaged among all samples, were 12.2 fish/5 min with E_{15}, 22.7 fish/5 min with E_{60}, 1.4 fish net night^{-1} with H_{61}, and 1.2 fish net night^{-1} with H_{122} (Table 4.1). Furthermore, electric fishing collected more species; a total of 40 species was collected with all four gears, 35 species with E_{15}, 33 species with E_{60}, 17 species with H_{61}, and 20 species with H_{122}. The substantial disparities in catch rates and numbers of species with the different gears presented problems developing a combined catch that equally reflected the catch with each gear. If all species were considered in developing the combined catch, the combined catch would include many species collected only with E_{15} and E_{60}; hence, E_{15} and E_{60} catches would appear more similar to the combined catch than would H_{61} and H_{122} catches. If the numbers of each species in the combined catch were the sum or arithmetic mean of catches with all four gears, the high catch rates with E_{15} and E_{60} would dominate the combined catch; again, E_{15} and E_{60} catches would appear more similar to the combined catch than would H_{61} and H_{122} catches. For these reasons, community resemblance indices that rely on measures of abundance of all species collected would not provide fair estimates of relative selectivity of the four gears.

Similar to Yeh (1977), rank correlation was chosen to quantify how similar the catch with each gear was to the combined catch with all four gears; i.e. the relative species selectivity of each gear. To eliminate bias from the higher catch rates with E_{15} and E_{60} than with H_{61} and H_{122}, the catch rates for each species with each gear was ranked (highest catch rate assigned rank = 1). The mean rank of catch rate for all gears for each species was the combined catch rate for that species. The bias resulting from the greater number of species collected with E_{15} and E_{60} was reduced by including in the correlation analyses species that were relatively abundant and consistently caught with all gears. Using the total catch at all sites and during all river stages for each gear, the average of the rank of catch rate with all four gears for each of the 40 species collected was calculated and the species was gain ranked (lowest mean rank assigned rank = 1). Species with low average catch ranks were generally caught with most or all gears, whereas species with high combined catch ranks were generally caught with one or two gears. The effect of species not collected with a gear was evaluated by comparing relative selectivity of each gear using the 6, 9, 12, and 15 lowest ranked species (i.e. species that were relatively abundant and most consistently caught with all four gears) in the combined catch. Spearman rank correlation coefficients were then calculated between the catch rates with each gear and the combined catch rate for all gears for each site (a habitat at a location) during each of the three river stages sampled.

The correlation analysis tested how well each gear represented the rank order of the relative abundance (rank order) of the prevalent species in the fish community at each site. A high positive correlation reflects concordance of the rank order of the species collected with one gear to the average rank order of the species collected with all four gears, and indicates low relative selectivity of that gear for individual species in the community. Conversely, a low or a negative correlation indicates high relative selectivity of that gear for individual species in the community. The rank correlation coefficients were used as an index of similarity between the relative abundance of the species collected with a single gear to the average catch with all four gears, not as a test of association; hence, no statistical significance was assigned to correlation coefficients.

Using locations as replicates, average correlations were calculated for each gear at steep bank habitat and sandbar habitat for each river stage; differences among these means were tested by Friedman's two-way analysis of variance by ranks and least significant difference test to separate means. Significant difference was set at $P = 0.05$.

If rank correlation is a meaningful metric of relative species selectivity, correlation coefficients should change when the catch rates of one or more species change for one or more gears. The sensitivity of rank correlation analyses was evaluated by measuring the changes in correlation coefficients that resulted from systematically altering catch rate ranks of two species at two randomly selected locations for each river stage. Six series of alterations were performed: (1) switched species ranked one and two for one gear; (2) switched species ranked one and two for two gears; (3) switched species ranked one and two for three gears; (4) switched species ranked one and three for one gear; (5) switched species ranked one and three for two gears; and (6) switched species ranked one and four for one gear. Sites and gears for rank alteration were randomly selected. Sensitivity analyses were done for fish communities with six and 12 species (i.e. the six highest ranking species in the combined catch and the 12 highest ranking species in the combined catch). Using sites as replicates, the mean differences (absolute values) between rank correlation coefficients was calculated from actual data and each series of systematically altered data.

4.3   Results

Using total catch rates at all sites sampled during three river stages, mean rank catch rate for E_{15}, E_{60}, H_{61} and H_{122} (Table 4.1) was obtained. Based on these ranks (highest rank = 1), the 15 most abundant and consistently collected species were freshwater drum, blue catfish, flathead catfish, white bass, channel catfish, gizzard shad, river carpsucker, longnose gar, common carp, shortnose gar, blue sucker, smallmouth buffalo, river shiner, bluegill and threadfin shad.

Rank correlation coefficients measuring the relative selectivity of E_{15}, E_{60}, H_{61}, and H_{122} for estimating the relative abundance of prevalent species were variable among locations, river stages and habitats, as exemplified by analysis for 12 prevalent species (Table 4.2), and as demonstrated by the relatively high standard errors for most gears for analyses of 6, 9, 12 and 15 prevalent species (Table 4.3). Mean rank correlation coefficients significantly differed among gears for six comparisons in the steep bank habitat and for six comparisons in the sandbar habitat (Table 4.3). In one of these 12 comparisons (steep bank habitat, low river stage, nine species), E_{15} and H_{122} had higher mean correlations with the average

Table 4.1 Number of fishes collected with 15 Hz pulsed DC electric fishing (E_{15}), 60 Hz pulsed DC electric fishing (E_{60}), 61 cm diameter hoop nets (H_{61}), and 122 cm diameter hoop nets (H_{122}) in lotic habitats in the Lower Mississippi River, August–December 1994. In parentheses are the number of 5-min samples for electric fishing and the number of net-night samples for hoopnetting. Rank is the rank of the average ranks of the catch with each gear.

Species	E_{15} (155)	E_{60} (165)	H_{61} (241)	H_{122} (233)	Rank
Shovelnose sturgeon, *Scaphirhynchus platorynchus* (Rafinesque)	1	2	1	8	19
Paddlefish, *Polyodon spathula* (Walbaum)	2	0	0	0	34
Longnose gar, *Lepisosteus osseus* (Linnaeus)	19	52	3	10	8
Shortnose gar, *Lepisosteus platostomus* Rafinesque	11	50	3	3	10
Bowfin, *Amia calva* Linnaeus	2	1	0	0	32
American eel, *Anguilla rostrata* (Lesueur)	0	2	1	2	25
Skipjack herring, *Alosa chrysochloris* (Rafinesque)	36	102	0	0	16.5
Gizzard shad, *Dorosoma cepedianum* (Lesueur)	594	2060	1	7	6
Threadfin shad, *Dorosoma petenense* (Gunther)	119	336	0	0	15
Goldeye, *Hiodon alosoides* (Rafinesque)	24	14	0	0	23
Mooneye, *Hiodon tergisus* (Lesueur)	0	1	0	0	38
Grass carp, *Ctenopharyngodon idella* (Valenciennes)	1	4	0	0	29
Common carp, *Cyprinus carpio* Linnaeus	45	86	1	1	9
Bighead carp, *Hypophthalmichthys nobilis* (Richardson)	0	0	0	1	35
Emerald shiner, *Notropis atherinoides* Rafinesque	22	152	0	0	18
Blacktail shiner, *Cyprinella venusta* Girard	1	0	0	0	38
River shiner, *Notropis blennius* Girard	16	24	1	1	13
River carpsucker, *Carpiodes carpio* (Rafinesque)	72	69	1	16	7
Quillback, *Carpiodes cyprinus* (Lesueur)	3	1	0	0	21
Blue sucker, *Cycleptus elongatus* (Lesueur)	16	19	2	7	11
Smallmouth buffalo, *Ictiobus bubalus* (Rafinesque)	30	29	0	19	12
Bigmouth buffalo, *Ictiobus cyprinellus* (Valenciennes)	11	24	0	1	20
Blue catfish, *Ictalurus furcatus* (Lesueur)	204	51	67	47	2
Channel catfish, *Ictalurus punctatus* (Rafinesque)	104	32	137	33	5
Flathead catfish, *Pylodictis olivaris* (Rafinesque)	238	20	80	71	3.5
Madtom, *Noturus* sp.	1	0	0	0	38
Inland silverside, *Menidia berylina* (Cope)	4	2	0	0	28
Yellow bass, *Morone mississippiensis* Jordan & Eigenmann	1	3	0	0	30.5
White bass, *Morone chrysops* (Rafinesque)	144	214	2	20	3.5
Striped bass, *Morone saxatilis* (Walbaum)	5	3	0	0	27
Green sunfish, *Lepomis cyanellus* Rafinesque	1	0	0	0	33
Orange spotted sunfish, *Lepomis humilis* (Girard)	0	1	0	0	38
Bluegill, *Lepomis macrochirus* Rafinesque	2	9	11	2	14
Largemouth bass, *Micropterus salmoides* (Lacépède)	1	3	0	0	30.5
White crappie, *Pomoxis nigromaculatus* Rafinesque	0	0	2	2	24
Black crappie, *Pomoxis nigromaculatus* (Lesueur)	0	0	2	2	26
Darter, *Percina* sp.	1	0	0	0	38
Sauger, *Stizostedion canadense* (Smith)	25	17	1	0	16.5
Freshwater drum, *Aplodinotus grunniens* Rafinesque	124	312	15	35	1
Striped mullet, *Mugil cephalus* Linnaeus	10	52	0	0	22
Total, all species	1892	3749	329	291	

catch rate than E_{60} or H_{61}. In all other comparisons, E_{15} or both E_{15} and E_{60} had higher positive mean correlations with the average catch rate than H_{61} or H_{122}.

Comparison of correlation coefficients among the four gears at each location indicated E_{15} and E_{60} most frequently had the highest correlation with the average catch of the four gears (i.e. had the lowest relative selectivity; Table 4.4). Of a total of 116 comparisons, E_{15} had the highest correlation coefficient in 66 comparisons (including six ties), E_{60} had the highest correlation coefficient in 41 comparisons (including five ties), H_{61} had the highest correlation coefficient in two comparisons (including one tie), and H_{122} had the highest correlation in 15 comparisons (including two ties).

If rank correlation is a useful metric of relative selectivity, correlation coefficients should change when estimates of relative abundance by a gear change. Sensitivity analyses conducted with six and 12 common species indicated that the correlation

Table 4.2 Rank correlation of catch rates of 12 common species in steep bank and sandbar habitats with 15 Hz electric fishing (E_{15}), 60 Hz electric fishing (E_{60}), 61 cm diameter hoop nets (H_{61}), 122 cm diameter hoop nets (H_{122}) and the average of the ranks of catch rates with E_{15}, E_{60}, H_{61}, and H_{122} at different locations in the Lower Mississippi River

Location	Steep bank habitat				Sandbar habitat			
	E_{15}	E_{60}	H_{61}	H_{122}	E_{15}	E_{60}	H_{61}	H_{122}
Falling river stage								
Victoria Bend	0.76	0.37	0.83	0.87	0.78	0.70	0.34	0.54
Yucatan Cutoff	0.72	0.65	0.56	0.43	0.76	0.56	0.48	0.46
Bondurant Towhead	0.89	0.83	0.45	0.71	0.77	−0.11	0.69	0.73
Low river stage								
Wolf Island	0.86	0.80	a	0.72	0.75	0.65	0.53	a
Island 8	0.91	0.73	0.57	0.58	0.76	0.84	0.48	0.67
Victoria Bend	0.71	0.44	0.58	0.76	0.78	0.28	0.35	0.66
Rosedale Bend	0.55	0.72	0.31	0.57	0.61	0.89	0.40	0.17
Yucatan Cutoff	0.76	0.48	0.28	0.26	0.73	0.73	a	0.75
Bondurant Towhead	0.97	0.49	0.61	0.48	0.79	0.91	0.16	−0.19
Rising river stage								
Wolf Island	0.83	0.82	0.19	0.43	b	b	b	b
Island 8	0.87	0.89	−0.04	−0.26	0.66	0.75	0.09	0.18
Victoria Bend	0.81	0.62	0.30	0.60	0.83	0.77	0.49	0.42
Rosedale Bend	0.84	0.77	0.31	0.27	0.89	0.93	a	a
Yucatan Cutoff	0.93	0.96	–	−0.05	0.68	0.59	0.44	0.44
Bondurant Towhead	0.81	0.94	0.23	a	0.85	0.54	0.20	0.44

[a]No fish caught.
[b]Habitat not sampled.

Table 4.3 Mean rank correlation of catch rates in steep bank and sandbar habitats with 15 Hz electric fishing (E_{15}), 60 Hz electric fishing (E_{60}), 61 cm diameter hoop nets (H_{61}), 122 cm diameter hoop nets (H_{122}) and the average of the ranks of catch rates with E_{15}, E_{60}, H_{61}, and H_{122} at different river stages in the Lower Mississippi River. Correlations were calculated for 6, 9, 12 and 15 common species. In parentheses are the number of locations used to calculate the mean and the standard error

No. species	Steep bank habitat				Sandbar habitat			
	E_{15}	E_{60}	H_{61}	H_{122}	E_{15}	E_{60}	H_{61}	H_{122}
Falling river stage								
6	0.50	−0.09	0.40	0.54	0.66	0.34	0.26	0.22
	(3, 0.11)	(3, 0.30)	(3, 0.21)	3, 0.14)	(3, 0.05)	(3, 0.21)	(3, 0.15)	(3, 0.41)
9	0.74	0.34	0.57	0.65	0.73	0.27	0.43	0.41
	(3, 0.09)	(3, 0.29)	(3, 0.16)	(3, 0.15)	(3, 0.02)	(3, 0.23)	(3, 0.17)	3, 0.19)
12	0.79	0.62	0.61	0.67	0.77^w	0.38^x	0.50^x	0.58x
	(3, 0.05)	(3, 0.13)	(3, 0.11)	(3, 0.13)	(3, 0.01)	(3, 0.25)	(3, 0.10)	3, 0.08)
15	0.81	0.64	0.55	0.62	0.80^w	0.42^x	0.47^x	0.56^x
	(3, 0.04)	(3, 0.11)	(3, 0.12)	(3, 0.13)	(3, 0.02)	(3, 0.16)	(3, 0.06)	3, 0.07)
Low river stage								
6	0.59	0.38	0.06	0.70	0.59	0.50	0.40	0.41
	(6, 0.13)	(6, 0.14)	(6, 0.21)	(6, 0.08)	(6, 0.10)	(6, 0.23)	(6, 0.17)	(6, 0.17)
9	0.76^y	0.53^z	0.41^z	0.61^y	0.69	0.68	0.38	0.37
	(6, 0.07)	(6, 0.03)	(6, 0.10)	(6, 0.12)	(6, 0.06)	(6, 0.13)	(6, 0.12)	(6, 0.19)
12	0.79	0.61	0.47	0.56	0.74	0.72	0.38	0.41
	(6, 0.06)	(6, 0.06)	(6, 0.07)	(6, 0.07)	(6, 0.03)	(6, 0.10)	(6, 0.06)	(6, 0.18)
15	0.83^y	0.66^{yz}	0.49^z	0.56^z	0.79^w	0.74^w	0.34^x	0.40^x
	(6, 0.05)	(6, 0.07)	(6, 0.08)	(6, 0.08)	(6, 0.02)	(6, 0.07)	(6, 0.07)	(6, 0.13)
Rising river stage								
6	0.89^y	0.81^y	$−0.09^z$	0.01^z	0.65	0.77	0.31	0.46
	(6, 0.03)	(6, 0.10)	(6, 0.17)	(6, 0.24)	(5, 0.10)	(5, 0.07)	(5, 0.18)	(5, 0.12)
9	0.86^y	0.83^y	0.13^z	0.18^z	0.75^w	0.74^w	0.30^x	0.42^x
	(6, 0.03)	(6, 0.08)	(6, 0.09)	(6, 0.20)	(5, 0.08)	(5, 0.07)	(5, 0.12)	(5, 0.07)
12	0.85^y	0.84^y	0.20^z	0.20^z	0.78^w	0.72^w	0.30^x	0.37^x
	(6, 0.02)	(6, 0.05)	(6, 0.06)	(6, 0.16)	(5, 0.05)	(5, 0.07)	(5, 0.10)	(5, 0.06)
15	0.82^y	0.83^y	0.28^z	0.28^z	0.78^w	0.72^w	0.28^x	0.37^x
	(6, 0.03)	(6, 0.05)	(6, 0.06)	(6, 0.14)	(5, 0.04)	(5, 0.06)	(5, 0.07)	(5, 0.04)

[y,z]Values in same row with different letters significantly different ($P < 0.05$) by Friedman's two-way analysis of variance by ranks and least significant difference test to separate means.
[w,x]Values in same row with different letters significantly different ($P < 0.05$) by Friedman's two-way analysis of variance by ranks and least significant difference test to separate means.

coefficients changed when the estimates of relative abundance with one or more gears changed (Table 4.5). Mean differences in rank correlation coefficient between actual catch rates and systematically altered catch rates increased with the amount of change in relative abundance. Switching species ranked one and three changed the correlation coefficients more than switching species ranked one and two, and switching species ranked one and four changed the correlation coefficients more than switching species ranked one and three. Switching species ranked one and two for three gears changed the correlation coefficients more than switching species ranked one and two for two gears, and switching species ranked one and two for two gears changed the correlation coefficients more than switching species ranked one and two for one gear. Similarly, switching species ranked one and three for two gears changed the correlation coefficients more than switching species ranked one and three for one gear. The magnitude of the effect of changes in relative abundance on correlation coefficients was greater when correlations were based on the relative abundance of six common species than when the correlations were based on the relative abundance of 12 common species.

Table 4.4 Number of locations where 15 Hz electric fishing (E_{15}), 60 Hz electric fishing (E_{60}), 61 cm diameter hoop nets (H_{61}), or 122 cm diameter hoop nets (H_{122}) had the highest rank correlation with the average of the ranks of catch rates with E_{15}, E_{60}, H_{61}, and H_{122} at different river stages in the Lower Mississippi River. Correlations were calculated for 6, 9, 12 and 15 common species. N is the number of locations compared at each river stage

| No. species | Steep bank habitat | | | | | Sandbar habitat | | | | |
	N	E_{15}	E_{60}	H_{61}	H_{122}	N	E_{15}	E_{60}	H_{61}	H_{122}
Falling river stage										
6	3	2	1[a]	1	0	3	3	0	0	1[a]
9	3	2	0	0	1	3	2	0	0	1
12	3	2	0	0	1	3	2	0	0	1
15	3	2	0	0	1	3	3	0	0	0
Low river stage										
6	6	2	1	0	3	6	2	3	0	1
9	6	4	0	0	2	6	3	4[a]	0	0
12	6	4	1	0	1	6	3	4[a]	0	0
15	6	4	1	0	1	6	4	2	0	0
Rising river stage										
6	6	2	4	0	0	5	2	3	1[b]	1[b]
9	6	3	3	0	0	5	3	2	0	0
12	6	3	3	0	0	5	3	2	0	0
15	6	3	4[a]	0	0	5	3	3[a]	0	0

[a]Tied for highest correlation with E_{15}.
[b]Tied for highest correlation with E_{60}.

Table 4.5 Average change in rank correlation for fish communities of six or 12 prevalent species when the species ranked first and second were switched for one gear (D1), two gears (D2), or three gears (D3); when the species ranked first and third were switched for one gear (D4) or two gears (D5); and when the species ranked first and fourth were switched for one gear (D6). N is the number of sites (habitats at locations) used to calculate the mean. Values in parentheses are standard errors

No. of prevalent species	River stage	N	D1	D2	D3	D4	D5	D6
6	Falling	16	0.08	0.14	0.18	0.14	0.16	0.38
			(0.03)	(0.03)	(0.03)	(0.03)	(0.03)	(0.07)
6	Low	15	0.15	0.19	0.29	0.19	0.25	0.23
			(0.05)	(0.05)	(0.06)	(0.04)	(0.06)	(0.07)
6	Rising	14	0.10	0.18	0.23	0.21	0.35	0.29
			(0.03)	(0.03)	(0.04)	(0.06)	(0.08)	(0.06)
12	Falling	16	0.01	0.02	0.03	0.04	0.04	0.08
			(0.00)	(0.00)	(0.01)	(0.01)	(0.01)	(0.03)
12	Low	15	0.01	0.04	0.08	0.04	0.10	0.04
			(0.00)	(0.01)	(0.01)	(0.01)	(0.03)	(0.01)
12	Rising	21	0.02	0.03	0.05	0.05	0.06	0.06
			(0.01)	(0.01)	(0.01)	(0.01)	(0.02)	(0.01)

4.4   Discussion

The ability to manage and gain insight into the ecology of large river fish populations and communities depends on safe and efficient sampling methods. Hoop nets have been one of the standard sampling devices for flowing waters and are one of the few gears that can be used to sample fish in swiftly flowing channel habitats of large rivers (Hubert and Schmitt 1982). Pugh and Schramm (1998) demonstrated that high and low frequency electric fishing were efficient sampling methods in Lower Mississippi River lotic habitats. However, useful sampling gears should also provide accurate assessments (i.e. representative samples) of fish communities. Hoop nets (e.g. Starrett & Barnickol 1955; Hubert & Schmitt 1982; Michaels & Williamson 1982) and electric fishing (e.g. Michaels & Williamson 1982; Nelson & Little 1986; Zalewski & Cowx 1990; Reynolds 1996) are both species selective. Although knowledge of the absolute selectivity (Hamley 1975) of these gears is desired, measuring selectivity by comparing the catch of a gear with a known population or community is difficult, if not impossible, with the current technology in the open-population conditions in large, free-flowing rivers. Relative selectivity provides a way to compare the selectivity of a suite of gears and select the gear that provides the best estimate of the community described by all gears.

Following Yeh (1977), rank correlation was used to measure relative selectivity and it was found to be sensitive to changes in catch rates. Methods were provided to reduce or eliminate bias resulting from greater catch rates and numbers of species caught by some gears. The method explored in this chapter can be used to calculate relative selectivity among any number of gears concomitantly used in the same habitat. Relative selectivity is not a replacement for absolute selectivity; however, comparison of the catch with a single gear to the combined catch of all useable gears is intuitively appealing. It is common practice for fishery workers to sample with a variety of gears, each selective for different species, to most thoroughly describe the community; hence, the combined catch with all usable gears, adjusted to prevent a single gear from dominating the combined catch, may be considered as the 'best estimate' of the community. Indeed, this best estimate may misrepresent the fish community, but the probability of this occurring should decline as the number of wisely selected sampling gears increases.

Based on comparisons of fish catches with two sizes of hoop nets and two electric fishing outputs in LMR riverine habitats, low pulse frequency electric fishing generally was found to have the lowest relative selectivity, and therefore provided the most representative samples in the habitats tested. However, measures of relative selectivity were variable among river stages, habitats and locations; and high pulse frequency electric fishing often had the lowest relative selectivity or relative selectivity not significantly different from low pulse frequency electric fishing. The low relative selectivity supports the preferred use of electric fishing if sampling constraints, such as sampling time, prevent more extensive sampling with diverse gears.

Few fish sampling gears are effective in both standing and flowing waters. Use of different gears provides comprehensive assessment of fisheries variables, but catches with different gears are, at best, weakly comparable. Our limited ability to collect comparable samples in different habitats, in particular habitats with different current velocities and different bottom substrates, has constrained our understanding of the ecology and management of the LMR (Fremling *et al.* 1989). Electric fishing is considered an efficient sampling method for many species in standing water. Pending evaluation of relative selectivity of electric fishing in LMR standing water habitats, electric fishing may prove a valuable method for estimating fish relative abundance and community composition in diverse habitats in the Lower Mississippi River and other large rivers (Fig. 4.1).

Fish Communities

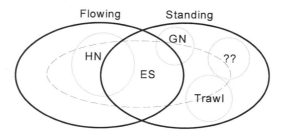

Figure 4.1 The solid-line ovals represent the fish communities in standing- and flowing-water habitats in large rivers. A sampling method (or methods) that obtains representative samples in standing- and flowing-water habitats will provide comparable estimates of fish abundance necessary for assessing the diverse fish communities in large rivers. ES = electric fishing, GN = gill nets, HN = hoop nets, and ?? = any other gear

Acknowledgements

Todd Driscoll, Ken Cash, and Sekou Sangare assisted with fish collection. Mark Boone, Steve Gutreuter and L. Esteban Miranda provided helpful comments on earlier versions of the manuscript. Funding for this study was provided by the US Army Corps of Engineers, Mississippi Valley Division, and the Mississippi Co-operative Fish and Wildlife Research Unit.

References

Baker J.A., Pennington C.H., Bingham C.R. & Winfield L.E. (1987) An ecological evaluation of five secondary channel habitats in the Lower Mississippi River, *Lower Mississippi River Environmental Program, Report 7*. Vicksburg, MI: US Army Engineer Waterways Experiment Station, 45 pp.

Beckett D.C. & Pennington C.H. (1986) Water quality, macroinvertebrates, larval fishes, and fishes of the Lower Mississippi River – a synthesis, *Technical Report E-86–12*. Vicksburg, MI: US Army Engineer Waterways Experiment Station, 136 pp.

Casselman J.M., Penczak T., Carl L., Mann R.H.K., Holcik J. & Woitowich W.A. (1990) An evaluation of fish sampling methodologies for large river systems. *Polish Archives for Hydrobiology* **37**, 521–555.

Cobb S.P. & Clark J.R. (1981) Aquatic habitat studies on the Lower Mississippi River, river mile 480 to 530; Report 2, aquatic habitat mapping, *Miscellaneous Paper E-80–1*. Vicksburg, MI: US Army Engineer Waterways Experiment Station, 24 pp.

Fremling C. R., Rasmussen J.L., Sparks R.E., Cobb S.P., Bryan C.F. & Claflin T.O. (1989) Mississippi River fisheries: a case history. *Canadian Special Publication of Fisheries and Aquatic Sciences* **106**, 205–219.

Hamley J. M. (1975) Review of gill net selectivity. *Journal of the Fisheries Research Board of Canada* **32**, 1943–1969.

Hovgard H. (1996) A two-step approach to estimating selectivity and fishing power of research gill nets used in Greenland waters. *Canadian Journal of Fisheries and Aquatic Sciences* **53**, 1007–1013.

Hubert W. A. (1996) Passive capture techniques. In B.R. Murphy & D.W. Willis (eds) *Fisheries Techniques*, second edition. Bethesda, MA: American Fisheries Society, pp. 157–181.

Hubert W.A. & Schmitt D.N. (1982) Factors influencing hoop net catches in channel habitats of Pool 9, Upper Mississippi River. *Proceedings of the Iowa Academy of Science* **89**, 84–88.

Michaels R. & Williamson P. (1982) Southcentral Region fisheries investigations: an evaluation of techniques for capturing channel catfish from the Altamaha River, *Final Report, Project GA F-029-9*, Georgia Department of Natural Resources, 24 pp.

Nelson K.L. & Little A.E. (1986) Evaluation of the relative selectivity of sampling gear on ictalurid populations in the Neuse River. *Proceedings of the Annual Conference of the Southeastern Association of Fish and Wildlife Agencies* **40**, 72–78.

Pierce R.B., Tomko C.M. & Kolander T.D. (1994) Indirect and direct estimates of gill-net size selectivity for northern pike. *North American Journal of Fisheries Management* **14**, 170–177.

Pugh L.L. & Schramm H.L. Jr (1998) Comparison of electric fishing and hoopnetting in the Lower Mississippi River. *North American Journal of Fisheries Management* **18**, 649–656.

Reynolds J.B. (1996) Electric fishing. In B.R. Murphy & D.W. Willis (eds) *Fisheries Techniques*, second edition. Bethesda, MA: American Fisheries Society, pp. 221–253.

Starrett W.C. & Barnickol P.G. (1955) Efficiency and selectivity of commercial fishing devices used on the Mississippi River. *Illinois Natural History Survey Bulletin* **26**, 325–366.

Yeh C. F. (1977) Relative selectivity of fishing gear used in a large reservoir in Texas. *Transactions of the American Fisheries Society* **106**, 309–313.

Zalewski M. & Cowx I.G. (1990) Factors affecting the efficiency of electric fishing. In I.G. Cowx and P. Lamarque (eds) *Fishing with Electricity: Applications in Freshwater Fisheries Management*. Oxford: Fishing News Books, pp. 89–111.

Section II
Large river fisheries

Chapter 5
The fish community of the River Thames: status, pressures and management

S.N. HUGHES
National Coarse Fisheries Centre, Environment Agency, Arthur Drive, Hoo Farm Industrial Estate, Worcester Road, Kidderminster DY11 7RA, UK (e-mail: Simon.hughes@environment-agency.gov.uk)

D.J. WILLIS
Environment Agency, Thames Region, Isis House, Howbery Park, Wallingford, Oxon. OX10 8BD, UK

Abstract

The River Thames is a highly regulated lowland river supporting a diverse fish community including recovering populations of eel, *Anguilla anguilla* (L.), and Atlantic salmon, *Salmo salar* L., forming a valuable ecological and recreational resource. An overview of the current fish community, including aspects of species composition, abundance and distribution is provided. Strategies for monitoring the fish populations in challenging sampling environments are presented, including hydroacoustic and 0-group surveys. Coarse fish angling is the principal fishery and its status is characterised. The main pressures are identified and considered in relation to current management approaches and the benefits of integrated and holistic management are emphasised. A vision of the fisheries of the River Thames in the twenty-first century is proposed.

Keywords: River Thames, fish community, management, pressures, regulated lowland river.

5.1 Introduction

The River Thames (Fig. 5.1) is the largest gauged catchment in the British Isles, covering 9950 km² upstream of the tidal limit at Teddington (Ward 1981). The river is 354 km from the source in the limestone Cotswolds to the North Sea. It flows east across chalk and tertiary rocks in deeply incised valleys, their shape depending on the underlying bedrock. The chalk valleys are relatively narrow and steep sided, whereas the clay valleys are wide with gently sloping sides. This chapter focuses on the freshwater river from source to London Bridge, a distance of 263 km with a mean gradient of 0.32 m km⁻¹.

The river has been regulated for almost 2000 years to facilitate its use as a navigation channel. Most of the freshwater river from the tidal limit to the town of Lechlade nearly 200 km upstream is extensively modified by 44 low head weirs operated to maintain the river for navigation. It forms a series of impounded sections of relatively uniform habitat, except where vestiges of the original channel remain. The impoundments can effectively inhibit upstream fish migration between reaches.

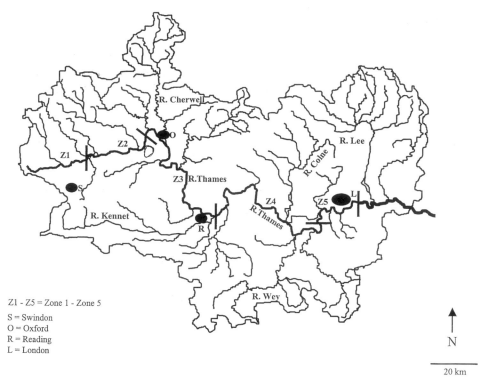

Figure 5.1   The catchment, tributaries and five zones of the River Thames, UK

The catchment has a human population of almost 12 million, which has remained relatively stable since the mid-1940s. The Thames catchment is intensively managed, with 3078 licenced abstractions for water supply and over 11 000 consented effluent discharges.

Mean daily river flow is about 78 m^3 s^{-1} at the tidal limit (Teddington), based on records since 1883. In the twentieth century, gauged flow has varied between 6.5 m^3 s^{-1} and 714 m^3 s^{-1}. The annual mean rainfall in the catchment upstream of Teddington is 710 mm, with a slight autumn maximum; about 460 mm of this is lost through evaporation. January river discharge is usually about four times the summer discharge, when most of the flow is made up of groundwater input (Littlewood & Marsh 1996).

Angling has been recorded on the River Thames since the publication of Izaak Walton's *The Compleat Angler* in 1653. The activity gained popularity in the mid-nineteenth century, when 187 angling clubs were listed in the London area. Anglers were principally responsible for documenting the decline of a number of important species, notably Atlantic salmon, *Salmo salar* L., shad, *Alosa fallax* (Lacépède), and eel, *Anguilla anguilla* (L.) (Francis & Urwin 1991).

Management of the River Thames and its tributaries is the responsibility of the Environment Agency. The primary fisheries duty is to 'maintain, improve and develop' the freshwater fisheries resource and the Agency also has responsibilities for other relevant activities such as water quality, water resources, flood defence, recreation, navigation and a wider conservation role.

5.2 Fisheries studies

Despite increasing pressures and management of the River Thames, scientific studies of fish populations were not carried out until the late 1950s, and only then near Reading (Mann 1964, 1965, 1972; Williams 1965, 1967; Mackay & Mann 1969; Mathews 1971; Berrie 1972; Mann *et al.* 1972). Mann (1989) reviewed some of the issues relating to fishery management in the River Thames.

Because of the complex nature of large lowland rivers and the harsh, variable sampling environments they represent, it is difficult to obtain good quality stock assessment in a cost effective way (Hickley 1996; Jackson 1996). Fishery managers therefore have a comparatively poor understanding of factors that influence coarse fish populations in large lowland rivers. Ideally, they should have robust qualitative and quantitative information for fish communities throughout the length of a river system to aid decision making.

Conventional techniques (electric fishing, seine netting) have not achieved this in sampling environments like the Thames that comprise some of the UK's major recreational fisheries. Other methods have been attempted (trawling, angler catch assessment, trapping and complex netting techniques), which are either too expensive, too site specific, too damaging or just inaccurate (Hickley 1996). Sampling fish communities in the River Thames is difficult and expensive, and until recently has not been possible for most of the river. In response to the growing need for appropriate information to inform fisheries management, an integrated approach has been developed by the Environment Agency using complementary techniques to provide a measure of fish community composition and size.

Sonar (hydroacoustic) survey methods developed principally by the Environment Agency for mobile use in shallow waters allow quantitative fish stock assessment (Duncan & Kubecka 1993, 1994). The method produces relatively precise assessments of riverine fish abundance, longitudinal distribution and population length structure over long lengths of river, but has a number of significant limitations. In general, limitations include distinguishing fish close to boundaries or other features in a watercourse, and differentiation between species. The equipment is relatively expensive and requires skilled and trained operators working for long periods at night, which have both cost and safety implications. Despite these issues the method is now successfully used for routine monitoring of the Thames and other large lowland rivers (e.g. Hughes 1998; Lyons 1998).

Developments in electric fishing sampling techniques have made it more suitable for use in large, deep rivers by UK standards. Electric fishing has a number of limitations related to river width and depth, water conductivity, clarity and velocity. Purpose-built electric fishing boats with large (> 1 m diameter) sequentially firing Wisconsin ring electrode arrays are an effective method for sampling fish in water up to 1.5 m deep, and associated with the river bed, aquatic plants and other features (Harvey & Cowx 1996). In rivers like the Thames, however, achieving the goal of quantitative sampling using this method alone remains difficult.

Information on 0-group recruitment in lowland rivers is an important management requirement given that spawning and fry recruitment are critical factors influencing subsequent year class strength and hence the sustainability of fish communities (Cowx *et al.* 1995). Micromesh seining is an effective method for qualitative and semi-quantitative sampling of appropriate juvenile fish habitats and has provided data on this key life stage.

The integrated approach to fisheries studies outlined above is currently being implemented by the Environment Agency within a wider multi-disciplinary strategy encompassing studies of phytoplankton, zooplankton, water quality and hydrology to continue to develop better understanding on which to base future management decisions.

5.3   Fish community status

5.3.1   *Species occurrence and distribution*

Longitudinal change in environmental factors along the river influences the distribution of fish species (e.g. Vannote *et al.* 1980). To simplify description of fish communities the river has been divided into five zones based on general habitat characteristics, as shown in Fig. 5.1 and described in Table 5.1.

The river has a diverse fish fauna by UK standards, with 28 of the 35 native riverine freshwater and diadromous fish species defined by Winfield *et al.* (1994) (Table 5.2). The main River Thames contains all fish species found in the catchment as a whole. The diversity of the Thames fish fauna includes a range of species that contribute to its high fisheries conservation value.

Brook lamprey, *Lampetra planeri* (Bloch), and bullhead, *Cottus gobio* L. are both present and are listed in Annex II of the EC directive on the Conservation of Natural Habitats and of Flora and Fauna (92/43/EEC). Although these species are not considered rare in the UK (Winfield *et al.* 1994), EC member countries have a duty to conserve viable populations.

Smelt, *Osmerus eperlanus* (L.), are nationally rare (Winfield *et al.* 1994) and have been lost from several other British estuaries in recent years, although they have made a return to Zone 5 of the River Thames. The group of species comprising silver bream, *Blicca bjoerkna* (L.), crucian carp, *Carassius carassius* (L.), and rudd, *Scardinius erythrophthalmus* (L.), have a restricted British distribution and now may be becoming rarer.

There is no evidence of breeding populations of non-native fish species in the River Thames, although rare captures of zander, *Stizostedion lucioperca* (L.), are made. Rainbow trout, *Oncorhynchus mykiss* (Walbaum), are occasionally found as a result of stocking of recreational fisheries in some tributaries.

With respect to fish distribution, Zone 1, the first 36 km of river, holds populations of 17 species. This can be explained by the high diversity of habitat present, resulting from historical river drainage schemes that have produced slow flowing, deep sections in addition to natural channel morphology.

Zones 2, 3 and 4 are similar in species occurrence. The current management of the river for navigation is a dominant factor in determining physical habitat

Table 5.1 General habitat characteristics of five zones in the River Thames

Zone	Begins	Ends	Length (km)	General habitat description
1	Source	Confluence of River Coln	36	Natural channel with pool riffle sequence, erosion and deposition banks. No weirs; largely rural land use. Extensive submerged and emergent aquatic plants, diverse habitat.
2	Confluence of River Coln	Farmoor Reservoir	38	Level control by 8 weirs. Dredging, channel modification and some bank protection. Natural course remains in many places, submerged and emergent plants are common, but less diverse habitat.
3	Farmoor Reservoir	Confluence of River Kennet	79	Heavy channel regulation by 16 weirs, some in-channel and bankside habitat modified by dredging and hard bank protection and some urban reaches. Moderate habitat diversity. Some areas with multiple channels with some natural pool riffle character. Flow regulation by large abstractions at Farmoor Reservoir and Didcot power station.
4	Confluence of River Kennet	Teddington Weir	87	Heavy channel regulation by 20 weirs. Riverine and bankside habitat heavily modified by dredging, straightening and reaches with low habitat diversity; aquatic vegetation is rare. Increased flow due to significant input from River Kennet. Large surface water abstractions in the lower section.
5	Teddington Weir	London Bridge	23	Tidal, much less level regulation. Extensive urban encroachment into foreshore. Low flow regime determined by abstraction in lower Zone 4. Sporadic and significant storm sewage input.

characteristics and the resulting fish community. One notable feature is the presence of Atlantic salmon in Zone 4 due to an active restoration programme and provision of fish passage facilities in this section.

Zone 5, the freshwater tidal zone, is similar to Zone 4, but notable differences include the presence of the anadromous smelt, and one record of the rare migrant Allis shad, *Alosa alosa* (L.).

5.3.2 *Abundance*

Gaining information on fish abundance on a large scale along a river such as the Thames remains a considerable challenge. For such large-scale assessments in this

Table 5.2 Native riverine freshwater and diadromous fish species and their status in the River Thames

Species	Common name	Presence/absence				
		Zone 1	Zone 2	Zone 3	Zone 4	Zone 5
PETROMYZONIDAE						
Petromyzon marinus L.	Sea lamprey	×	×	×	×	✓
Lampetra fluviatilis (L.)	River lamprey			Absent		
Lampetra planeri (Bloch)	Brook lamprey	✓	×	×	×	×
ACIPENSERIDAE						
Acipenser sturio L.	Sturgeon			Absent		
CLUPEIDAE						
Alosa alosa (L.)	Allis shad		One record only in Zone 5.			
Alosa fallax (Lacépède)	Twaite shad		Absent, but present downstream of Zone 5.			
SALMONIDAE						
Salmo salar[1] L.	Atlantic salmon	×	×	✓[1]	✓[1]	✓[1]
Salmo trutta trutta L.	Sea trout	×	×	×	✓	✓
Salmo trutta fario L.	Brown trout	✓	✓	✓	✓	✓
COREGONIDAE						
Coregonus autumnalis (Pallas)	Houting			Absent		
THYMALLIDAE						
Thymallus thymallus (L.)	Grayling	✓	×	×	×	×
OSMERIDAE						
Osmerus eperlanus (L.)	Smelt	×	×	×	×	✓
ESOCIDAE						
Esox lucius L.	Pike	✓	✓	✓	✓	✓
CYPRINIDAE						
Rutilus rutilus (L.)	Roach	✓	✓	✓	✓	✓
Leuciscus leuciscus (L.)	Dace	✓	✓	✓	✓	✓
Leuciscus cephalus (L.)	Chub	✓	✓	✓	✓	✓
Phoxinus phoxinus (L.)	Minnow	✓	✓	✓	✓	✓
Scardinius erythrophthalmus (L.)	Rudd	×	×	✓	✓	✓
Tinca tinca (L.)	Tench	✓	✓	✓	✓	✓
Gobio gobio (L.)	Gudgeon	✓	✓	✓	✓	✓
Barbus barbus (L.)	Barbel	✓	✓	✓	✓	✓
Alburnus alburnus (L.)	Bleak	✓	✓	✓	✓	✓
Blicca bjoerkna (L.)	Silver bream	×	✓	✓	✓	✓
Abramis brama (L.)	Common bream	✓	✓	✓	✓	✓
Carassius carassius (L.)	Crucian carp	×	×	×	×	✓
Cyprinus carpio L.	Common carp	✓	✓	✓	✓	✓
COBITIDAE						
Barbatula barbatulus (L.)	Stoneloach	✓	✓	✓	✓	✓
Cobitis taenia L.	Spined loach			Absent		

Table 5.2   continued

| Species | Common name | Presence/absence | | | | |
		Zone 1	Zone 2	Zone 3	Zone 4	Zone 5
ANGUILLIDAE						
Anguilla anguilla (L.)	Eel	✓	✓	✓	✓	✓
GADIDAE						
Lota lota (L.)	Burbot			Absent		
GASTEROSTEIDAE						
Gasterosteus aculeatus L.	Three spined stickleback	✓	✓	✓	✓	✓
Pungitius pungitius (L.)	Ten spined stickleback	✓	✓	✓	✓	✓
PERCIDAE						
Perca fluviatilis L.	Perch	✓	✓	✓	✓	✓
Gymnocephalus cernuus (L.)	Ruffe	×	✓	✓	✓	✓
COTTIDAE						
Cottus gobio L.	Bullhead	✓	✓	✓	✓	✓

[1]Active restoration programme in place.
✓ = Present. × = Absent.

river type, hydroacoustics is the most cost-effective tool available. Figure 5.2 (after Hughes 1998) presents information collected in 1995 and 1996 on the Thames and is the first insight into spatial abundance for most of the non-tidal river. Fish densities (15–230 individuals 1000 m^{-3}) for the Thames are generally greater than in other lowland rivers such as the River Ouse in Yorkshire (A. Duncan, personal communication), and the River Trent (Lyons 1998).

The pattern of fish densities (Fig. 5.2) shows higher values in Zone 2, associated with a natural, sinuous river channel habitat type, with well-developed and diverse instream and bankside physical habitat and vegetation. In the zones further downstream, where fish abundance was lower, the river passes through increasingly urban and developed areas and has extensive regions with engineered bank protection unfavourable for spawning and survival of juvenile fish. Results within individual reaches offer evidence for patchiness in continuous longitudinal distributions of fish targets, some of which can be attributed to river features or particular events (Duncan & Kubecka 1996).

5.3.3   *Status of diadromous fish*

In the past, the River Thames supported healthy populations of Atlantic salmon, smelt and eels with associated fisheries. Stocks declined in the early 1800s, mainly because of deteriorating water quality in the tidal river due to the rapid expansion of the City of London.

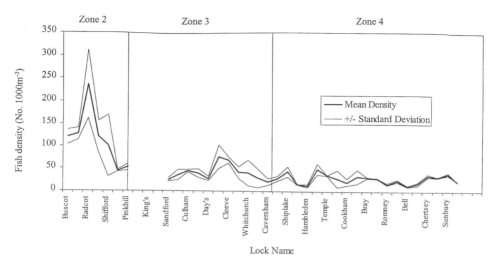

Figure 5.2 Hydroacoustic fish density in three zones of the River Thames (after Hughes 1998)

Following substantial improvements to the water quality of the tidal river, a programme to restore Atlantic salmon commenced in 1979 (Alabaster & Gough 1986). Its aim is to restore a self-sustaining run with an annual target of 1000 adults. It is primarily a fisheries conservation initiative that has associated benefits of the role of salmon as a biological indicator of river quality.

The programme is based on introducing juvenile salmon and has sustained a run of adults since 1982. A total of over 2500 adult returns, monitored by two fish traps on lower non-tidal weirs, has been confirmed; the average annual return is 160, although results of mark, release, recapture studies (Alabaster & Gough 1986) indicate that the total estimated run is 50–100% higher.

An important component of the programme is the construction of a combination of Denil, Larinier and pool/traverse fish passes at 20 low head weirs upstream of the tidal limit at Teddington. These allow access to spawning and nursery habitat in key tributaries of the middle river, particularly the River Kennet; a chalk stream tributary that offers the best juvenile production potential. A fish trap on the lower River Kennet has recorded adult returns to this tributary but no productive recruitment has been confirmed. The fish passes were constructed with funding from a successful collaboration with an independent charity – the Thames Salmon Trust (Willis & Armstrong 1995).

The current phase is near completion and is to be reviewed in the year 2000, but interim results suggest that the longer-term objective of a self-sustaining run remains very ambitious. Remaining constraints relate to water quality and flow management in the lower river; poor fish pass performance and weak juvenile production in selected tributaries. Further progress will depend on the outcome of the continuing technical assessment and the continued success of the Thames Salmon Trust in providing supporting funds.

Recent reviews of the status of eels in the catchment (Naismith & Knights 1997; Naismith 1998) found that there is a widespread population of eels in the tidal river.

Upstream of Teddington, eel density and the proportion of sites containing eels decreases markedly. The present status is one of partial and restricted recolonisation due to low natural recruitment and barriers to upstream migration. Current management policy is to improve upstream fish passage for elvers and small eels to increase recruitment to the freshwater sector of the river; some eel passes are already in place on weirs in the lower river.

Smelt has also made a natural recovery, with a good population migrating into Zone 5 to spawn during the spring. Current management includes monitoring of the smelt population, particularly with respect to consideration of its potential use as an environmental indicator of water quality.

5.4   Fisheries status

Fishing on the main river Thames is almost exclusively a recreational fishery for coarse fish: very few fish are removed and anglers return their catch to the river. There are no major commercial or other fisheries in the freshwater sector. The river is a very important recreational resource for both organised matches and casual pleasure anglers, with over 200 angling associations who lease fishing rights (McKeever 1995). It is a venue for national angling competitions and it is known that anglers travel considerable distances to fish the river.

A socio-economic review of angling showed that the Thames catchment contains 21% of all anglers in England and Wales with 9% of households containing at least one angler. This study estimated a figure of 2 300 000 anglers in England and Wales (Moon & Souter 1995), 91% of whom are coarse anglers; this represents over 430 000 coarse anglers within the Thames catchment, the highest number in a single catchment in England and Wales.

Approximately 36% of coarse anglers nationally indicated a preference for fishing in rivers; they travel a mean distance of over 30 km per fishing trip and spend around £1200 a year on angling (Moon & Souter 1995). Based on these figures, about 158 000 anglers with a preference for river fishing live within the Thames catchment, spending approximately £190 million a year on coarse fishing.

Studies by the Environment Agency based on results from monthly creel surveys on the Thames in two fishing seasons (1993 and 1994) indicate a mean density of around four anglers per km of available river bank at more popular times, such as summer week-ends. With respect to angling success, mean annual catch per unit effort was between 92 and 411 g rod hour^{-1}, in the higher range of results from similar rivers in England (Axford 1991).

Angling effort and equipment is regulated by the Salmon and Freshwater Fisheries Act 1975 and local byelaws. Fishing effort is controlled and promoted by rod fishing licence; up to four rods can be used at any time, and strict byelaws regulate their use. Nets, traps and other methods are not allowed except with the consent of the Environment Agency. A close season on the river between 15 March and 15 June aims to protect the fish community during peak spawning times. In addition, many angling clubs enforce their own rules, for example restricting the use of barbed hooks, keepnets and particular types of bait.

5.5   Pressures and management

5.5.1   *Water resources*

Approximately 5000 ML are abstracted from the river and groundwater sources daily. The majority is for public supply (80%), most of this (60%) coming from major surface water abstractions from lower Zone 4. Water resources are supported by significant re-use of water, taking advantage of highly treated effluent from treatment works and the natural purification capacity of the river.

Water abstraction introduces a range of impacts on the fisheries resource, and the recent series of drought years from 1989 has heightened concern over them. The reduction of flows produced by abstraction may represent a significant threat to flow sensitive life stages of the resident coarse fish population (e.g. rheophilic spawning habitat). In contrast, there is evidence that recruitment to limnophilic species may be enhanced by river conditions associated with reduced flow (e.g. Mills & Mann 1985). The potential impact on the objectives of the Atlantic salmon restoration programme is also significant. A critical issue is environmental quality of the tidal river, which affects downstream passage of smolts and upstream passage of adults (Alabaster & Gough 1986; Alabaster *et al.* 1991).

Current management determines a residual target flow at head of tide as a function of the time of year and the amount of water stored in the reservoirs which supply London (Sexton 1988). The normal residual target flow is 9.26 m s^{-3}, and recent drought years have demonstrated that the protection of target flow, particularly while adult salmon are returning (April–October) is unfeasible. The operating agreement allows river flows to be reduced to as low as 2.32 m s^{-3} if reservoir storage is low, representing an abstraction of up to 80% of average summer flows.

Alabaster (1990), and Alabaster, Gough and Brooker (1991) found that adult salmon migration in the Thames is positively related to flow and implies that the current operating agreement may not support the effective passage of salmon through the tidal river. The need to balance abstraction with environmental objectives has been highlighted as an important issue in the Local Environment Agency Plan (LEAP) for the Thames Tideway (Environment Agency 1997).

Entrapment of resident fish and migrating salmon smolts at surface water intakes feeding storage reservoirs can be very large. Studies undertaken in the 1980s by one of the Environment Agency's predecessor organisations demonstrated that a high proportion of salmon smolts may be lost in years with low spring/early summer flows and high abstraction rates. The work also demonstrated that juvenile coarse fish were entrapped, but was not clear what the subsequent impacts would be at population level. Migrating salmonids and coarse fish now receive protection from entrapment by physical and behavioural screens deployed at major abstraction intakes.

Between 1960 and 1988, water demand increased annually by approximately 2.0% per year. This trend was temporarily reduced in the late 1980s and early 1990s by economic decline and leakage reduction by water companies. Current demand is close to a high growth forecast and the Environment Agency's Thames Region

Water Resources Strategy indicates that resource deficits will occur around 2010–2015. Early planning of new strategic water resource schemes is therefore necessary.

Strategic options to meet the increase in demand include:

(1) a new pumped storage reservoir with a capacity of approximately 150 million cubic metres in Oxfordshire to supply the upper catchment and augment river flow for major abstraction sites downstream in London;
(2) an inter-catchment transfer from the lower River Severn to the upper River Thames to augment flow for downstream abstractions, primarily in London.

Continuing studies by the Environment Agency and the water company commenced in the 1990s to assess the viability and potential impacts of these developments. Both have common aims and require additional river regulation. The development of an ecologically acceptable operating regime is a challenging task in such a complex system where the ability to predict and mitigate impacts is still developing.

Impacts identified by studies to date include changes in water quality and water temperature, impacts on trophic interactions and physical habitat caused by changes in regulation. The developments could provide advantages such as use of the resource to gain environmental benefits at locations where flow dependent habitats may already be affected by existing abstractions.

5.5.2   *Water quality*

Appropriate water quality is essential to support resident fish populations and enable migratory species to move freely. The river is designated as a cyprinid fishery under the European Community Freshwater Fisheries Directive (79/659/EEC) for 230 km upstream of Teddington Weir (Zone 4, 3, 2 and part of Zone 1). The freshwater sector of the river receives a considerable pollution load with 67 direct and indirect discharges from sewage treatment works with populations of over 10 000, and many smaller sources.

Large sewage treatment works occasionally cause problems, particularly in the summer months when river flows are low and temperatures high. Localised problems, including fish mortalities, have been experienced in the recent past. The majority of the river from Zone 1 to Zone 4 is either proposed or designated as eutrophic under the Urban Waste Water Treatment Directive (91/271/EEC). Mean nitrate concentrations are in excess of the world averages for flowing rivers, and was the only nutrient measure related to phytoplankton distribution in the Thames (Ruse & Love 1997).

The Environment Agency sets water quality objectives (WQO) as targets based on the uses of watercourses. The standards defining the River Ecosystem classes (RE1–5) address the chemical quality requirements of fish communities (Department of the Environment 1994). A large proportion of the River Thames is classified as RE2: 'water of good quality suitable for all fish species'. Other sections are classified as RE3: 'water of fair quality suitable for high class coarse fish populations'. The whole of the non-tidal river is of sufficient quality to support high quality coarse fish populations and the passage of migratory fish.

The tidal freshwater stretch from Teddington Weir to London Bridge, is not subject to the WQO scheme, and estuary specific standards are set which recognise that low dissolved oxygen levels will occur and do not guarantee protection during critical summer periods. This is particularly relevant to the current aim of restoring a salmon population as the passage of migratory fish cannot be guaranteed. Combined storm sewage discharges into the tidal freshwater stretch following heavy rain over London can have a major impact. Current management initiatives include the use of two vessels which inject oxygen into the river (the 'Thames Bubblers') and the dosing of storm overflows with hydrogen peroxide.

5.5.3 *Navigation and flood defence*

The principal type of navigation on the Thames is pleasure craft (McKeever 1995). Boat traffic declined from 1978 to 1993, although a slight rise is expected in future with more boats visiting from other waterways. In 1996, 28 644 vessels were registered for use on the freshwater sector of the Thames; about 50% of these were powered craft. In the same year 775 471 lock passages were recorded (E. McKeever personal communication).

Habitat modification associated with river flow and level regulation for navigation is a major factor in determining the character and quality of the fish community. The management issues associated with operating a navigation also have a modifying impact on fish communities (Pinder 1997). Considerable lengths of sheet piled bank erosion protection, frequent dredging of gravel shoals and blocking off anastomosed channels all contribute to an overall loss of habitat diversity.

Industrial and residential development in the Thames valley has encroached onto the flood plain in many places. As a result of the need to protect homes and industry from flooding, a number of management activities take place. These involve activities that increase channel capacity and conveyance such as dredging, trimming bankside trees and vegetation, and building bank protection against scour and erosion, all of which can have a marked impact on critical habitats for key life stages of coarse fish. Other workers have stressed the importance of these critical habitats (Pilcher & Copp 1997) and the need for their protection (Pinder 1997).

The Environment Agency promotes reinstatement for more natural and diverse margin and bankside habitats. Old meander loops, backwaters, distributaries and other off-stream habitats are being more actively restored, created and managed for the benefit of the wider aquatic environment.

By influencing the design of schemes such as weir and lock replacement, managers have been able to incorporate solutions such as purpose-built spawning channels for rheophilic species. Appropriate development of off-river marinas has been demonstrated to benefit many components of the fish communities found in large lowland rivers (Copp 1997). On major new schemes such as the Maidenhead, Eton and Windsor Flood Alleviation Scheme, new channels and flood plains are being designed to meet the needs of both flood defence and the environment.

More recently, there is evidence of a positive approach to the management of river control structures to benefit the wider environment. A recent project to assess fish communities in the many anastomosed channels around Oxford demonstrated their value as spawning habitat for rheophilic fish. The project seeks to manage flow allocation to protect flow-dependent spawning habitat, whilst continuing to fulfil the needs of flood protection and the navigation.

5.6   Prospects for the twenty-first century

Future management of the fish community of the River Thames will continue to be coordinated by the Environment Agency but will depend increasingly on partnership with the angling community and other stakeholders. A strategic plan to implement these four themes successfully has the potential to protect and increase the ecological and recreational value of the fish community of the River Thames.

(1) *Protection of existing resource*: the fundamental role of future managers is to protect the existing fisheries resource and the pre-requisites known to be important for the balanced development of fish communities. These include water quality and quantity, physical habitat and responsible use of the resource.

(2) *Integrated strategy*: current management of the river is undertaken in a fragmented manner. Effective management of the fish community will increasingly require integration on both a geographical and disciplinary basis. Issues such as restoration of diadromous fish populations and proposed strategic water resource developments require holistic river basin management to provide effective solutions. For such a strategy to work it is essential that scientists, engineers and decision makers build effective relationships based on mutual understanding of the whole aquatic resource and all its needs and sensitivities.

(3) *Research*: managers must continue to build on knowledge of the fish community and the wider aquatic ecosystem. New challenges continue to emerge, such as recent evidence of the impact of endocrine disrupters on coarse fish populations (Jobling *et al.* 1998). It is recognised that we have much to learn in the management of large lowland rivers and some fundamental integrated research is still required to develop management capabilities. The importance of learning from the opportunities provided by applied river management initiatives is fundamental to continue progress in this area.

(4) *Promoting use of the resource*: anglers are key stakeholders in the aquatic environment and contribute significant resources towards its management. Future management options must address the needs and aspirations of these groups whilst encompassing wider conservation issues.

(5) *Opportunity to enhance*: returning the river to a more diverse habitat of increased ecological value is not necessarily at odds with the pressures on it. A range of options exist for the restoration of riverine habitats that meet both environ-

mental and other needs, many of which have been successfully applied to other rivers (Cowx & Welcomme 1998; Williams *et al.* 1997).

Based on current knowledge, some anticipated outcomes in the twenty-first century are outlined below.

(1) Fish species composition may remain stable, with no anticipated losses, but possible breeding additions by natural dispersion *via* inter-catchment canals may occur, especially the introduction of zander.

(2) There may be evidence of successful salmon recruitment in key tributaries, but aspirations for the restoration of a self-sustaining run may remain highly ambitious. Multi-species fish passage facilities should be included in all river control structures as they are refurbished, supporting a slow recovery upstream of eel population.

(3) Uniform good quality water capable of supporting high quality coarse fish populations will prevail. Considerable long-term investment to enable sustainable solutions for tidal storm sewage overflows may occur, dependant on the future economic environment.

(4) There will be further strategic water resource regulation of the middle and lower river, that may influence the fish community towards more rheophilic species, but without changing species composition. The quest for higher residual flow to the estuary remains ambitious, given the cost of supporting an increased target flow and demand for water for consumption.

(5) The dominant effect of the impounded navigation will remain but individual river habitat restoration schemes, based on methods currently used, will link together to represent significant change towards more natural in-channel and river corridor habitats. Sections of the river with low habitat diversity may develop into a mosaic of more diverse and ecologically valuable habitats.

(6) There will be an increase in the demand for recreational angling in line with a general increase in leisure time available to individuals. The river will remain an important and popular coarse fishery.

Acknowledgements

We gratefully acknowledge the assistance of our colleagues in the Environment Agency who have contributed data and information to this work. Dr R.H.K. Mann, MBE and Mr W. Yeomans made valuable comments on an early draft of the chapter. The views expressed in this chapter are entirely those of the authors.

References

Alabaster J.S. (1990) *Water Quality and Freshwater Flow in the Thames Tideway in Relation to Salmon Migration, with Particular Reference to the Effect of Low Flow in 1990.* Reading: National Rivers Authority, 32 pp.

Alabaster J.S. & Gough P.J. (1986) The dissolved oxygen and temperature requirements of Atlantic Salmon, *Salmo salar* L., in the Thames estuary. *Journal of Fish Biology* **29**, 613–621.

Alabaster J.S., Gough P.J. & Brooker W.J. (1991) The environmental requirements of Atlantic salmon, *Salmo salar* (L.), during their passage through the Thames Estuary, 1982–1989. *Journal of Fish Biology* **38**, 741–762.

Axford S. (1991) Some factors affecting angling catches in Yorkshire rivers. In I.G. Cowx (ed.) *Catch Effort Sampling Strategies: Their Application in Freshwater Fisheries Management.* Oxford: Fishing News Books, Blackwell Science, pp. 14–153.

Berrie A.D. (1972) Productivity of the River Thames at Reading. In R.W. Edwards & D.J. Garrod (eds) Conservation and productivity of natural waters. *Symposium of the Zoological Society of London* **29**, 69–86.

Copp G.H. (1997) Importance of marinas and off-channel water bodies as refuges for young fishes in a regulated lowland river. *Regulated Rivers: Research and Management* **13**, 303–307.

Cowx I.G. & Welcomme R.L. (eds) (1998) *Rehabilitation of Rivers for Fish.* Oxford: Fishing News Books, Blackwell Science, 260 pp.

Cowx I.G., Pitts C.S., Smith K.L., Hayward P.J. & van Breukelen S.W.F. (1995) Factors influencing coarse fish populations in rivers, *R&D Note 460.* Bristol: National River Authority, 125 pp.

Department of the Environment (1994) *The Surface Waters (River Ecosystem) (Classification) Regulations.* London: HMSO, 24 pp.

Duncan A. & Kubecka J. (1993) *Hydroacoustic Methods of Fish Surveys, R&D Note 196.* Bristol: National Rivers Authority, 163 pp.

Duncan A. & Kubecka J. (1994) *Hydroacoustic Methods of Fish Surveys, A Field Manual, R&D Note 329.* Bristol: National Rivers Authority, 52 pp.

Duncan A. & Kubecka J. (1996) Patchiness of longitudinal fish distributions in a river as revealed by continuous hydroacoustic survey. *ICES Journal of Marine Science* **53**, 161–165.

Environment Agency (1997) *Thames Tideway (Teddington to Rower Bridge) LEAP Consultation Report.* Frimley: Environment Agency, 137 pp.

Francis J.M. & Urwin A.C.B. (1991) Francis Francis, 1822–1866. Angling and fish culture in the Twickenham, Teddington and Hampton reaches of the River Thames. *Borough of Twickenham Local History Society, Paper Number* **65**, 27 pp.

Harvey J. & Cowx I.G. (1996) Electric fishing for the assessment of fish stocks in large rivers. In I.G. Cowx (ed.) *Stock Assessment in Inland Fisheries.* Oxford: Fishing News Books, Blackwell Science, pp. 11–26.

Hickley P. (1996) Fish population survey methods: a synthesis. In I.G. Cowx (ed.) *Stock Assessment in Inland Fisheries.* Oxford: Fishing News Books, Blackwell Science, pp. 3–10.

Hughes S. (1998) A mobile horizontal hydroacoustic fisheries survey of the River Thames, United Kingdom. *Fisheries Research* **35**, 91–97.

Jackson D.C. (1996) Stock assessment considerations for riverine fisheries. In I.G. Cowx (ed.) *Stock Assessment in Inland Fisheries,* Oxford: Fishing News Books, Blackwell Science, pp. 407–408.

Jobling S., Tyler C., Nolan M. & Sumpter J.P. (1998) The identification of oestrogenic effects in wild fish, *R&D Technical Report W119.* Bristol: Environment Agency, 35 pp.

Littlewood I.G. & Marsh T.J. (1996) Re-assessment of the monthly naturalised flow record for the River Thames at Kingston since 1883, and the implications for relative severity of historical droughts. *Regulated Rivers: Research and Management* **12**, 13–26.

Lyons J. (1998) A hydroacoustic assessment of fish stocks in the River Trent, England. *Fisheries Research* **35**, 83–90.

Mackay I. & Mann K.H. (1969) Fecundity of two cyprinid fishes in the River Thames, Reading, England. *Journal Fisheries Research Board of Canada* **26**, 2795–2805.

Mann K.H. (1964) The pattern of energy flow in the fish and invertebrate fauna of the River Thames. *Verhandlungen der Internationalen Vereinigung für theoretische und angewandte Limnologie* **15**, 485–495.

Mann K.H. (1965) Energy transformations by a population of fish in the River Thames. *Journal of Animal Ecology* **34**, 253–275.

Mann K.H. (1972) Case history: the River Thames. In R.T. Ogelsby, C.A. Carlson & J.A. McCann (eds) *River Ecology and Man*. New York: Academic Press, pp. 215–232.

Mann R.H.K. (1989) The management problems and fisheries of three major British rivers: the Thames, Trent and Wye. In D.P. Dodge (ed.) *Proceedings of the International Large River Symposium. Canadian Special Publication on Fisheries and Aquatic Science* **106**, 444–454.

Mann K.H., Britton R.H., Kowalczewski A., Lack T.J., Mathews C.P. & McDonald I. (1972) Productivity and energy flow at all trophic levels in the River Thames, England. In Z. Kajak & A. Hillbricht-Ilkowska (eds) Productivity problems in freshwaters. *Proceedings of the IBP/ UNESCO Symposium, Kazimierz Dolny, Poland, 1970*, pp. 579–596.

Mann, K.H., Britton, R.H., Kowalczewski, A., Lack, T.J., Mathews, C.P. & McDonald, I. (1972) Productivity and energy flow at all trophic levels in the River Thames, England. In Z. Kajak & A. Hillbricht-Ilkowska (eds) Productivity problems in freshwaters. *Proceedings of the IBP/ UNESCO Symposium, Kazimierz Dolny, Poland, 1970*, pp. 579–596.

Mathews C.P. (1971) Contribution of young fish to total production of fish in the River Thames near Reading. *Journal of Fisheries Biology* **3**, 157–180.

Mckeever E. (1995) *A Recreational Strategy for the River Thames*. Reading: National Rivers Authority, Thames Region, 65 pp.

Mills, C.A. & Mann, R.H.K. (1985) Environmentally induced fluctuations in year class strength and their implications for management. *Journal of Fish Biology* **27** (Suppl. A): 209–226.

Moon N. & Souter G. (1995) Socio-economic review of angling 1994, *R&D Note 385*. Bristol: National Rivers Authority, 32 pp.

Naismith I.A. (1998) *Status of Eel in the Thames Catchment, EA Contract 11096*. Copies available from WRC Medmenham, Henley Road, Medmenham, Bucks, SL7 2HD, UK.

Naismith I. A. & Knights B. (1997) The distribution, density and growth of the eel (*Anguilla anguilla* L.) in the freshwater catchment of the River Thames. *Journal of Fish Biology* **42**, 217–226.

Pilcher M.W. & Copp G.H. (1997) Winter distribution an habitat use by fish in a regulated lowland river system of south-east England. *Fisheries Management and Ecology* **4**, 199–215.

Pinder L.C.V. (1997) Research on the Great Ouse: Overview and implications for management. *Regulated Rivers: Research and Management* **13**, 309–315.

Ruse L. & Love A. (1997) Predicting phytoplankton composition in the River Thames, England. *Regulated Rivers: Research and Management* **13**, 171–183.

Sexton J.R. (1988) Regulation of the River Thames a case study on the Teddington flow proposal. *Regulated Rivers: Research and Management* **2**, 323–333.

Vannote R.L., Minshall G.W., Cummins K.W., Sedell J.R. & Cushing C.E. (1980) The river continuum concept. *Canadian Journal of Fisheries and Aquatic Sciences* **37**, 130–137.

Walton I. (1653) *The Compleat Angler*. London: The Folio Society, 272 pp.

Ward R.C. (1981) River systems and river regimes. In R.I. Ferguson (ed.) *British Rivers*. London: Allen and Unwin, pp. 1–33.

Williams W.P. (1965) The population density of four species of freshwater fish, roach (*Rutilus rutilus* L.), bleak (*Alburnus alburnus* L.), dace (*Leuciscus leuciscus* L.) and perch (*Perca fluviatilis* L.) in the River Thames at Reading. *Journal of Animal Ecology* **34**, 173–185.

Williams W.P. (1967) The growth and mortality of four species of fish in the River Thames at Reading. *Journal of Animal Ecology* **36**, 695–720.

Williams J.E., Wood C.A. & Dombeck M.P. (eds) (1997) *Watershed Restoration: Principles and Practices*. Bethesda, MA: American Fisheries Society, 561 pp.

Willis D.J. & Armstrong G.S. (1995) Collaborative promotional strategies; Fisheries trusts – a case study. *Proceedings of the Institute of Fisheries Management 26th Annual Study Course*. Nottingham: Institute of Fisheries Management, pp. 67–76.

Winfield I.J., Fletcher J.M. & Cragg-Hine D. (1994) *Status of Rare Fish, Vol. 2, R&D Project Record 249/8/NW*. Bristol: National Rivers Authority, 108 pp.

Chapter 6
Cyclic behaviour of potamodromous fish in large rivers

R. QUIROS and J.C. VIDAL

Facultad de Agronomía, Universidad de Buenos Aires, Av. San Martín 4453, CF 1417, Buenos Aires, Argentina

Abstract

Potamodromous fish species have evolved under natural river conditions, and have usually keyed migrations and spawning behaviour to coincide with high-water events. Knowledge of riverine fish populations has increased significantly over the past decade, but advancement in the study of potamodromy is lagging behind. However, potamodromy is strongly represented in large South American rivers. Over the past 30 years the upper Parana basin has been highly regulated, and damming has altered flow regimes, especially between 1985 and 1990. To study the effects of river regulation on fish movements, a site situated in the main channel of the lower middle Parana River was studied for fish abundance twice a week during that period. The fish species have changed their abundance cyclically in the main channel. Moreover, the potamodromous fishes maintained their seasonal position in the main channel, despite huge flow regulation by upper basin dams. Most fish species maintained seasonal cycles similar to those evident in the pre-regulation period. Potamodromous fishes have retained the migration patterns evolved in the pristine riverine system. Consequently, it may be that factors other than water level and temperature trigger upstream migration of potamodromous fish in large river–floodplain systems.

6.1 Introduction

Because the best breeding habitat rarely coincides with the best feeding habitat, most fish species inhabiting large river–floodplain systems have two distinct centres of concentration and fish migrate between the two (Welcomme 1985). Longitudinal migrations take place within the main river channel, sometimes over long distances. Potamodromy is strongly represented in large river–floodplain systems from the warm temperate zone and the tropics. In South America, a number of medium-sized river–floodplain systems seem to have migratory fish populations showing similar characteristics, with a single seasonal movement to and from a downstream feeding zone and an upstream breeding zone (e.g. Bayley 1973; Godoy 1975; Paiva & Bastos 1982). This pattern is presumably repeated throughout most moderate size river–floodplain systems of the continent which carry characin and siluroid populations (Welcomme 1985). However, for the middle Parana river the migration patterns obtained using tagging methods are more complex (see Espinach Ros & Delfino (1993) for an overview).

Oscillating cycles of abundance between spawning and feeding areas have been shown for potamodromous fishes in the lower Rio de la Plata basin (Bonetto & Pignalberi 1964; Bonetto *et al.* 1971). Because of the close relationship between water level fluctuations and rainfall, cyclic fish behaviour in seasonal comparisons was assigned to water level variations in the Parana River (Quiros & Cuch 1989). However, river regulation through damming in the headwaters has changed the flow regime, especially for the period 1985–1990 (Fig. 6.1) and provided the opportunity to test the effects of flooding as an onset for fish upstream movements.

Potamodromous fish species have coevolved with the system under natural river conditions and usually keyed their movements to coincide their spawning with high-water events. Some examples from natural undeveloped river–floodplain systems support this statement (Welcomme 1985). But what happened when the flood pulse changed its seasonal variability by human intervention, assuming all the other characteristics remained constant? To respond to this question would help to clarify the main determinants of upstream potamodromous fish movements. Therefore, the main objective of this chapter was to study patterns in riverine fish abundance within and between years in a regulated, but highly variable, river–floodplain system, and to compare cyclic fish patterns for years with moderate to unusually high discharge manipulations. The aim was to test the hypothesis that potamodromous fish move into the main channel for spawning or feeding according to the river discharge,

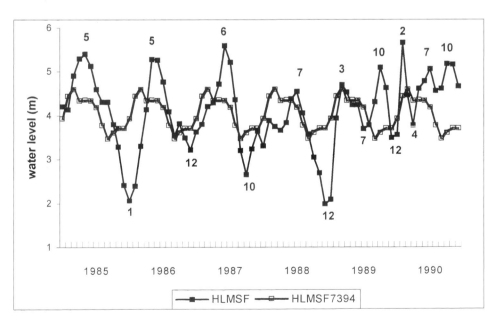

Figure 6.1 Water level: monthly mean of daily averages for the period 1985–1990 (HLSFM) and monthly for the 1973–1994 post-dam period (HLMSF7394). Figure numbers indicate month of the year

independent of the cyclic patterns they have inherited through coevolution with the unaltered environment.

6.2   Description of the system and the main channel fishery

The Parana River flows 4000 km southwards from its sources in the Brazilian Shield to its mouth in the Pampa Plain, discharging 20 000 $m^3 s^{-1}$ in the Rio de la Plata river. From its confluence with the Paraguay River, the Parana River has extensive floodplains which widen downstream and cover more than 20 000 km^2. These floodplains reduce the variability and distribute the flow evenly throughout the year, suppressing sharp and pronounced flood peaks.

Before most of the upstream dams were built, the lower middle Parana River showed a regular annual cycle, usually reaching its maximum in March or April (late summer–early autumn) and its minimum flow in September. The volume of water in reservoirs in the upper basin has been increasing since the early 1950s but a sharp increase occurred from 1972–1973 up to the present. It represents more than 50% of the mean annual discharge at the mouth (Rio de la Plata) (Quiros 1990). Damming has contributed to the disappearance of large migratory fish species, mainly in the dammed upper reaches of the Parana River (Agostinho *et al.* 1994).

The flood pulse now tends to occur earlier but is more attenuated than in the unregulated river. Moreover, water remains in the floodplain for most of the year (Quiros 1990). The upper dams do not stop floods; the attenuation occurs because the water management policies of hydroelectric companies retain the water in reservoirs during high rainfall and river discharge periods and release it during low flow conditions.

However, for the 1985–1990 period, upper basin dams situated in Brazil induced huge changes in discharge, seemingly during turbine and spillway probing (Fig. 6.1). Flow variability was distinct and dramatic, involving changes of several thousands of cumecs during the years 1989 and 1990.

For the lower Parana fishery, active fishing methods are usually used, although some static gear is also used within the floodplain during high water periods. The lower middle Parana fishery takes place in the main channel all year round using drift gillnets, with panels with stretched mesh sizes ranging from 26 cm to over 34 cm. The nets are set on the bottom of the channel facing downstream and positioned in a manner considered to maximise catch efficiency. The current drags the net on the bottom in areas especially prepared for fishing (canchas). This fishing gear usually catches large fish more than four years old (Quiros & Cuch 1989).

6.3   Materials and methods

To study fish movements, changes of fish abundance over time was assessed. The fishing site was located in the main channel of the lower middle Parana River

(Rosario City). The draft gill nets catches were sampled twice a week during 1985–1990. On each sample date, the catch of 6–7 fishermen and 10–30 hauls was sampled at the landing site nearby. All fish were identified to species (Ringuelet, Aramburu & Alonso de Aramburu 1967) and individually measured for total length (mm) and weight (g). *Prochilodus lineatus* (Holmberg) (previously *platensis*) (sabalo 41.0%), *Leporinus obtusidens* (Val.) (boga 34.3%), *Pterodoras granulosus* (Bleeker) (armado 12.2%), *Pseudoplatystoma coruscans* (Eigenmann & Eigenmann) (surubi 5.2%), *Luciopimelodus pati* (Eigenmann & Eigenmann) (pati 4.8%), and *Salminus maxillosus* (Val.) (dorado 2.4%) were the principal species captured and contributed (by weight) the most to the total catch. Catch per unit effort (CPUE) was expressed as kg per net-haul for total catch and by species. Monthly mean CPUE for the large siluroids (surubi, pati, armando) and characins (sabalo, boga, dorado) are presented here.

Variability in fish abundance was large between years. For the Parana River, as for other river–floodplain systems (Welcomme 1985), fish species abundance depends on the flood conditions, flooding intensity and amount of water remaining in the system during the low flow period in the years when the fishes are born (Quiros & Cuch 1989). To analyse the relationship between the seasonal fish abundance by species and water level, the former variable was normalised, dividing it by the annual CPUE for the species. Consequently, mean monthly normalised fish abundance by species is expressed as the percentage of the annual catch per unit effort for the species.

Water level data for Santa Fe Harbor station were collected from the Nacional de Construcciones Portuarias y Vias Navegables. The monthly mean of daily averages was considered as the water level variable (HLSF). To analyse relationships between CPUE and flood regime for pre-dam (1925–1972) and post-dam (1973–1994) periods, the monthly mean water level data are presented as HLMSF25–72 and HLMSF73–94, respectively. Water temperature data for the main channel at Rosario City (TEMP) were obtained for 1985 through 1990 from the Instituto Nacional de Aqua y del Ambiente.

6.4   Results

The analyses of the fish abundance series for the 1985–1990 period indicated regular cyclic changes in abundance of fish in the main channel for most of the species studied (Figs 6.2 and 6.3). The large potamodromous fishes appear to retain seasonal movements in the main channel despite large variations in the intensity and timing of water discharges by the dams situated in the upper basin.

Normalised fish abundance by species in the main channel was slightly related to water level for the studied period. The abundance values of only two species (*Leporinus* and *Luciopimelodus*) were positively related to water level ($P < 0.01$; $n = 72$). Similar results were obtained for fish abundance in each year studied (Table 6.1). For both the pre-dam and the post-dam periods, potamodromous fishes were in the

Table 6.1 Seasonal relationships between fish abundance by species in the main channel and water temperature (climate), the mean monthly water level for the 1925–1971 period (pre-dam flood pulse), and the mean monthly water level for the studied period (flood pulse for the 1985–1990 period)

Conceptual variable	Fish species					
	Prochilodus	Leporinum	Luciopimelodus	Salminus	Pterodoras	Pseudoplatystoma
Climate	Winter	Late autumn	Late autumn/early winter	Late summer/early autumn	Late summer/autumn	Summer/Early autumn
Pre-dam flood pulse	Falling	Falling	High falling	High	High	High
Flood pulse for the 1985–1990 period	High increasing, high, and high falling	High increasing, high, and high falling	High, high falling, and low	Low, low and high increasing, high and falling	Low increasing, high increasing, high, and low	Low ansd high increasing, falling, and low
Fish abundance in the main channel						
Spawning period[a, b]						
Peaking in	Increasing	Increasing?	High increasing??	Flood peak?	High increasing?	High increasing?
Peaking in	Spring–summer	Spring–summer??	Summer??	Summer?	Summer?	Summer?
Larvae drift[a,b,c]						
Peaking in	Increasing	Increasing	High increasing??	Flood peak?	High increasing	High increasing
Peaking in	Spring–summer	Spring–summer	Summer??	Summer?	Summer	Summer

[a]Oldani and Oliveros (1984). [b]Cordiviola de Yuan et al. (1984). [c]Fuentes (1998). [d]Fuentes & Espinach Ross (1998).

main channel most of the year (R. Quiros unpublished data), but for the post-dam period, fish abundance varied drastically during the flood cycle (Fig. 6.2). However, for the undeveloped system, the abundance of main-channel spawners increased substantially in the main channel during the time of maximum flood (*Pseudopla-tystoma, Pterodoras,* and *Salminus*) and during high falling waters (*Prochilodus, Leporinus,* and *Luciopimelodus*). From these species, the top predators *Salminus maxillosus* and *Pseudoplatystoma coruscans* have reportedly decreased in abundance in the lower middle Parana during the last twenty years. On the other hand, detritivorous and omnivorous fishes increased their relative abundance in the system (R. Quiros unpublished data).

The relationship between normalised fish abundance and water temperature describes the sequence of fish species abundance in the main channel (Tables 6.1 and 6.2). As expected, for the three species that were more abundant in the main channel during summers (*Salminus, Pterodoras,* and *Pseudoplatystoma*) fish abundance was positively related to water temperatures (n = 72, P <0.01). Conversely, for the two species that were more abundant in the main channel during autumn and winter (*Leporinus* and *Prochilodus*), abundances were inversely related to water tempera-tures (n = 72, P < 0.01). Similar results were obtained when fish abundances were analysed for each studied year, although statistical significance was lower (Table 6.2).

However, simple correlation between these time-related variables is merely a simplified view of more complex processes. Potamodromous fishes were in the main channel when the water was both increasing and falling (Figs 6.3–6.6), indicating that they followed seasonal patterns more than flood patterns (Tables 1 and 2). During the period 1988–1990, when more dramatic changes in water level timing were expressed, fish species retained their seasonal patterns. Moreover, for the 1988–1990 period, fish cycling patterns relative to water level variability were highly chaotic, while there were highly regular cycles related water temperature variation (Figs 6.3–6.6).

Different fish species peaked in the main channel at different times, however, there was little overlap in the timing of fish movements in the main channel (Fig. 6.2). This suggests that adult potamodromous fish species maintain an independent existence in the main channel, possibly moving in distinct groups or schools.

In previous studies, particularly for *Prochilodus lineatus,* the reproductive period was estimated using gonad maturity of adult fish (Bonetto 1963; Pignalberi 1965; Oldani & Oliveros 1984; Tablado, Oldani, Ulibarrie & Pignalberi de Hassan 1988). These studies showed that the reproductive period for *Prochilodus* ranges from October to April, but the months with more intense spawning activity are November and December. Larval abundance of *Prochilodus* and almost every other migratory species peaked when the water level of Parana River peaked. *Prochilodus* and *Leporinus* larvae abundance usually peak during November, while siluroids peak during February (Fuentes 1998; Fuentes & Espinach Ross 1998) (Table 6.2). It is apparent that fish larvae go into the floodplain ponds and sloughs when water rises. These areas also provide fish nursery and wintering areas when waters recede. It has

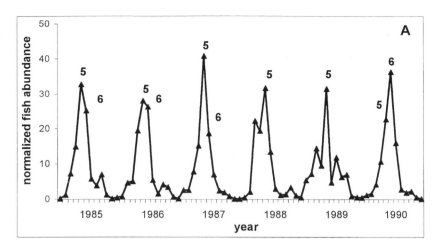

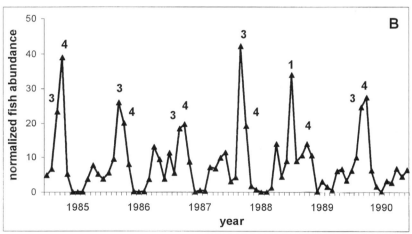

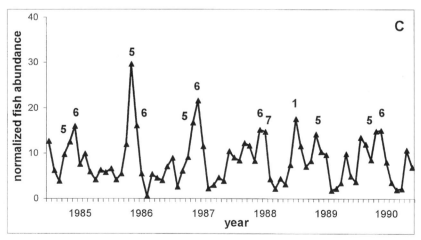

Figure 6.2 Fish abundance (normalised) for selected fish species during the period 1985–1990. A, *Leporinus obtusidens*; B, *Pterodoras granulosus*; C, *Luciopimelodus pati*.

Table 6.2 Cycling of fish abundance in the main channel and correlation coefficients[a] between fish abundance by species and mean monthly water level (HLSF) and water temperatures (T_{water}) for the studied years. Direction of cycling trajectory was included between brackets (see text).

Year	Fish species											
	Prochilodus		Leporinus		Luciopimelodus		Salminus		Pterodoras		Pseudoplatystoma	
	HLSF	T_{water}	HLSF	T_{water}	HLSF	T_{water}	HLSF	T_{water}	HLSF	T_{water}	HLSF	T_{water}
1985	0.03 (−)	−0.72 (−+)	0.74 (−)	−0.48 (−)	0.51 (+−)	−0.36 (−)	0.54 (+)	0.20 (−)	0.36 (+)	0.19 (−)	0.36 (+)	0.40 (−)
1986	0.79 (−)	−0.89 (−+)	0.73 (+)	−0.37 (−)	0.61 (+)	−0.19 (−)	−0.17 (+)	0.28 (−)	−0.23 (+)	0.36 (−)	−0.68 (+)	0.74 (−)
1987	0.13 (−)	−0.76 (+)	0.59 (+)	−0.32 (−)	0.73 (+)	−0.35 (−)	0.21 (+)	0.53 (−)	−0.33 (+)	0.68 (−)	−0.18 (+)	0.66 (−)
1988	−0.03 (−+)	−0.58 (−+)	0.31 (+)	−0.03 (−)	0.54 (+)	−0.15 (−)	0.11 (+−)	0.43 (−)	−0.14 (chaotic)	0.53 (−)	0.07 (+)	0.57 (−)
1989	−0.13 (chaotic)	0.06 (+−)	0.07 (chaotic)	−0.23 (−)	−0.50 (chaotic)	0.36 (−)	0.04 (chaotic)	0.70 (+−)	−0.59 (chaotic)	0.59 (−)	−0.21 (chaotic)	0.33 (−)
1990	0.20 (chaotic)	−0.76 (+)	−0.01 (chaotic)	−0.63 (−)	0.27 (chaotic)	0.03 (−)	0.01 (chaotic)	0.06 (−)	−0.39 (chaotic)	0.46 (−)	−0.25 (chaotic)	0.42 (−)

[a] $r = 0.71$ ($P<0.01$, $n=12$), $r = 0.58$ ($P<0.05$, $n=12$).

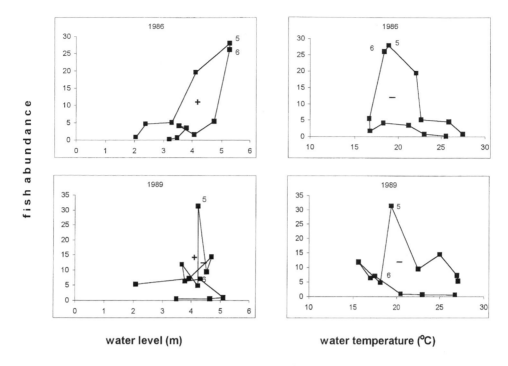

Figure 6.3 Cycling behaviour for *Leporinus obtusidens* in selected years. Figure numbers, month of the year; (+) and (−), positive and inverse direction cycling, respectively; (+ −), chaotic cycling

also been suggested that fish wait for a favourable second water level increase before spawning (Fuentes & Bonetto, unpublished data).

For the studied period, fish abundance was rarely related to the flood pulse. However, fish abundances for the majority of species studied were positively related to mean monthly water levels for both the pre-dam and post-dam periods ($P > 0.01$). *Prochilodus* abundance was an exception. It was inversely related to water level for the pre-dam period (R. Quiros unpublished data). During the years when more dramatic changes to flood timing occurred the fishes approximated their pre-dam seasonal patterns in the main channel. For the study period, the main difference between *Prochilodus* and the other migratory fish species was that *Prochilodus* was present in the main channel most of all year.

6.5 Discussion

The abundance of potamodromous fish species in the main channel shifted in a cyclical manner with season. Although the flood pulses in the 1985–1990 period were

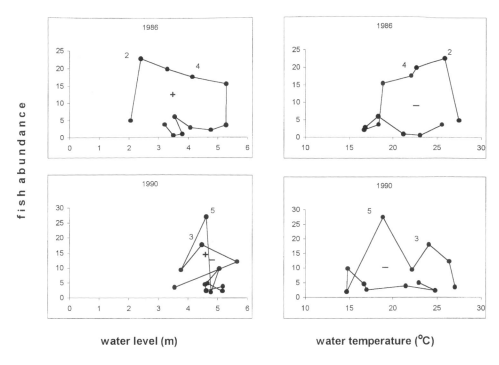

Figure 6.4 Cycling behaviour for *Salminus maxillosus* in selected years. Figure numbers, month of the year; (+) and (−), positive and inverse direction cycling, respectively; (+ −), chaotic cycling

suppressed, the peak in fish abundance in the main channel was during the winter similar to before regulation of the upper river. Despite changes in the frequency and intensity of floods, fish movement has also remained similar to that observed before regulation of the upper river. Among year comparisons suggest the fishes maintained their movements on a climate basis, but not on a flood-event basis. For all the potamodromous fish species studied, fish cyclical behaviour in the main channel was more consistent when compared to water temperature than when compared to water level variability. However, the flood pattern for 1985 and 1986 were different from the flood patterns for the pre-dam period (R. Quiros unpublished data). Those results support the conclusion that flooding does not appear to be the factor determining fish movements on a seasonal basis. Furthermore, for the periods 1941–1968 and 1985–1990, changes in the time of maximum abundance have not occurred. Therefore, potamodromous fishes have retained their original migration patterns.

For many years, individuals of the large fish species in the lower Rio de la Plata basin moved upstream more than 400 km from the lower reaches of the lower basin (Rio de la Plata, lower Uruguay River, and the Parana delta) to the middle Parana during late autumn and winter to spawn in the main channel during late spring and

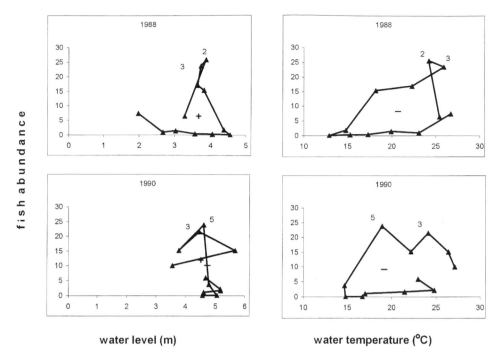

Figure 6.5 Cycling behaviour for *Pseudoplatystoma coruscans* in selected years. Figure numbers, month of the year; (+) and (−), positive and inverse direction cycling, respectively; (+ −), chaotic cycling

summer (Bonetto *et al.* 1971; Espinach Ros & Delfino 1993). Moreover, fish catches for both the Rio de la Plata and the middle Parana are inversely related (Quiros & Cuch 1989). However, other group of species appear to move from upstream areas to the middle Parana during the summers (Bonetto, Canon Veron & Roldan 1981; Oldani & Oliveros 1984). Little information exists about the migratory behaviour of individual species, but this study shows that some species, like *Prochilodus*, *Leporinus* and *Luciopimelodus* move into the lower middle Parana in winter, while other fish species such as *Pseudoplatystoma*, *Pterodoras*, and *Salminus*, move in summer. Most of those species spawn in the main channel during late spring and summer (Table 6.1). Both *Prochilodus* and *Leporinus*, and probably *Luciopimelodus*, exhibit time lags ranging from three to five months between upstream migration and spawning. Tablado *et al.* (1988) mentioned mature *Prochilodus* in a floodplain pond during late spring before spawning. For some migratory species, to rest in floodplain ponds after upstream migration and before spawning may be an explanation for the time lags between upstream movements and spawning in the main channel (C. Fuentes, personal communication). For example, *Prochilodus reticulatus* in the Sinú River move

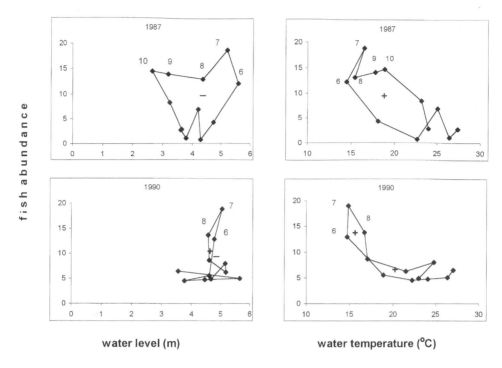

Figure 6.6 Cycling behaviour for *Prochilodus lineatus* in selected years. Figure numbers, month of the year; ($+$) and ($-$), positive and inverse direction cycling, respectively; ($+-$) chaotic cycling

into the upper watershed during their migration, rest for a period of time in the floodplain, and then move downstream to spawn (E. Theiss, personal communication).

Fish tagging (Bonetto *et al.* 1981; Sverlij, Espinach Ros & Ort 1993) suggests the existence of different fish stocks north and south of the site where the river–floodplain geomorphology changes abruptly (Quiros & Cuch 1989). This region is situated 150 km north from the study site. Therefore, homogeneous fish stocks for the studied site can be assumed, but different overlapping stocks cannot be dismissed.

Fish movement remains similar to that observed before regulation of the upper river although changes in the intensity and timing of the flood are apparent. Moreover, cyclic patterns for water temperature were usually more regular than for water level. For the lower middle Parana River, it was suggested that higher water levels trigger reproduction of principal migratory species (Bonetto *et al.* 1981; Tablado *et al.* 1988). However, results from fish larvae studies suggest that the maturation cycle may be controlled by factors such as water temperature and photoperiod (Fuentes 1998). This study does not answer the question of why potamodromous fishes migrate. However, it supports the hypothesis that time of migration for potamodromous fishes, and by analogy with diadromous fishes (Gross 1987), must have a genetic component. However, the influence of environmental factors cannot be rejected

It is not known what overall effects the changes in river flow regime will have on the fish populations. In other large systems for the pre-dam period most of riverine fish species anticipate flood conditions by spawning before or during the water rise (Bayley 1995). However, it is highly probable that changes in timing of discharge have produced a disconnection between spawning period and flood conditions. This may explain the shift of potamodromous fish assemblage composition in the regulated river when compared with the unregulated river (R. Quiros unpublished data). However, it is difficult to assign a causal relationship between river regulation and change in fish assemblage (Petts, Imhof, Manny, Maher & Weisberg 1989) since other factors related to development activities may be involved. For example, previous results have supported river water pollution as a primary negative factor on fish abundance for the lower basin (Quiros 1990).

The southern portion of the Parana River has a warm temperate thermal regime, and in the middle reaches, where the current study was conducted, water temperature varies by more than 15°C over the year (Drago 1984). However, the potamodromous fish species studied and related genera are also potamodromous in tropical systems, where annual water temperature usually ranges just a few degrees (Welcomme 1985). Therefore, water temperature is probably not an important factor in triggering the start of upstream fish movement.

Fish abundance for all the species studied, with exception of *Prochilodus*, were positively related to water levels for the pre-dam period (Quiros 1990). Moreover, for the lower Parana basin, there were no noticeable climatic changes during the last centuries. Therefore, it may be supposed that the thermal regime for the 1985–1990 period was similar to that of the pre-dam river period. This is an important conclusion to support the hypothesis that factors other than flood and water temperature, trigger upstream migration for potamodromous fishes in large river–floodplain systems. As suggested for diadromy (Gross 1987), potamodromy may be a complex assortment of life history traits under competitive and environmental selection pressure.

The output of this study is important for the restoration of river–floodplain systems, and recovery and management of riverine fish populations. Potamodromous fishes have maintained their seasonal position in the main channel despite huge flow regulation. The effects of change in frequency and intensity of floods on spawning and recruitment cannot be addressed here, but some negative effects may be suspected. Periodic flooding is critical for maintaining a floodplain river's ecological integrity and biological productivity. For the maintenance of riverine fish populations however, sustaining the timing of floods appears to be as important as maintaining river-floodplain connectivity. Commonly, the spatial scale will be related to total biomass but flood timing will be basically related to healthy population life cycles. The lack of baseline data for this kind of system in developed countries presents an enormous problem to those who attempt to restore them (Bayley 1991). However, some wide generalisations, obtained from natural fluctuations in unmodified rivers or large manipulations in relatively less developed systems, could be easily tested for developed systems. If the aim is to restore large rivers for riverine fish population, then natural flood pulses will have to be reimposed both in temporal and spatial scales (Bayley 1991). Most of the world's large rivers are greatly affected by human activity (Welcomme, Ryder & Sedell 1989). The necessity for a clear theoretical basis for how large river operate has been stressed before (Johnson, Richardson & Naimo 1995). This study also suggests that to uphold the widespread consensus on statements like 'most common fish species in river–floodplain systems are fluvial generalists well adapted to highly variable environments' would be misleading for the management of riverine populations in both developed and undeveloped systems.

Potamodromous migrations appear to have several advantages for the fish species undertaking them. In those species that are obligate main channel spawners, upstream migration by adult fish prior to spawning must have a role in counteracting downstream drift of eggs, larvae and fry (Welcomme 1985). However, this study suggests the proposed hypothesis that potamodromous fishes move in the main channel according to river discharge independent of cyclical patterns evolved within the undeveloped environment must be rejected. Potamodromous fishes retained their seasonal patterns, evolved with the pristine riverine system, despite of huge changes in water discharge.

Acknowledgements

R. Quiros acknowledges research support from the Consejo Nacional de Investigaciones Cientificas y Tecnologicas. We thank Maria Boveri and Eric Theiss for insightful comments and Hugo T. von Bernard for technical assistance.

References

Agostinho A.A., Julio H.F. & Petrere M. (1994) Itaipu reservoir (Brazil): impacts of the impoundment on the fish fauna and fisheries. In I.G. Cowx (ed.) *Rehabilitation of Freshwater Fisheries*. Oxford: Fishing News Books, Blackwell Science, pp. 171–184.

Bayley P.B. (1973) Studies on the migratory characin, *Prochilodus platensis* Holmberg 1988 (Pisces, Characoidei) in the River Pilcomayo, South America. *Journal of Fish Biology* **5**, 25–40.

Bayley P.B. (1991) The flood pulse advantage and the restoration of river-floodplain systems. *Regulated Rivers: Research and Management* **6**, 75–86.

Bayley P.B. (1995) Understanding large river-floodplain ecosystems. *BioScience* **45**, 153–158.

Bonetto A.A. (1963) Investigaciones sobre migraciones de peces en los ríos de la Cuenca del Plata. *Ciencia e Investigaçion* **19**, 12–26.

Bonetto A.A. & Pignalberi C. (1964) Nuevos aportes al conocimiento de las migraciones de peces en los Ríos Mesopotámicos de la República Argentina. Santo Tome, Sante Fe: *Comunicaciones del Instituto Nacional de Limnologia* **1**, 19 pp.

Bonetto A.A., Pignalberi C., Cordiviola de Yuan E. & Oliveros, O. (1971) Informaciones complementarias sobre migraciones de peces en la cuenca del Plata. *Physis (Buenos Aires)* **30**, 305–320.

Bonetto A.A., Canon Veron M. & Roldan D. (1981) Nuevos aportes al conocimiento de las migraciones de peces en el río Paraná. *Ecosur* **8**, 29–40.

Cordiviola de Yuan E., Oldani N., Oliveros O. & Pignalberi de Hassan C. (1984) Aspectos limnologicos de ambientes proximos a la ciudad de Santa Fe (Parana Medio). Poblaciones de peces ligadas a la vegetacion. *Neotropica* **30**, 127–129.

Drago E.C. (1984) Estudios limnologicos en una seccion transversal del tramo medio del rio Parana. VI: Temperatura del agua. *Revista de la Asociacion Ciencias Naturales del Litoral* **15**, 79–92.

Espinach Ros A. & Delfino R. (1993) Las pesquerias de la Cuenca del Plata en Bolivia, Paraguay, Argentina y Uruguay. COPESCAL. Informe de la Sexta Reunion del Grupo de Trabajo sobre Recursos Pesqueros. Montevideo, Uruguay, 10–13 de mayo de 1993, *FAO Informe de Pesca 490, Anexo IV: 36–51*. Rome: FAO, 80 pp.

Fuentes C.M. (1998) Deriva de larvas de sabalo, Prochilus lineatus, y otras especies de peces de interés comercial en el río Paraná Inferior. PhD dissertation, Universidad de Buenos Aires, Argentina, 136 pp.

Fuentes C.M. & Espinach Ros A. (1998) Distribución espacial y temporal del ictioplancton en un punto de delta del río Paraná. Revista. de Museo Argentino de Ciencias Naturales. *Hydrobiologia* TomoVIII, No. **6**, 51–61.

Godoy M.P. (1975) *Peixes do Brasil. Suborden Characoidei. Bacia do Rio Moggi Guassu.* Piracicaba: Editora Franciscana, 4 vols, 30 pp.

Gross M.R. (1987) Evolution of diadromy in fishes. *American Fisheries Society Symposium* **1**, 14–25.

Johnson B.L., Richardson W.B. & Naimo T.J. (1995) Past, present, and future concepts in large river ecology. *BioScience* **45**, 134–141.

Oldani N.O. & Oliveros O.B. (1984) Dinamica temporal de peces de importancia economica. *Revista de la Asociacion Ciencias Naturales del Litoral* **15**, 175–183.

Paiva M.P. & Bastos S.A. (1982) Marcacao de peixes nas regioes do alto e medio Sao Francisco (Brasil). *Ciencia e Cultura* **34**, 1362–1365.

Petts G.E., Imhof J.G., Manny B.A., Maher J.F.B. & Weisberg S.B. (1989) Management of fish populations in large rivers: a review of tools and approaches. In D.P. Dodge (ed.) *Proceedings of the International Large River Symposium. Canadian Special Publication of Fisheries and Aquatic Sciences* **106**, 578–588.

Pignalberi C. (1965) Evolucion de las gonadas en *Prochilodus platensis* y ensayo de clasificacion de los estados sexuales (Pisces, Characoidei). *Anales II Congreso Latino Americano do Zooligia, Sao Paulo* **2**, 203–208.

Quiros R. (1990) The Parana river basin development and the changes in the lower basin fisheries. *Interciencia* (Venezuela) **15**, 42–451.

Quiros R. & Cuch E. (1989) The fisheries and limnology of the lower La Plata Basin. In D.P. Dodge (ed.) *Proceedings of the International Large River Symposium. Canadian Special Publication of Fisheries and Aquatic Sciences* **106**, 429–443.

Ringuelet R.A, Aramburu R.H., & A. Alonso de Aramburu. 1967. *Los Peces Argentinos de Agua Dulce.* La Plata, Argentina: Comision de Investigacion Cientifica. Gobernacion de la Provincia de Buenos Aires, 602 pp.

Sverlij S.B., Espinach Ros A. & Orti G. (1993) Sinopsis de los datos biologicos y pesqueros del sabalo *Prochilodus lineatus* (Valenciennes, 1847). *FAO Sinopsis sobre la Pesca* **154**. Rome: FAO, 64 pp. (in Spanish).

Tablado A., Oldani N., Ulibarrie L. & Pignalberi de Hassan C. (1988) Dinamica temporal de la taxocenosis de peces en una laguna del valle aluvial del río Parana (Argentina). *Revue de Hydrobiologie Tropicale* **21**, 335–348.

Welcomme R.L. (1985) River fisheries, *FAO Fisheries Technical Paper 262*. Rome: FAO, 330 pp.
Welcomme R.L., Ryder R.A. & Sedell J.A. (1989) Dynamics of fish assemblages in river systems – a synthesis. In D.P. Dodge (ed.) *Proceedings of the International Large River Symposium. Canadian Special Publication of Fisheries and Aquatic Sciences* **106**, 569–577.

Chapter 7
Seasonal movements of coarse fish in lowland rivers and their relevance to fisheries management

M.C. LUCAS and T. MERCER

Department of Biological Sciences, University of Durham, South Road, Durham DH1 3LE, UK (e-mail: m.c.lucas@durham.ac.uk)

G. PEIRSON and P.A. FREAR

Environment Agency North East Region, Coverdale House, Aviator Court, Amy Johnson Way, Clifton Moor, York YO3 4UZ, UK

Abstract

Telemetry of fish behaviour has provided a wealth of information on the migratory behaviour of salmonids, and the impacts of potential obstructions to their riverine migrations. Although coarse fish species may also move considerable distances over a range of temporal scales, detailed quantitative knowledge is very limited, despite the importance of these fish communities in lowland rivers. The aim of this work was to obtain information on the magnitude and timing of movements of coarse fish in the Nidd/ Yorkshire Ouse system in Northern England by the use of radio tracking. Studies since 1993 have shown that considerable movements – up to 30 km – were exhibited by several cyprinid species, including adult barbel, *Barbus barbus*, chub, *Leuciscus cephalus*, dace, *Leuciscus leuciscus* and roach, *Rutilus rutilus*, although the range of movements between individuals was substantial.

Examples of movement patterns are provided, illustrating upstream spawning migrations, movement to refuge areas and homing responses. The influence of a flat-V gauging weir on the behaviour of tracked fish is considered in relation to environmental parameters and visual observations. It was concluded that migration by coarse fish in lowland rivers such as the Yorkshire Ouse system is an important component of the life cycle, and that to sustain stocks in a moderately natural state, river management must be sensitive to these requirements.

Keywords: Cyprinidae, telemetry, migration, homing, river obstructions, efficiency of passage.

7.1 Introduction

Coarse fish, especially cyprinids, are often the major component of fish communities in the middle and lower reaches of temperate rivers and often provide valuable recreational fisheries. In representing the most mobile fraction of biomass in lowland river ecosystems, coarse fish may play an important role in energy and nutrient fluxes (Lucas *et al.* 1998). These river systems are increasingly subject to impoundment and weir construction (Welcomme 1994). Some riverine cyprinids are recognised as being migratory (Smith 1991), but in Europe, and in the UK in particular, the possible importance of migration and other movements in the life cycle of coarse fish species has received relatively little attention. O'Hara (1986) argued that such information, on which to base sound fisheries management, was urgently needed, but only recently has such work begun in earnest.

Mark–recapture work on fish such as barbel, *Barbus barbus* (L.) (Hunt & Jones 1974), and bream, *Abramis brama* (L.) (Whelan 1983), has shown that these fish can travel tens of kilometres, and in the case of bream, home to discrete spawning areas. Linfield (1985) found that the mean length of cyprinids such as dace, *Leuciscus leuciscus* (L.), decreased away from the river source, and interpreted this as being due to active upstream migration by adult fish and downstream dispersal of juveniles. Even for those species recognised as being clearly migratory, such as barbel, the extent of our knowledge has been limited, until recently, to data obtained from methods such as mark–recapture which, although useful, are constrained by infrequency of location and by sampling problems in lowland environments. Tracking is a better method of supplying more detailed information on fish movements and for elucidating the reasons associated with fish movement. For European coarse fishes, tracking studies have considered a limited range of species such as pike, *Esox lucius* L. (Malinin 1970; Diana *et al.* 1977; Langford 1981; Lucas 1992), bream (Malinin 1970; Langford 1981; Malinin *et al.* 1992), eel, *Anguilla anguilla* (L.), (LaBarr *et al.* 1977; McGovern & McCarthy 1992) and barbel (Baras & Cherry 1990; Baras 1992; Baras 1995; Lucas & Batley 1996; Lucas & Frear 1997). However, research is currently expanding to other species including dace (Clough & Ladle 1997; Lucas 1999), chub *Leuciscus cephalus* (L.) (Fredrich 1996; Lucas *et al.* 1998), tench, *Tinca tinca* (L.) (Perrow *et al.* 1996) and perch, *Perca fluviatilis* L. (M. Perrow, personal communication). The earlier studies mentioned were generally limited in their usefulness by the low numbers of fish tracked.

The aim of this ongoing work, begun in 1993, was to obtain information on the magnitude and timing of coarse fish movements in the Nidd/Yorkshire Ouse system, and to examine fish behaviour in relation to weir obstructions.

7.2   Materials and methods

7.2.1   *Study site*

The study area comprises the Yorkshire Ouse catchment, N.E. England, which contains substantial areas of relatively unpolluted upland and lowland river habitat, but no longer maintains important anadromous salmonid populations due to chronic pollution of the Humber estuary into which the Ouse flows. There are important recreational fisheries for brown trout, *Salmo trutta* L. and grayling *Thymallus thymallus* (L.) in the upper reaches and for coarse fishes (mainly cyprinids) in the lower reaches. More detailed descriptions of the catchment and its biological and fisheries characteristics are given in Whitton and Lucas (1997) and Lucas *et al.* (1998).

The study concentrated on the lower Nidd and Ouse systems (Fig. 7.1). The lower Nidd is highly convoluted, with an average width of 12 m and an average depth of 1.2 m. The substratum is mainly sand and silt overlying clay. Gravel riffles are rare in the Ouse and lower 10 km of the Nidd, but are increasingly common further upstream. The Ouse has a mean width of 35 m, and a depth of 3 m. The Ouse has a higher conductivity than the Nidd ($\sim$500 μS vs $\sim$250 μS), but these are low compared to many more eutrophic lowland rivers, and make radio-tracking on the Ouse system feasible.

The Ouse and Nidd are obstructed by several weirs, with navigation passes on the Ouse, but not on the Nidd. Hunsingore weir (HW on Fig. 7.1) appears to be impassable, and a flat-V flow-gauging weir at Skip Bridge (SBW) built in 1978 by Yorkshire Water Authority was assumed to be of no consequence to fish populations. Since this time there has been concern from anglers that the weir substantially inhibits the natural upstream migration of coarse fish. Objective evidence of inhibition of upstream movements by coarse fish, has accumulated from angler catch–effort records, with reduced catch rates upstream of Skip Bridge weir (Axford 1991). A subsidiary weir built 50 m downstream of the gauging weir in 1982, to increase tailwater levels immediately below the weir, did not appear to improve passage past the weir (Axford 1991).

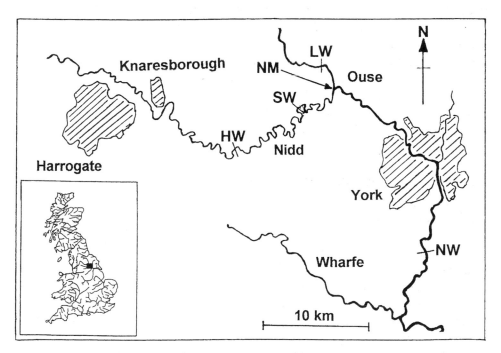

Figure 7.1 Map of the Yorkshire Ouse and Nidd and its position within the UK. HW – Hunsingore weir, SW – Skip Bridge weir, NM – Nidd Mouth, LW – Linton weir, NW – Naburn weir

The fish which are potentially most affected are principally rheophilic, lithophilic spawners (*sensu* Balon 1975), such as barbel, chub and dace, which utilise gravel areas to spawn (Philippart *et al.* 1988). The lack of gravel spawning habitat in the Nidd below SBW (Lucas & Batley 1996) currently necessitates their movement past the weir. However, a variety of smaller species, such as bleak, *Alburnus alburnus* (L.), roach, *Rutilus rutilus* (L.), and gudgeon, *Gobio gobio* (L.), also appear to have been influenced (Axford 1991) with reduced catch rates of these species upstream of SBW. This has been interpreted as an effect of the weir limiting upstream movement, following downstream movement in winter. Certainly, passage of small fish is likely to be more inhibited since maximum absolute swimming speed increases with fish length (Wardle 1975). In the summer of 1995, six PVC baffle units of a Larinier-type superactive design (Larinier 1992) were added to the downstream weir face to reduce average water velocity in an attempt to improve upstream traversal of the weir. Reduction in mean water velocity is likely to be especially beneficial to smaller species due to the influence of body length on swimming speed.

7.2.2   *Fish tracking*

Since 1993 over 100 adult riverine cyprinids have been tracked in the Nidd/Ouse system – mainly barbel ($N = 31$, fork length 47–64 cm), roach ($N = 24$, FL 18–27 cm), chub ($N = 30$, FL 37–46 cm) and dace ($N = 18$, FL 18–23 cm) – using varying frequency and pulse rate combinations for identification of individual fish. Fish for tagging were obtained from the lower 12 km of the Nidd, mainly be electric fishing using pulsed direct current apparatus, or in some cases by angling. In spring 1993 barbel were tagged and released at their capture sites upstream of SBW. In spring 1994 all barbel were tagged and released at their capture sites downstream of SBW. All dace, captured at the onset of the spawning migration in 1996 were caught and released below SBW. All chub, tagged in batches in the spring, summer and autumn of 1996, and late winter of 1997 were captured and released downstream of SBW. All roach were captured 1–3 km upstream of SBW, tagged and translocated to the pool of SBW. This latter experiment was based on observations of homing with cyprinid species (Stott *et al.* 1963; Smith 1991; Clough & Ladle 1997) and the hypothesis that they would seek to home back to their capture site upstream of the weir, and hence be highly motivated to pass the weir.

Most of the fish tracked were tagged by surgical placement of the transmitter. This is the preferred technique of attachment for long-term studies (Winter 1983) and long-term survival of tagged individuals has been demonstrated (Lucas & Batley 1996; Lucas *et al.* 1998). Identification of sex was possible for those fish caught during spring by careful internal inspection prior to transmitter implantation. A variety of standard radio transmitters produced by Argus Electronics (Lowestoft, UK), Biotrack (Wareham, UK) and Mariner Radar Ltd. (Lowestoft, UK) ranging in weight from 2 g in air to 16 g in air, and giving useful lives of 8 weeks to 15 months were used. The ratio of transmitter/fish weight (in air) did not exceed 2% for any fish tagged. Depending on tag power output (generally proportional to tag size) ranges to a hand-held 3-element Yagi antenna were typically 80–400 m in the Nidd, and 20–300 m in the Ouse. The smallest tags were often very difficult, and in some cases impossible, to locate in the Ouse, particularly at Nidd Mouth where depths approach 10 m. Normally the tags incorporated an integral antenna, and were sealed in polycarbonate, epoxy or dental acrylic. Some roach were tagged using externally attached transmitters, with whip antennae, adjacent to the dorsal fin, using neoprene foam backing mounts and coated vicryl sutures. All tag attachment was carried out under general anaesthesia (0.1 mg L^{-1} neutralised MS222 or benzocaine) on site in a mobile field laboratory, using specially designed stainless steel trays, under as near aseptic conditions as were possible, and prophylactic antimicrobial applications (Lucas & Batley 1996).

Tracking used a combination of passive and active techniques; with programmable scanning and data logging stations (Argus Electronics, Mariner Radar Ltd.) at key locations, and active tracking using a modified Yaesu FT-290R receiver (Argus Electronics) and a 3-element Yagi antenna, mainly on foot or by two-man canoe. Poor road access, the convoluted nature of the river and high flood banks made tracking by vehicle impossible. Low tag range for smaller tags (< 100 m to a hand-held Yagi), low to moderate search distances (typically 5–25 km) and low budget, made tracking by aerial survey impracticable.

7.2.3   Observations of fish at Skip Bridge weir

In 1996, counts of fish attempting to leap/swim over the weir were also made periodically between the middle of April and the middle of June, to complement tracking studies of fish behaviour in relation to the weir. Times and observation duration varied between 06.00–19.00 h and 15–60 min respectively, since these observations were carried out opportunistically. Success or failure of ascent was noted, as was species and size of fish where possible.

7.2.4   Environmental parameters

Water temperature and flow were recorded during the study. Temperature was recorded hourly using a data logger deployed at SBW at a depth of 1 m. Mean daily discharge rates were obtained from the Environment Agency, North East region.

7.3   Results

7.3.1   Barbel and chub

The results show that riverine cyprinid fishes in the Ouse system are much more seasonally mobile than was generally assumed in the past. Barbel and chub, tracked throughout the whole year, tended to exhibit upstream movements in spring, prior to spawning, and in general a net downstream movement in autumn and winter. Both chub and barbel moved substantial distances, up to 30 km, over the course of a year, but there was substantial individual variability in the distances involved (Fig. 7.2). The movements followed a seasonal pattern with upstream movements in spring and

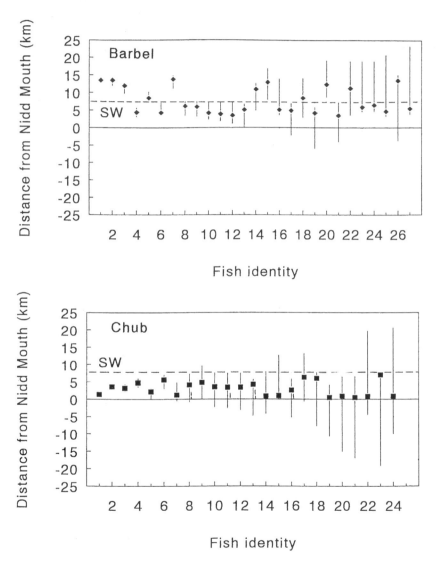

Figure 7.2 Ranges of movement of barbel tracked for more than 3 months and chub tracked for more than 5 months. Capture and release point is identified by a closed symbol. The horizontal dashed line, marked SW, indicates the position of Skip Bridge flow-gauging weir. Negative *y*-axis values reflect the position in the Ouse downstream of Nidd Mouth. The vertical dashed lines indicate the range of movement within the Ouse, upstream of Nidd Mouth

downstream movements in summer and autumn. Some fish utilised both the River Nidd and Yorkshire Ouse, demonstrating that at least a component of the stocks are common to both rivers. High flow events often resulted in downstream displacement of both species. In summer this was usually brief, and was followed by a subsequent upstream homing movement to the location occupied prior to the high flow (Lucas & Batley 1996; Lucas *et al.* 1998). In autumn and winter, successive downstream movements, often associated with high flow, resulted in a step-wise pattern of downstream movement (Fig. 7.3).

In a detailed study on the ability of barbel to ascend SBW to reach spawning areas upstream in 1994, it was shown that some 40% of fish considered to have attempted passage succeeded in passing, but that others would not or could not, and that the weir had a highly significant delaying effect on rates of movement (Lucas & Frear 1997).

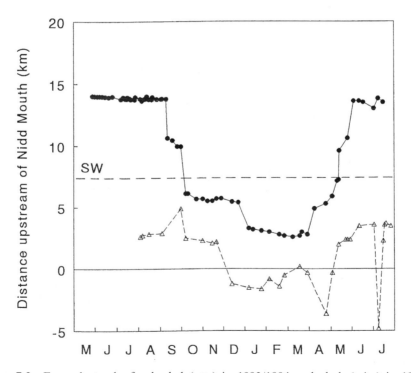

Figure 7.3 Example tracks for barbel (–●–) in 1993/1994 and chub (–△–) in 1996–1997 tagged in the River Nidd, illustrating seasonal migratory behaviour and, for chub displacement from spawning area by a large flood, and subsequent rapid homing. The horizontal dashed line, marked SW, indicates the position of Skip Bridge flow-gauging weir. Negative *y*-axis values reflect the position in the Ouse downstream of Nidd Mouth

7.3.2   *Dace and roach*

In the tracking experiments with dace and roach, five dace (27.8% of tagged dace) and nine roach (37.5%) were recorded as having ascended the weir. True figures could be somewhat higher, especially for dace, since the positions of some tagged fish could not be located, through lack of range and/or tag failure. Most dace initially moved downstream towards the River Ouse where tagged fish were very difficult to locate. Once past the weir, dace moved 3.5–14 km upstream to gravel spawning areas. Seven roach surmounted the weir within 10 days of release, and two more ascended after 4 weeks, moving upstream in groups to spawning areas 0.1–4.5 km upstream of the weir. Several roach, although passing the weir on different dates, rejoined the same group, initially in the locality of capture, but subsequently they moved upstream to spawning areas where pre-spawning aggregations of roach engaged in courtship were observed (Fig. 7.4). Seven fish remained within 150 m downstream of SBW, where they possibly spawned, and a group of five fish moved about 1 km downstream of the weir. Chi-square analysis showed no significant differences (all $P > 0.05$) between observed age and sex distributions of dace or roach which ascended the weir compared to the expected values.

From the tracked fish there was no clear evidence of utilisation of specific flow or temperature windows for passage. Data of adequate precision were not available for dace to make any useful analysis. Eight of the nine roach that ascended did so under low flow conditions of 2.3–2.8 $m^3 s^{-1}$, but one moved through at 4–5 $m^3 s^{-1}$. Temperatures at which roach passed over the weir varied between 10 and 18°C.

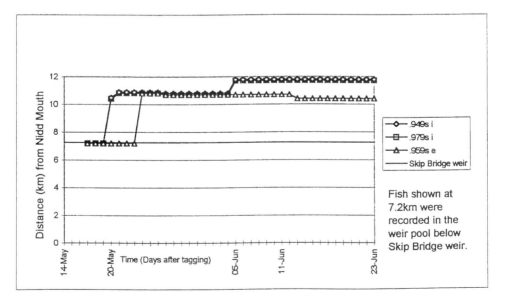

Figure 7.4   Upstream movements of several roach captured approximately 2 km above Skip Bridge weir, radio-tagged and released below Skip bridge weir

Activity of fish visually observed attempting to pass the weir increased with temperature, and was particularly evident above 12°C (Fig. 7.5). Most fish observed were juvenile roach and dace together with gudgeon, bleak and minnow, although

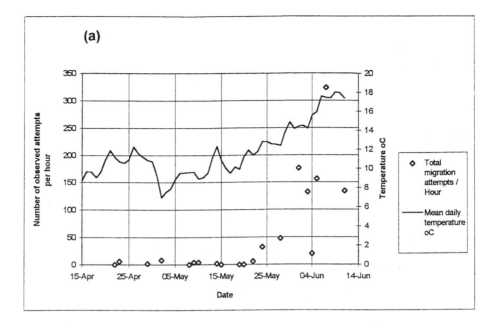

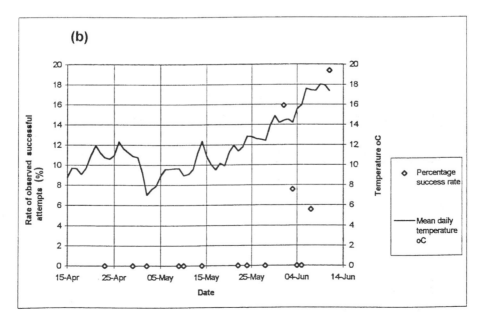

Figure 7.5 Rates of (a) total attempted passage, and (b) percentage successful passage by cyprinid fish at Skip Bridge weir in relation to water temperature

one adult barbel, one adult chub and approximately 30 adult roach and dace
(>15 cm) were seen attempting to pass the weir. Water velocities at the weir were
high, exceeding 2 m s^{-1} adjacent to the baffles. Under 5% of observed passage
attempts were successful. Most fish attempting to use the baffles to ascend were
swept out of them by the turbulent flow associated with the shallow water depth of
$\sim$20 cm, which barely covered several baffles. The baffles appeared to channel a
proportion of the flow through the central 'V' between the two symmetrical sets of
baffle units (three in each). Few fish were observed to succeed passing through the
central region, although this seemed to be a preferred area for attempts by larger
(>15 cm) fish, which on occasions frequently leapt or swam into this zone. Most fish
that successfully traversed the weir did so through the gaps between the baffle units,
although the correct orientation and positioning to enable them to do this appeared
to be a random event, after leaping at the protruding sill of the weir.

7.4   Discussion

An increasing number of studies using telemetry to study the movements of riverine
coarse fish (e.g. Baras 1992; Clough & Ladle 1997), as well as studies using other
methods such as acoustics (Duncan & Kubecka 1996; Lucas *et al.* 1998), particularly
in deeper lowland rivers, where radio-tracking is less effective, are beginning to
provide a consensus as to the extent and types of movement patterns exhibited by
these fish. Much less clear is an understanding of the way in which extrinsic stimuli,
such as flow and temperature, affect movement. An improved understanding of this
requires larger samples than are accommodated for in many tracking studies, as well
as careful experimental protocol, and in many cases extensive data analysis.
Furthermore, it is likely that river habitats may exert substantial influences on the
movement patterns discovered. For example, in degraded, channelised river habitat
the magnitude of movements of coarse fish appears to be greater than in natural,
more heterogeneous habitats, where under changing environmental conditions such
as increasing flow, suitable microhabitats for refuge are likely to be closer (Bruylants
et al. 1986). This is not perhaps surprising; after all, movement is one of the main
options available to river fish in responding to changes in their environment.

The proportion of fish successfully passing SBW under conditions of spawning or
homing movement ranged between 27% (minimum estimate) for dace and 40% for
barbel. This indicates that the weir presents a substantial obstacle for upstream
movement by a range of the more common fish species in the Nidd/Ouse, which does
not appear to have been greatly altered by the addition of the baffles. All recorded
successful traversals of SBW by roach and most chub (data for dace are insufficiently
precise) appear to have occurred during the hours of darkness or early morning. This
is a similar pattern to that found by Lucas and Frear (1997), who suggested that this
could be interpreted as evidence of behavioural inhibition of movement past the
weir, in addition to physical inhibition of traversal. Evidence from comparison of
telemetered passage of SBW with visual records suggests that most attempts of adult

roach to pass the weir occurred at night. Typically, during daytime, less than 5% of visually observed attempts to ascend the weir were successful (this is a conservative figure; the true value is likely to be rather less), while at least 27.8% of radiotagged dace and 37.5% of radiotagged roach are known to have ascended the weir.

Successful traversal of SBW by cyprinids was greater at water temperatures above 10°C. This can be explained in terms of the increase in physiological and swimming capabilities of fishes that occur with increasing temperature (Wardle 1975), and the higher temperature optima of cyprinids compared to salmonids (Wieser 1991). There is an inverse relationship between mean daily temperature and river flow during the spring–summer transition (Fig. 7.6), exaggerated in recent drought years, and although precise environmental requirements for migration and spawning may vary between coarse fish species and catchments, a genuine conflict is likely to exist for passage of coarse fish during spawning migrations at obstructions. It would appear that unlike upstream-migrating salmonids, cyprinids face the problems of ascending obstructions at higher early spring flows when their swimming performance and natural activity is low, or ascending obstructions when basal flows have declined greatly but their swimming performance is nearer its optimum (Fig. 7.6). A similar problem was reported for early spring salmon when water temperatures were 2–3°C, but when temperatures rose to 5–8°C, levels providing substantially improved salmonid swimming performance (Beamish 1978), basal flow over obstructions was still relatively high, thus aiding ascent. These differences in physiology and behaviour, and environmental conditions, suggest that a fundamental dichotomy in approach may be needed for maximising passage of cyprinids or salmonids past some obstacles.

Much attention has been paid to the potential and actual impact of obstructions on the passage of migratory salmonids and there is a substantial set of literature describing the behaviour of salmon in relation to obstructions and the design and

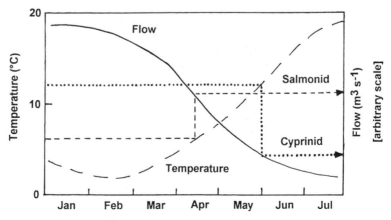

Figure 7.6 Schematic model illustrating the different flow conditions occurring for spring-migrating salmonids and cyprinids at temperatures for which moderate swimming performance (based on data from Beamish 1978) may be expected

provision of fish passage facilities to mitigate obstructing effects (e.g. Beach 1984). By contrast, the possible impact of river-obstructing devices on populations of coarse fish has received scant attention (it is, for example fully discounted by Beach), despite the ecological and economic importance of this species group. Little information regarding the behaviour of migratory non-salmonid fish in response to weirs and fish passes is available, although improvements in fish pass design and monitoring are now being made in continental Europe (Larinier 1983; Travade & Larinier 1992) and the UK.

There is little to be gained through the expensive installation of a fish pass in a lowland river dominated by coarse fish when the efficacy of fish passage has not been quantified for different species and life stages under varying flow and temperature conditions. The passage of several coarse fish through a fish pass does not prove it adequate; the number of successful passages must be related to the number of fish accumulated below the pass, and those attempting to traverse the obstacle. Repeated unsuccessful attempts of passage are likely to result in exhaustion. Use of *in situ* automated PIT systems (Castro-Santos *et al.* 1996) and physiological telemetry (Lucas *et al.* 1993) are proving to be instrumental in answering these problems in North America and the UK, principally for salmonids.

The significance of river obstruction in affecting natural movement patterns of some riverine fishes, even on a minor scale, has probably been underestimated. River obstruction was identified as a major cause in population declines of lithophilic and rheophilic cyprinids in the River Meuse, Belgium (Philippart *et al.* 1988; Baras *et al.* 1994), and is likely to be important in producing observed changes in the fish community in terms of ecological structure, availability to commercial and recreational fisheries, and conservation status (Welcomme 1994). This study has shown that migration by coarse fish in lowland rivers such as the Yorkshire Ouse system is an important component of the fish's life cycle, and that to sustain stocks in a moderately natural state, river management must be sensitive to these requirements.

Acknowledgements

We are grateful to local angling clubs, especially Leeds and District Amalgamated Society of Anglers and York and District Amalgamation of Anglers for their co-operation with this study. We thank the Environment Agency North East region fisheries inspectors for assistance during capture of fish. MCL thanks the following for financial assistance: the Environment Agency, Nuffield Foundation, Fisheries Society of the British Isles, Leeds and District Amalgamated Society of Anglers, York and District Amalgamation of Anglers, Wetherby Angling Association and Bradford No. 1 Angling Association.

References

Axford S. (1991) Some factors affecting catches in Yorkshire rivers. In I.G. Cowx (ed.) *Catch Effort Sampling Strategies*. Oxford: Fishing News Books, Blackwell Science, pp. 143–153.

Balon E.K. (1975) Reproductive guilds of fishes: a proposal and definition. *Journal of the Fisheries Research Board of Canada* **32**, 821–864.

Baras E. (1992) Etude des strategies d'occupation du temps et de l'espace chez le barbeau fluviatile, *Barbus barbus* (L.) *Cahiers d'Ethologie* **12**, 125–412.

Baras E. (1995) Seasonal activities of *Barbus barbus*: effect of temperature on time budgeting. *Journal of Fish Biology* **46**, 806–818.

Baras E. & Cherry B. (1990) Seasonal activities of female barbel *Barbus barbus* (L.) in the River Ourthe (southern Belgium) as revealed by radio-tracking. *Aquatic Living Resources* 7, 181–189.

Baras E., Lambert H. & Philippart J.C. (1994) A comprehensive assessment of the failure of *Barbus barbus* spawning migration through a fish pass in the canalized River Meuse (Belgium). *Aquatic Living Resources* 7, 181–189.

Beach M.H. (1984) Fish pass design criteria for the design and approval of fish passes and other structures to facilitate the passage of migratory fishes in rivers, *Fisheries Technical Report 78*. London: Ministry of Agriculture, Fisheries and Food, 46 pp.

Beamish F.W.H. (1978) Swimming capacity. In W.S. Hoar & D.J. Randall (eds) *Fish Physiology, Vol. VII*. New York: Academic Press, pp. 101–187.

Bruylants B., Vandelannoote A. & Verhayen R. (1986) The movement pattern and density distribution of perch, *Perca fluviatilis* L., in a channelized lowland river. *Aquaculture and Fisheries Management* **17**, 49–57.

Castro-Santos T., Haro A. & Walk S. (1996) A passive integrated transponder (PIT) tag system for monitoring fishways. *Fisheries Research* **28**, 253–261.

Clough S. & Ladle M. (1997) Diel migration and site fidelity in a stream-dwelling cyprinid, *Leuciscus leuciscus*. *Journal of Fish Biology* **50**, 1117–1119.

Diana J.S., Mackay W.C. & Ehrman M. (1977) Movements and habitat preference of northern pike (*Esox lucius*) in Lac Ste Anne, Alberta. *Transactions of the American Fisheries Society* **106**, 560–565.

Duncan A. & Kubecka J. (1996) Patchiness in longitudinal fish distribution of a river as revealed by continuous hydroacoustic survey. *ICES Journal of Marine Science* **53**, 161–165.

Fredrich F. (1996) Preliminary studies on daily migration of chub (*Leuciscus cephalus*) in the Spree River. In E. Baras & J.Cl. Philippart (eds) Underwater biotelemetry. *Proceedings of the First Conference and Workshop on Fish Telemetry in Europe*. Belgium: University of Liege, p. 66.

Hunt P.C. & Jones J.W. (1974) A population study of *Barbus barbus* L. in the River Severn, England. II. Movement. *Journal of Fish Biology* **6**, 269–278.

LaBarr G.W., Hernando Casal J.A. & Delgado C.F. (1987) Local movements and population size of European eels, *Anguilla anguilla*, in a small lake in southwestern Spain. *Environmental Biology of Fishes* **19**, 111–117.

Langford T.E. (1981) The movement and distribution of sonic-tagged coarse fish in two British rivers in relation to power station cooling outfalls. In F.M. Lang (ed.) *Proceedings of the 3rd International Conference on Biotelemetry*. Laramie, WY: University of Wyoming, pp. 197–232.

Larinier M. (1983) Guide pour la conception des dispositifs de franchissement des barrages pour les poissons migrateurs. *Bulletin Français de la Pêche et de la Pisciculture, numero special, Conseil Superieur de la Pêche*, 36 pp.

Larinier M. (1992) Facteurs biologiques a prendre en compte dans la conception des ouvrages de franchissement, notions d'obstacles a la migration. *Bulletin Français de la Pêche et de la Pisciculture* **326–327**, 20–29.

Linfield R.S.J. (1985) An alternative concept to home range theory with respects to populations of cyprinids in major river systems. *Journal of Fish Biology* **27** (Suppl. A), 187–196.

Lucas M.C. (1992) Spawning activity of male and female pike, *Esox lucius* L., determined by acoustic tracking. *Canadian Journal of Zoology* **70**, 191–196.

Lucas M.C. (1999) Recent advances in the use of telemetry and tracking applied to freshwater fishes. In L. Le Maho (ed.) *Proceedings of the Fifth European Wildlife Telemetry Conference, Strasbourg, Aug. 1996* (in press).

Lucas M.C. & Batley E. (1996) Seasonal movements and behaviour of adult barbel *Barbus barbus*, a riverine cyprinid fish: implications for river management. *Journal of Applied Ecology* **33**, 1345–1358.

Lucas M.C. & Frear P. (1997) Effects of a flow-gauging weir on the migratory behaviour of adult barbel, *Barbus barbus*, a riverine cyprinid. *Journal of Fish Biology* **50**, 382–396.

Lucas M.C., Johnstone A.D.F. & Priede I.G. (1993) Use of physiological telemetry as a method of estimating metabolism of fish in the natural environment. *Transactions of the American Fisheries Society* **122**, 822–833.

Lucas M.C., Mercer T., Batley E., Frear P.A., Peirson G., Duncan A. & Kubecka J. (1998) Spatio-temporal variations in the distribution and abundance of fishes in the Yorkshire Ouse system. *Science of the Total Environment* **210/211**, 437–455.

Malinin L.K. (1970) The use of ultrasonic transmitters for the marking of bream and pike. *Fisheries Research Board of Canada Translation Series* **2146**, 7 pp.

Malinin L.K., Kijasko V.I. & Vääränen P.L. (1992) Behaviour and distribution of bream (*Abramis brama*) in oxygen deficient regions. In I.G. Priede & S.M. Swift (eds) *Wildlife Telemetry: Remote Monitoring and Tracking of Animals*. Chichester: Ellis Horwood, pp. 297–306.

McGovern P. & McCarthy T.K. (1992) Local movements of freshwater eels (*Anguilla anguilla* L.) in western Ireland. In I.G. Priede & S.M. Swift (eds) *Wildlife Telemetry: Remote Monitoring and Tracking of Animals*. Chichester: Ellis Horwood, pp. 319–327.

O'Hara K. (1986) Fish behaviour and the management of freshwater fisheries. In T.J. Pitcher (ed.) *The Behaviour of Fishes*. London: Chapman & Hall, pp. 496–521.

Perrow M.R., Jowitt A.J.D. & Johnson S.R. (1996) Factors affecting the habitat selection of tench in a shallow, eutrophic lake. *Journal of Fish Biology* **48**, 859–870.

Philippart J.Cl., Gillet A. & Micha J.C. (1988) Fish and their environment in large European river ecosystems. The River Meuse. *Sciences de l'Eau* **7**, 115–154.

Smith R.J.F. (1991). Social behaviour, homing and migration. In I.J. Winfield & J.S. Nelson (eds) *Cyprinid Fishes: Systematics, Biology and Exploitation*. London: Chapman & Hall, pp. 509–529.

Stott B., Elsdon J.W.V. & Johnston J.A.A. (1963) Homing behaviour in gudgeon (*Gobio gobio* (L.)). *Animal Behaviour* **11**, 93–96.

Travade F. & Larinier M. (1992) Les techniques de controle des passes a poissons. *Bulletin Français de la Pêche et de la Pisciculture* **326–327**, 151–164.

Wardle C.S. (1975) Limit of fish swimming speed. *Nature* **255**, 725–727.

Welcomme R.L. (1994). The status of large river habitats. In I.G. Cowx (ed.) *Rehabilitation of Freshwater Fisheries*. Oxford: Fishing News Books, Blackwell Scientific Publications, pp. 11–20.

Whelan K.F. (1983) Migratory patterns of bream *Abramis brama* (L.) shoals in the River Suck system. *Irish Fisheries Investigations, Series A* **23**, 11–15.

Whitton B.A.W. & Lucas M.C. (1997) Biology of the Humber rivers. *Science of the Total Environment* **194/195**, 247–262.

Wieser W. (1991) Physiological energetics and ecophysiology. In I.J. Winfield & J.S. Nelson (eds) *Cyprinid Fishes: Systematics, Biology and Exploitation*. London: Chapman & Hall, pp. 426–455.

Winter J.D. (1983) Underwater biotelemetry. In L.A. Nielsen & D.L. Johnson (eds) *Fisheries Techniques*. Bethesda, MA: American Fisheries Society, pp. 372–395.

Chapter 8
Seasonal and diel changes of young-of-the-year fish in the channelised stretch of the Vltava River (Bohemia, Czech Republic)

O. SLAVÍK

TGM Water Research Institute, Podbabská 30, 160 62 Prague 6, Czech Republic
(e-mail: ondrej_slavik@vuv.cz)

L. BARTOŠ

Institute of Animal Production, Ethology Group, 104 00 Prague 10 – Uhříněves, Czech Republic

Abstract

In the littoral zone of the Vltava River, changes in the abundance of young-of-the-year fish (YOY) were linked to abiotic factors and the presence of potential predators. During the diel period, abundance of YOY varied. In early summer and autumn, YOY abundance was more affected by potential predators when compared with late summer. This suggests that diel changes of YOY abundance in the littoral zone were primarily related to ontogenetic development. No relationship was found between roach abundance in the period between June and September and their length. In the poorly structured littoral zone of the channelised river, no long-term refugia for juvenile fish were available and no shifts in abundance of juvenile fish with time were observed.

Keywords: Juvenile fish, abundance, length, diurnal period, seasons, habitat shift.

8.1 Introduction

Availability of refugia plays a major role in the spatial distribution of young-of-the-year (YOY) fishes. To avoid predation risk from piscivorous fish, small fish tend to occupy shallow habitats (Power 1984; Harvey 1991). However, as fish grow, the predation risk in shallow habitats from wading/diving predators increases, and bigger fish inhabit deeper habitats (e.g. Schlosser 1982; Mahon & Portt 1985; Harvey & Stewart 1991). The ontogenetic habitat shift must also take account of the microhabitat carrying capacity; e.g. roach, *Rutilus rutilus* (L.), larvae inhabit submerged vegetation in the deep water but juveniles shift to shallower water (Copp 1990).

During low light intensity, the predation risk for small fish is highest (Cerri 1983). With decreasing light intensity, the daytime refuges become less safe, and small fish move into shallow habitats (Schlosser 1988). The intensity and/or direction of movements between day and night positions differ according to quality of bottom substrate (Copp & Jurajda 1993), and the movements are also affected by variable physical conditions of the environment, e.g. concentration of dissolved oxygen (Suthers & Gee 1986). According to Post and McQueen (1988), the occurrence of juveniles in different habitats is determined ontogenetically, and responses on predation risk, physical conditions or food availability are inherent within this development.

The present study was carried out to test three hypotheses:

(1) in early summer, diel changes in YOY abundance are influenced more by abiotic factors than by predators;
(2) in late summer and autumn the influence of predation risk increases with increasing mobility of juveniles; and
(3) the growth of YOY is influenced by abundance (density) of the population.

8.2   Materials and methods

8.2.1   *Study area*

The Vltava River (Elbe catchment area) is 430 km long with a catchment area of about 2800 km^2. A cascade of five reservoirs is situated on the river, 43 km and more upstream from the study site. The release of hypolimnial water causes relatively low temperature during spring and summer, and relatively high temperature during autumn and winter (Kubecka & Vostradovský 1995). The study site was a pool in the littoral zone of the main channel. River banks were linked with stones and concrete, and the bottom substratum (an area of 3×12 m) was formed by panels set in concrete. Submerged vegetation was absent, and the maximum depth was 1.2 m. Water velocity was negligible. The average annual discharge at the site in 1992 was 104 m^3 s^{-1}. Mean dissolved oxygen, water temperature, BOD$_5$, and conductivity were 10.7 mg L^{-1}, 10.9°C, 5.3 mg L^{-1}, and 355 μS, respectively (Kult 1995). Below Prague, the river receives 600–800 L s^{-1} of poorly treated sewage from the Prague Sewage Works, which doubles the BOD$_5$, and increases the ammonia, alkalinity, nitrate and phosphate concentrations (Vostradovský 1994). The study site was situated 14 km downstream of the pollution source.

8.2.2   *Sampling procedures*

A 10-m long seine net, 1.8 m high and with a mesh size 1.7 mm was used for sampling. Juvenile fishes were collected monthly from June to October 1992 (15 June, 16 July, 14 August, 17 September and 16 October). Samples (60 in total) were collected in 2-h intervals during the 24-h period. The first sample was taken between 05:00 and 07:00 am, and belonged to the light phase of the diel period. Potential predators were detected within YOY patterns, and/or caught by setting gillnets (20 m long, 2 m high, mesh size 20 mm) parallel to the river bank 3 m apart (depth 2 m). In early summer, fish longer than 60 mm were considered as potential predators. During late summer and autumn, fish longer than 80 mm were considered as predators (Copp & Jurajda 1993). Fish caught were measured to the nearest 0.1 mm, fixed in 4% formaldehyde for further laboratory analysis, and identified according to Koblickaya (1966) and Mooij (1989). During each sampling, water temperature and dissolved oxygen concentration (microprocessor Oxi 196 – WTW) were also measured.

8.2.3 *Analysis*

Data were subjected to the General Linear Models Procedure (GLM) for unbalanced ANOVA. They were classified by month (June, July, August, September, October) and time of day (twelve 2-h intervals between 05:00 and 03:00 h). Time was nested to season. For season, June to July was considered early summer; August to September as late summer; and October as autumn. Least squares means (LSMEANs) were computed for each class, and differences between classes were subjected to *t*-tests. To calculate the relationships between variables, residual Spearman rank correlation coefficients were computed for each season.

8.3 Results

A total of 5853 0+ fish and 1714 fish >0+ were caught, of which 153 potential predators were used for the statistical analysis (Table 8.1). Roach, *Rutilus rutilus* (L.), and gudgeon, *Gobio gobio* (L), were the most numerous species. GLM showed significant relationships between abundance of YOY and potential predators (Table 8.2), and YOY length and temperature and dissolved oxygen (Table 8.3).

8.3.1 *Fish abundance changes during the day by season*

Roach abundance fluctuated during the day and this change shifted with season (Fig. 8.1a). In early summer, maximum roach abundance was registered during the night (23:00–01:00 h), then it gradually decreased. During light hours, maximum roach abundance was registered between 07:00 and 09:00 h. In late summer, maximum roach abundance was recorded after sunrise (05:00–07:00 h), reaching a daylight minimum between 17:00 and 19:00 h. During the night, the maximum was registered between 21:00 and 23:00 h, and the minimum between 23:00 and 01:00 h.

Gudgeon abundance fluctuated during the day and also changed with season (Fig. 8.1b). In early summer, the highest abundance of gudgeon was registered at night, during the same period as for roach (23:00–01:00 h). Abundance increased from sunrise (03:00–05:00 h) to noon (11:00–13:00 h), and then gradually declined to a minimum between 15:00 and 21:00 h (no significant differences between three sampling periods were found). In late summer, gudgeon abundance increased after

Table 8.1   Number of caught fish

Species	Number of individuals
0+ fish	
Gudgeon *Gobio gobio*	3344
Roach *Rutilus rutilus*	2509
Total	5853
Fish >1	
Roach *Rutilus rutilus*	1341
Dace *Leuciscus leuciscus* (L.)	128
Bleak *Alburnus alburnus* (L.)	85
Rainbow trout *Oncorhynchus mykiss* (Walbaum)	2
Vimba *Vimba vimba* (L.)	3
Ruffe *Gymnocephalus cernuus* (L.)	2
Total	1561
Potential predators	
Chub *Leuciscus cephalus* (L.)	103
Perch *Perca fluviatilis* L.	22
Asp *Aspius aspius*(L.)	12
Brown trout *Salmo trutta* L.	9
Pike *Esox lucius* L.	5
Pikeperch *Stizostedion lucioperca* (L.)	2
Total	153

sunrise, reaching a maximum between 07:00 and 09:00 h. The minimum abundance was detected between 17:00 and 21:00 h (no significant differences between two sampling periods were found). During autumn, differences in gudgeon abundance during the diel period were not significant.

Table 8.2   The result of GLM for roach, gudgeon and predators abundance

Variable	Class	F	d.f.	
Roach abundance	Model	347.38	15, 2509	$P < 0.0001$
	Months	188.98	4, 2509	$P < 0.0001$
	Time (season)	914.69	26, 2509	$P < 0.0001$
Gudgeon abundance	Model	332.92	15, 3344	$P < 0.0001$
	Months	581.15	4, 3344	$P < 0.0001$
	Time (season)	2176.54	26, 3344	$P < 0.0001$
Predator abundance	Model	105.4	15, 5853	$P < 0.0001$
	Months	298.57	4, 5853	$P < 0.0001$
	Time (season)	162.6	26, 5853	$P < 0.0001$

Table 8.3   Results of the GLM for roach and gudgeon length and physical characteristic

Variable	Class	F	d.f.	
Roach length	Model	298.05	15, 2509	*P* < 0.0001
	Months	530.91	4, 2509	*P* < 0.0001
Gudgeon length	Model	3950.9	15, 3344	*P* < 0.0001
	Months	7371.76	4, 3344	*P* < 0.0001
Temperature	Model	11300.15	15, 5853	*P* < 0.0001
	Months	28855.81	4, 5853	*P* < 0.0001
	Time(season)	154.94	26, 5853	*P* < 0.0001
Oxygen	Model	336.17	15, 5853	*P* < 0.0001
	Months	16.34	4, 5853	*P* < 0.0001
	Time(season)	1367.22	26, 5853	*P* < 0.0001

The abundance of potential predators fluctuated during the day and with season (Fig. 8.1c). In early summer, the maximum predator abundance was registered between 09:00 and 11:00 h, and predator abundance was higher during the day than the night. Conversely, in late summer, the situation was opposite, with maximum abundance between 23:00 and 01:00 h. During autumn, changes in abundance during the diel period were not significant.

8.3.3   *Fish abundance changes between months*

Abundance of roach and gudgeon fluctuated between months (Fig. 8.2a), and was highest in September. The lowest abundance of both species was recorded in October (*P* < 0.001). No significant differences in roach abundance were found between June and September.
    Abundance of potential predators also fluctuated between months (Fig. 8.2a) and reached a maximum in August.

8.3.4   *Fish length changes among months*

The size of roach gradually increased over the study period from June to October (Fig. 8.2b). Gudgeon size varied between months but the maximum length was attained in September and the minimum in August.

8.3.5   *Physical characteristics*

Water temperature differed significantly during the day and with seasons (Fig. 8.3a), and between months (Fig. 8.3b). Temperature increased from June to August, and

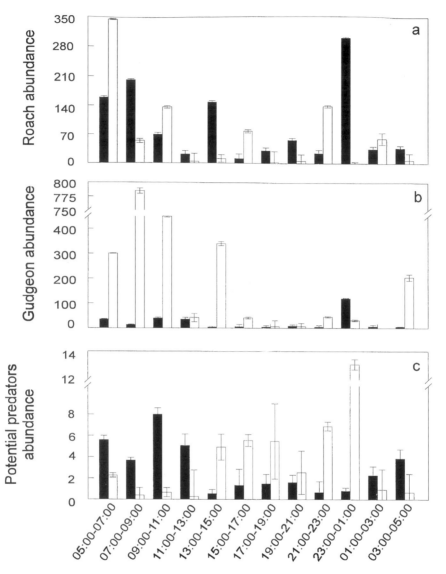

Figure 8.1 Roach (top), gudgeon (middle) and potential predators (bottom) abundance during the day by season: solid bars, early summer; open bars, late summer (LSMEAN; S.E.)

then gradually decreased to a minimum in October. In early summer, the maximum temperature was between 11:00 and 15:00 h (no significant differences were registered between sampling periods), and the minimum was recorded during the night until dawn (01:00–05:00 h). In late summer, the maximum was later in the afternoon (15:00–17:00 h), and the minimum was detected between 03:00 and 05:00 h.

Dissolved oxygen concentration varied significantly during the day and between seasons (Fig. 8.3c), and across months (Fig. 8.3d). Dissolved oxygen concentration

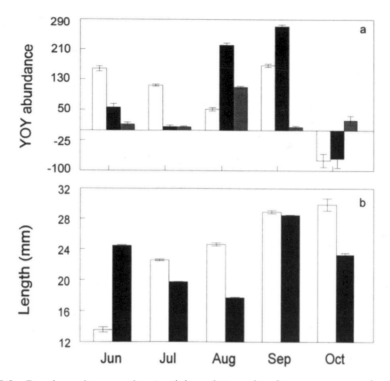

Figure 8.2 Roach, gudgeon and potential predators abundance across months (top) and roach and gudgeon length across months (bottom): open bars, roach; solid bars, gudgeon; shaded bars, potential predators (LSMEAN; S.E.)

increased from June to September, reaching the minimum in October. In early summer, oxygen content was higher during the night than the day. In late summer, the situation was opposite.

8.3.6 *Correlation between fish abundance, length and physical parameters*

Relationships for fish abundance changes during the day and with season between biotic and abiotic variables are shown in Table 8.4. No relationship was found between roach abundance and fish length across months ($r_s = 0.11$, $n = 2509$, n.s.). This relationship was negative for gudgeon ($r_s = -0.36$, $n = 3344$, $P < 0.0001$).

8.4 Discussion

This study supports the hypothesis that changes in YOY abundance in the littoral zone are primarily influenced by ontogenetic development (e.g. Copp 1990). YOY

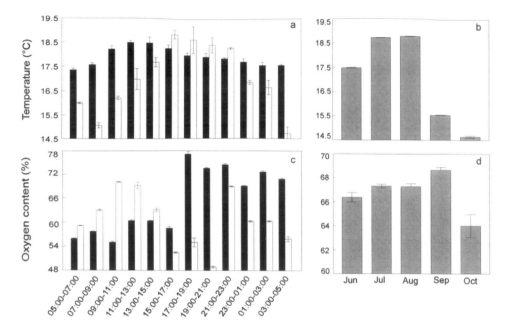

Figure 8.3 Water temperature (upper left) and oxygen content (lower left) during the day by season and water temperature (upper right) and oxygen content (lower right) across months: solid bars, early summer; open bars, late summer (LSMEAN; S.E.)

abundance reflected both abiotic factors and predation risk. In early summer and autumn, the influence of predation risk on YOY abundance was higher when compared with abiotic factors. In late summer, the situation was opposite, and the

Table 8.4 Residual Spearman rank correlation between YOY and potential predator abundance, dissolved oxygen concentration and temperature between seasons

	Early summer (n = 1284)	Late summer (n = 1208)	Autumn (n = 18)
Abundance of potential predators			
Roach abundance	0.77***	−0.64***	−0.86***
Gudgeon abundance	0.53***	0.69***	−0.86***
Dissolved oxygen			
Roach abundance	−0.13***	0.88***	−0.61**
Gudgeon abundance	−0.21***	−0.77***	−0.61**
Temperature			
Roach abundance	−0.54***	−0.73***	0.2 n.s.
Gudgeon abundance	−0.79***	0.77***	0.15 n.s.

$*P < 0.05$; $**P < 0.01$; $***P < 0.0001$; n.s.: not significant.

influence of abiotic factors predominated. The YOY abundance in the littoral zone was not as expected.

In early summer, maximum abundance of the roach in the littoral zone was recorded during the night, and the highest positive correlation of all observed factors was between abundance of roach and potential predators. The results are consistent with the theory that increasing predator activity causes movements of small fish into the shallow water (Schlosser 1988). No relationship was found between oxygen concentration and roach abundance (Table 8.4). However, the correlation between water temperature and roach abundance was negative, possibly linked to lower water temperatures as solar radiation raised the temperature in this zone during the day. High abundance of the roach during the dark period when water temperature was low shows the considerable influence of predation risk and dispersion of juveniles.

During late summer, maximum roach abundance was recorded during the daylight hours. Relationships between roach abundance and oxygen concentration, water temperature and potential predator abundance were found, but the effect of abiotic factors predominated over predation risk. The same trend was reported in lakes by Suthers and Gee (1986). They described movements of juvenile fish from habitats with high oxygen deficit into the deeper water where fish were exposed to higher predation risk.

By contrast, Copp and Jurajda (1993) found the maximum abundance of roach during the summer was at night, but they did not record any significant changes in the physical characteristics at their sites. The differences between the studies may be because the habitat used in the present study was more akin to a lake situation than a river.

The correlation between roach and potential predators abundance was negative, and when abundance of potential predators was high roach abundance was low (between 23:00 and 01:00 h). According to Christensen and Persson (1993), roach seek refuges only when predators are present. Unstructured habitat, as found at the study site, does not offer any long-term refugium, and to avoid instantaneous predation risk, roach probably move into the deeper water to gain a short-term benefit (Fraser & Emmons 1984).

In autumn, no significant variation in roach abundance was found during the diel cycle. Potential predation risk predominated over the influence of abiotic factors. This was confirmed by the strong negative correlation between predation and YOY abundance, and that predation pressure can still be high even at low prey density.

Similar trends in diel distribution of YOY individuals were recorded for gudgeon. In early summer, maximum abundance was found during the night, while in late summer it occurred during the day. Again the influence of potential predation risk during early summer was higher than from abiotic factors. In late summer, the influence of higher water temperature, and the associated lower dissolved oxygen concentration increased, while potential predation risk reduced. In autumn, a repeat of the situation in early summer was found.

In late summer, the relationship between YOY abundance and biotic/abiotic factors differed for roach and gudgeon, e.g. juvenile roach and perch differ in their antipredatory behaviour (Christensen & Persson 1993) and capacity to avoid predation (Eklöv & Persson 1995). It was assumed that the different correlations (negative/positive) for juvenile roach and gudgeon relate to different predator–prey relationships.

In the littoral zone, no relationship between roach abundance and length was found; thus this hypothesis could not be tested. However, body size of roach increased during the study period, as expected (Fig. 8.2c), but abundance remained high and stable. It is possible that this lack of change was related to the homogeneity of the structured habitat of the littoral zone, i.e. it did not offer any long-term refuge for the fish.

References

Cerri D.R. (1983) The effect of light intensity on predator and prey behaviour in cyprinid fish: Factors that influence prey risk. *Animal Behaviour* **31**, 736–742.

Copp G.H. (1990) Shift in the microhabitat of larval and juvenile roach, *Rutilus rutilus* (L.), in a floodplain channel. *Journal of Fish Biology* **36**, 683–692.

Copp G.H. & Jurajda P. (1993) Do small riverine fish move inshore at night? *Journal of Fish Biology* **43** (Suppl. A), 229–241.

Christensen B. & Persson L. (1993) Species-specific antipredatory behaviours: effects on prey choice in different habitats. *Behavioural Ecology and Sociobiology* **32**, 1–9.

Eklöv P. & Persson L. (1995) Species-specific antipredator capacites and prey refuges: interactions between piscivorous perch (*Perca fluviatilis*) and juvenile perch and roach (*Rutilus rutilus*). *Behavioural Ecology and Sociobiology* **37**, 169–178.

Fraser F.D. & Emmons E.E. (1984) Behavioral response of blacknose dace (*Rhinichthys atratulus*) to varying densities of predatory creek chub (*Semotilus atromaculatus*). *Canadian Journal of Fisheries and Aquatic Sciences* **41**, 364–370.

Harvey B.C. (1991) Interaction of abiotic and biotic factors influences larval fish survival in a Oklahoma streams. *Canadian Journal of Fisheries and Aquatic Sciences* **48**, 1476–1480.

Harvey B.C. & Stewart A.J. (1991) Fish size and habitat depth relationships in headwater streams. *Oecologia* **87**, 336–342.

Koblickaya A. F. (1966) Opredelitel molodi presnovodnych ryb (Identification keys for young of freshwater fishes). Moskva, Izd: Legkaja i piscevaja promyslennost, 208 pp.

Kubecka J. & Vostradovsky J. (1995) Effect of dams, regulation and pollution on fish stocks in the Vltava river in Prague. *Regulated Rivers: Research and Management* **10**, 93–98.

Kult A. (1995) *Water Quality in Rivers at 1992–1993.* Annual report of the Czech Hydrometeorological Institute, Prague, 315 pp.

Mahon R. & Portt C.B. (1985) Local size related segregation of fishes in streams. *Archiv für Hydrobiologia* **103**, 267–271.

Mathews C.P. (1971) Contribution of young fish to total production of fish in the River Thames near Reading. *Journal of Fish Biology* **3**, 157–187.

Mooij M.W. (1989) A key to the identification of larval bream, *Abramis brama*, white bream, *Blicca bjoerkna*, and roach, *Rutilus rutilus*. *Journal of Fish Biology* **34**, 111–118.

Post J.R. & McQueen D.J. (1988) Ontogenetic changes in the distribution of larval and juvenile yellow perch (*Perca flavescens*): A response to a prey or predators? *Canadian Journal of Fisheries and Aquatic Sciences* **45**, 1820–1826.

Power M.W. (1984) Depth distributions of armoured catfish: predator-induced resource avoidance? *Ecology* **65**, 523–528.

Schlosser I.J. (1982) Fish community structure and function along two habitat gradients in a headwater stream. *Ecological Monographs* **52**, 395–414

Schlosser I.J. (1988) Predation risk and habitat selection by two size classes of a stream cyprinid: experimental test of a hypothesis. *Oikos* **52**, 36–40.

Suthers T.M. & Gee J.H. (1986) Role of hypoxia in limiting dial spring and winter distribution of juvenile yellow perch in a praire marsh. *Canadian Journal of Fisheries and Aquatic Sciences* **43**, 1562–1570.

Vostradovský J. (1994) Impact of urbanisation on the fish community of River Vltava upstream and downstream of Prague, Czechoslovakia. In I.G. Cowx (ed.) *Rehabilitation of Freshwater Fisheries*. Oxford: Fishing News Books, Blackwell Science, pp. 458–466.

Section III
Habitat requirements

Chapter 9
Trout, summer flows and irrigation canals: a study of habitat condition and trout populations within a complex system

D.J. WALKS* and H.W. LI

Oregon Cooperative Fisheries Research Unit, Department of Fisheries and Wildlife, Oregon State University, Corvallis, OR 97331, USA

G.H. REEVES

United States Forest Service, Forestry Sciences Laboratory, Corvallis, OR 97331, USA

Abstract

The severely reduced summer flows (98% discharge diverted for irrigation) in the Upper Deschutes River of Central Oregon contributes to water temperatures of up to 29°C, which harms resident trout populations. An increase in discharge is currently under consideration, but will this increase trout numbers? Standard instream flow habitat models often assume that increased habitat quantity equals improved habitat quality. However, some important physiological variables such as water temperature are lacking from these models. To assess the rehabilitation potential of the trout fishery on the Deschutes, the instream relationship between discharge and water temperatures, and the influence of discharge on trout populations at a habitat unit level were examined. There was an inverse relationship between discharge and water temperature. Trout populations were affected by thermal habitat conditions; however, native rainbow trout, *Oncorhynchus mykiss* (Walbaum) were more limited by higher water temperatures than nonnative brown trout, *Salmo trutta* L.

Keywords: Trout, water temperature, discharge, habitat, stream ecology.

9.1 Introduction

Traditional techniques used to assess flow requirements for salmonids address issues of habitat quantity, but not habitat quality. The three most prominent instream flow methodologies in use today, Instream Flow Incremental Methodology (IFIM: Bovee & Cochnaur 1977; Bovee & Milhous 1978; Bovee 1982; Estes & Osborn 1986), Weighted Useable Area (WUA: Newcombe 1981), and the Tennant Method (Tennant 1976), have limited applicability, and frequently fail to meet tests of their validity (see Annear & Conder 1984; Mathur *et al.* 1985; Mosley 1985; Conder &

---

*Address for correspondence: University of Toronto, Department of Zoology, Ramsey Wright Room 401, Toronto, ON, Canada M55 3G5 (e-mail: dwalks@zoo.utoronto.ca)

Annear 1987; Scott & Shirvell 1987; Fausch *et al.* 1988 for reviews). In particular, no available technique addresses the issue of water temperature and how it can be linked to discharge changes.

The relationship between water temperature and discharge and how discharge and water temperature influenced trout densities in the Deschutes River of Central Oregon, USA were examined. The Upper Deschutes River currently experiences extremely low summer flows: irrigation withdrawals remove 98% of the regulated flow. Over the next 50 km water temperature increases, reaching up to 29°C, well above the upper incipient lethal temperature for trout (recorded summer 1994, Oregon Department of Fisheries and Wildlife (ODFW), unpublished data). One restorative measure that has been suggested to improve water quality and trout habitat in the upper river is to increase summer low flows by decreasing the amount of water diverted for irrigation. Whether an increase in discharge will help to increase trout populations is probably determined more by the effect of discharge on water temperature in this system than by a simple increase in the water level or habitat quantity.

9.2   Materials and methods

9.2.1   *Study site*

The Deschutes River flows north through Central Oregon, ending at the Columbia River, and drains most of the eastern flank of the Central Cascade Mountains (Fig. 9.1). Its drainage area is approximately 27 200 km^2, making it the second largest watershed in Oregon (US Bureau of Reclamation 1993). In the Upper Deschutes River, impoundments and irrigation withdrawals markedly modify the river, and have been in place since the early 1900s (Nehlsen 1995). Present-day trout numbers are much reduced from historical abundances (Nehlsen 1995).

Dams at two reservoirs upstream of Bend (Crane Prairie and Wickiup) control flow of the main stem. Hydrographical data from 1966 to 1994 for the regulated section of river downstream of Bend show higher winter and spring discharges and lower summer flows (Fig. 9.2a). Upstream of Bend, hydrographs are reversed from historical flows, with summer flows higher than winter flows (Fig. 9.2b). These highly regulated flows create relatively stable discharge in the Deschutes River study area between mid-April and October. Upstream of Bend, after the confluence with the warm water river, Little Deschutes (Fig. 9.1), the summer discharge of the Deschutes is approximately 51.0 m^3 s^{-1} with a summer near-surface mean water temperature of 16°C. Whilst passing through Bend, almost all water is diverted into irrigation canals, leaving 0.9–1.7 m^3 s^{-1} in the channel. Approximately 50 km downstream there is a marked increase in flow from underground aquifers, and this increase in discharge coincides with a decrease in temperature from over 29 to around 12°C.

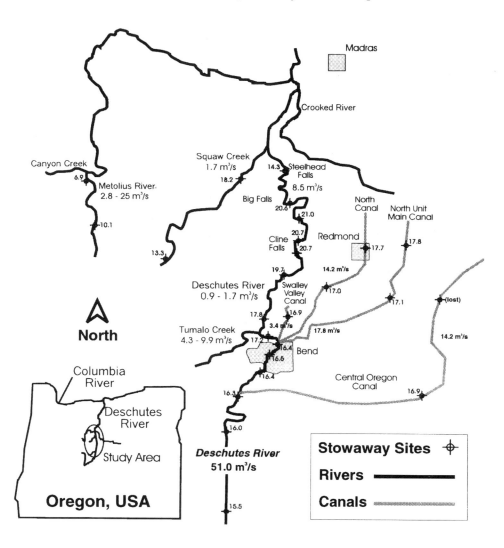

Canyon Creek
6.9
Metolius River.
2.8 - 25 m³/s
-10.1
13.3

Squaw Creek
1.7 m³/s
18.2

Madras

Crooked River

14.3 Steelhead
Falls
8.5 m³/s

Big Falls
20.6
21.0

North
Canal   North Unit
Main Canal

Cline
Falls
20.7
20.7

Redmond
17.7   17.8

19.7   14.2 m³/s

Deschutes River
0.9 - 1.7 m³/s

Swalley
Valley
Canal
17.0
17.1   (lost)

17.8
16.9
3.4 m³/s   17.8 m³/s   14.2 m³/s

Tumalo Creek
4.3 - 9.9 m³/s   17.2   16.4   Bend
16.6

16.4   Central Oregon
Canal   16.9
16.3

16.0

North

Columbia
River

Deschutes
River

Study Area

Oregon, USA

Deschutes River
51.0 m³/s

15.5

Stowaway Sites	⬦
Rivers	▬
Canals	▭

Figure 9.1   Map of the Deschutes River study area in Central Oregon showing the location of Stowaway™ thermographs in the rivers, and in four of the irrigation canals. Summer average discharges are given for rivers and canals

9.2.2   *Water temperature response to discharge changes*

Effect of discharge on water temperature was tested by comparing summer water temperatures in the main study area (flows between 0.9–1.7 $m^3 s^{-1}$), and upstream of the main study area (34.0–51.0 $m^3 s^{-1}$), with water temperatures in the irrigation canals (flows ranging from 3.1–19.8 $m^3 s^{-1}$). Twenty-six Stowaway™ water temperature data loggers were placed in the Deschutes River and in the canals (Fig. 9.1) (see Walks (1997) for details), and water temperatures were recorded every 15 min from June to late August, 1996. All sites were similar, except that the canals generally have less or no riparian cover so there is less shading of the water.

9.2.3   *Fish surveys and physical habitat data collection*

Fish surveys were conducted in the summers of 1995 and 1996. Sites were selected along the Deschutes River, Squaw Creek, Tumalo Creek and the Metolius River (Walks 1997). At 44 sites physical habitat measurements were taken and trout surveys conducted (see Walks (1997) for details). To examine the effects of water temperature, care was taken to select sites where discharge and other physical habitat characteristics were held as constant as possible. Discharge at all sites ranged from 1.7 to 3.4 m^3 s^{-1}. Sampling was carried out in water with low turbidity, some instream cover and some riparian cover; substrate type was typically basalt (lava) with gravel, cobble and rubble. Physical habitat was estimated qualitatively (see Walks (1997)). Pools were the primary habitat sampled because they were the preferred habitat for trout (Scott & Crossman 1973; Fausch 1984; Bowlby & Roff 1986; Bisson *et al.* 1988; Gelwick 1990; Decker & Erman 1992; Gowan *et al.* 1994).

Snorkel surveys were used to estimate trout density (number per unit volume). Snorkel surveys have been shown to provide an accurate count of trout, especially in clear waters (Northcote & Wilkie 1963; Bozek & Rahel 1991; Hillman *et al.* 1992). Electric fishing was not used because trout populations in the study area are already severely affected by low flows and high water temperatures (ODFW 1993; Nehlsen 1995).

9.2.4   *Analysis of trout population data*

One-way analysis of variance (ANOVA) was used to examine variance of trout density due to water temperature. Sites were characterised using maximum and mean daily water temperatures recorded for the summer as having low (mean $<10°$C, maximum $<14°$C), medium (mean 10–$19°$C, maximum 14–$24°$C) and high (mean $>19°$C, maximum $>24°$C) water temperature regimes. These correspond with recorded water temperature preferences for rainbow trout, *Oncorhynchus mykiss* (Walbaum), and brown trout, *Salmo trutta* L. (Preall & Rigler 1989; Weatherley *et al.* 1991; Elliott 1994; Diana 1995; Filbert & Hawkins 1995). Trout density data were a square-root transformed to normalise the data and equalise variances.

9.3   Results

9.3.1   *Changes in water temperature*

Water temperature on the Deschutes River was highly associated with discharge, with lower water temperatures being found in higher discharge and water temperature increasing with decreasing discharge. Maximum and mean water temperatures were higher in the river after the irrigation diversions than before the diversions or in any of the irrigation canals (Fig. 9.3). The main river after the

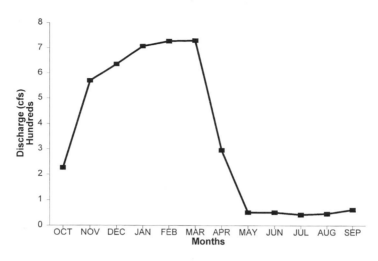

(a) Deschutes River Below Bend
Mean Daily Flows: 1966-1994

(b) Deschutes River at Wickiup
Mean Daily Flows: 1966-1990

Figure 9.2 Hydrographs for the Deschutes River (a) downstream, and (b) upstream of main irrigation diversions

diversions had a mean discharge of 1.7 $m^3 s^{-1}$, whereas mean discharge within the canals varied between 3.1 $m^3 s^{-1}$ (Swalley Valley Canal), 11.3 $m^3 s^{-1}$ (North Unit Main Canal) and 14.2 $m^3 s^{-1}$ (Pilot Butte Canal, Central Oregon Canal). The highest water temperature recorded in the river was 27.6°C, whereas the highest water temperature recorded in the canals was 22.4°C. Water temperatures rise at a relatively constant rate in the large canals, change slightly in the smaller canal, and change drastically in the river at the point of the main diversion. Based on these data, the river without any diversion could potentially have had a maximum water

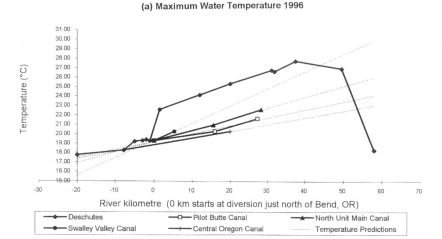

(a) Maximum Water Temperature 1996

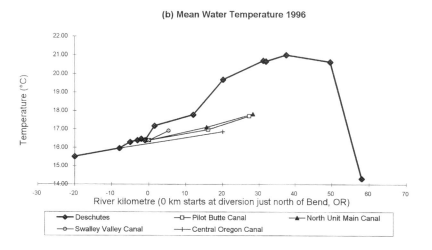

(b) Mean Water Temperature 1996

Figure 9.3 (a) Maximum, and (b) mean daily water temperatures for the Deschutes River, and four irrigation canals which divert water from the river, as recorded in summer 1996

temperature below 23°C (Fig. 9.3a). The Swalley Valley Canal, which in the summer of 1996 had a mean discharge of 3.1 m^3 s^{-1}, showed a maximum water temperature of just over 20°C when the river was up to 23°C at the same distance from the diversion (Fig. 9.3a).

A regression of mean and maximum daily water temperature against discharge, as recorded by the data loggers placed in the Deschutes River, found:

$$\text{Maximum WT} = 26.04 - 0.196 \ (Q) \qquad P < 0.0001, \ r^2 = 0.65, \ \text{and}$$
$$\text{Mean WT} = 19.33 - 0.086 \ (Q) \qquad P < 0.0001, \ r^2 = 0.42$$

where, $n = 13$ for both relationships, WT is daily water temperature in °C and Q is discharge in $m^3 s^{-1}$. These equations show that the variation in water temperature can, in part, be explained by variation in discharge. Maximum water temperature differences were highly correlated with discharge ($r^2 = 0.65$), as was mean water temperature ($r^2 = 0.42$).

9.4.2 *Water temperature effects on trout*

Trout population densities changed among the different water temperature regimes in the Deschutes River basin, and the relationship was different for rainbow trout than for brown trout. Lower rainbow trout densities were associated with water temperatures above 24°C and below 14°C ($P = 0.0001$) (Fig. 9.4). Highest rainbow trout densities were found in sites with their preferred mean water temperature, between 14–19°C. Brown trout density increased somewhat with an increase in water temperature, but not significantly ($P = 0.1205$) (Fig. 9.4).

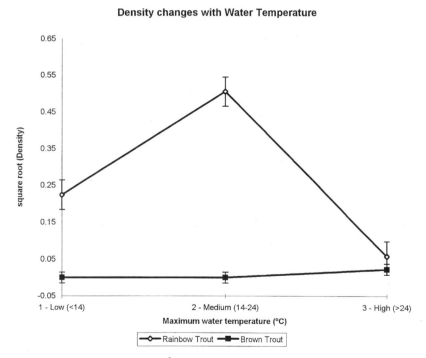

Figure 9.4 Density (abundance m^{-3}) of trout in study sites relative to changes in water maximum temperature (low: maximum <14°C, medium: 14–24°C, high: >24°C)

9.5   Discussion

The Upper Deschutes River maximum daily water temperature, under its current summer flow regime, exceeds the incipient lethal level for trout of approximately 25–26°C (see Preall & Rigler 1989; Jensen 1990; Weatherley *et al.* 1991; Elliott 1994), and the mean water temperatures are far above those preferred by trout or which benefit trout growth (Papoutsoglou & Papaparaskeva-Papoutsoglou 1978; Filbert & Hawkins 1995). Because of the direct physiological effects of high water temperature, the extreme temperature conditions of the Deschutes River probably denote critical habitat limitations for rainbow and brown trouts (Christie & Regier 1988; Regier *et al.* 1990; Bailey *et al.* 1991; Vigg & Burley 1991). The results indicate that: (i) the trout in the Deschutes River study basin are found in higher densities in their preferred thermal habitat; (ii) water temperatures in the Deschutes River are directly related to discharge levels; and (iii) the amount of thermal habitat would be likely to increase with an increase in discharge, overall having the effect of increasing both habitat quality and the quantity of habitat available to trout. If a typical study had been conducted, using one of the traditional techniques to determine instream flow minimums, it is likely that the importance of water temperature to this system would have been ignored.

The response of rainbow trout populations to the different water temperature regimes was expected from their temperature preferences (Papoutsoglou & Papaparaskeva-Papoutsoglou 1978; Diana 1995; Filbert & Hawkins 1995). Rainbow trout population density was highest at sites with maximum water temperatures between 14°C and 24°C (mean 10–19°C), and decreased significantly at lower and higher water temperatures. These low and high temperature habitats probably represent a range of temperatures that are limiting to trout density. Therefore, high summer water temperature in the Upper Deschutes River appears to be limiting to rainbow trout populations in this system. Brown trout densities did not respond to changing water temperature in the same manner as did rainbow trout densities. This difference could represent a difference in preferences for water temperature (Reeves *et al.* 1993; Reeves *et al.* 1987; Lin 1995; Regier *et al.* 1996), with the introduced brown trout adapted to, or less sensitive to, a broader range of water temperatures.

Water temperature has been shown to be influenced by discharge (e.g. Voelz & Ward 1990; Adams *et al.* 1993; Li *et al.* 1994). The canals 20 km from the irrigation withdrawals were 20–21°C, whereas the main river reached 25.2°C. The main difference was the amount of water. If anything, one would expect the water temperature in the canals to be influenced by their lack of shade (neither bankside vegetation nor topographic), whereas the river runs through a valley canyon and also has some riparian shading. The relatively lower summer water temperatures in the canals demonstrate that an increase in discharge will probably lower water temperatures in the main river, and that even a doubling of the mean summer flow, from $1.7 \text{ m}^3 \text{ s}^{-1}$ to $2.8 \text{ m}^3 \text{ s}^{-1}$, could result in a biologically significant reduction in water temperature in the main river.

Acknowledgements

Support for this project comes from the US Department of Agriculture CSREES grant 995–37102–2336. We would also like to thank Dan Sabath and Renee Ripley, our field-assistants, who also spent countless hours entering and double-checking data.

References

Adams R.M., Berrens R.P., Cerda A., Li H.W. & Klingeman P.C. (1993) Developing a Bioeconomic Model for Riverine Management: Case of the John Day River, Oregon. *Rivers* **4**, 213–226.

Annear T.C. & Conder A.L. (1984) Relative bias of several fisheries instream flow methods. *North American Journal of Fisheries Management* **4**, 531–539.

Bailey, J., Sephtoon, D. & Driedzic, W.R. (1991) Input of an acute temperature change on performance and metabolism of pickerel (*Esox niger*) and eel (*Anguilla rostrata*) hearts. *Physiology and Zoology* **64**, 697–716.

Bisson P.A., Sullivan K. & Nielsen J.L. (1988) Channel Hydraulics, habitat use, and body form of juvenile coho salmon, steelhead, and cutthroat trout in streams. *Transactions of the American Fisheries Society* **117**, 262–273.

Bovee K. (1982) A guide to stream habitat analysis using the Instream Flow Incremental Metholodogy, *Instream Flow Information Paper 12, FWS/OBS 82/26*. Fort Collins, CO: Cooperative Instream Flow Service Group, US Fish and Wildlife Service, 26 pp.

Bovee K. & Cochnaur T. (1977) Development and evaluation of weighted criteria probablility-of-use curves for instream flow assessments, *Fisheries Instream Flow Information Paper 3, FWS/OBS 77/63*. Fort Collins, CO: Cooperative Instream Flow Service Group, US Fish and Wildlife Service, 53 pp.

Bovee K. & Milhous R. (1978) Hydraulic simulation in instream flow studies: theory and technique, *Instream Flow Information Paper 5, FWS/OBS 78/33*. Fort Collins, CO: Cooperative Instream Flow Service Group, US Fish and Wildlife Service, 33 pp.

Bowlby J.N. & Roff J.C. (1986) Trout biomass and habitat relationships in Southern Ontario streams. *Transactions of the American Fisheries Society* **115**, 503–514.

Bozek M.A. & Rahel F.J. (1991) Comparison of streamside visual counts to electrofishing estimates of Colorado River cutthroat trout fry and adults. *North American Journal of Fisheries Management* **11**, 38–42.

Christie G.C. & Regier H.A. (1988) Measures of optimal thermal habitat and their relationship to yields for four commercial fish species. *Canadian Journal of Fisheries and Aquatic Sciences* **45**, 301–314.

Conder A.L. & Annear T.C. (1987) Test of weighted usable area estimates derived from a PHABSIM model for instream flow studies on trout streams. *North American Journal of Fisheries Management* **7**, 339–350.

Decker L.M. & Erman D.C. (1992) Short-term seasonal-changes in composition and abundance of fish in Sagehen creek, California. *Transactions of the American Fisheries Society* **121**, 297–306.

Diana J.S. (1995) *Biology and Ecology of Fishes*. LCC, IN: Cooper Publishing, 441 pp.

Elliott J.M. (1994) *Quantitative Ecology and the Brown Trout*. Oxford: Oxford University Press, 286 pp.

Estes C.C. & Osborn J.F. (1986) Review and analysis of methods for quantifying instream flow requirements. *Water Resources Bulletin* **22**, 389–397.

Fausch K.D. (1984) Profitable stream positions for salmonids: relating specific growth rate to net energy gain. *Canadian Journal of Zoology* **62**, 441–451.

Fausch K.D., Hawkes C.L. & Parsons M.G. (1988) Models that predict standing crop of stream fish from habitat variables: 1950–85, *General Technical Report PNW-GTR-213*. USDA Forest Service, 52 pp.

Filbert R.B. & Hawkins C.P. (1995) Variation in condition of rainbow trout in relation to food, temperature, and individual length in Green River, Utah. *Transactions of the American Fisheries Society* **124**, 824–835.

Gelwick F.P. (1990) Longitudinal and temporal comparisons of riffle and pool fish assemblages in a Northeastern Oklahoma Ozark stream. *Copeia* **4**, 1072–1082.

Gowan C., Young M.K., Fausch K.D. & Riley S.C. (1994) Restricted movement in resident stream salmonids: A paradigm lost? *Canadian Journal of Fisheries and Aquatic Sciences* **51**, 2626–2637.

Hillman T.W., Mullan J.W. & Griffith J.S. (1992) Accuracy of underwater counts of juvenile chinook and coho salmon and steelhead. *North American Journal of Fisheries Management* **21**, 590–603.

Jensen A.J. (1990) Growth of young migratory brown trout (*Salmo trutta*) correlated with water temperature in Norwegian rivers. *Journal of Animal Ecology* **59**, 603–614.

Li H.W., Lamberti G.A., Pearsons T.N., Tait C.K., Li J.L. & Buckhouse J.C. (1994) Cumulative effects of riparian disturbances along high desert trout streams of the John Day basin, Oregon. *Transactions of the American Fisheries Society* **123**, 627–640.

Lin P. (1995) Adaptations to temperature in fish: Salmonids, Centrarchids, and Percids. PhD thesis, University of Toronto, 171 pp.

Mathur D., Bason W.H., Purdy E.J. & Silver C.A. (1985) A critique of the instream flow incremental methodology. *Canadian Journal of Fisheries and Aquatic Sciences* **42**, 825–831.

Mosley M.P. (1985) River channel inventory, habitat and instream flow assessment. *Progress in Physical Geography* **9**, 494–523.

Nehlsen W. (1995) *Historical Salmon and Steelhead Runs of the Upper Deschutes River and their Environments*. Portland, OR: Pacific Rivers Council, 71 pp.

Newcombe C. (1981) A procedure to estimate changes in fish population caused by changes in stream discharge. *Transactions of the American Fisheries Society* **110**, 382–390.

Northcote T.G. & Wilkie D.W. (1963) Underwater census of stream fish populations. *Transactions of the American Fisheries Society* **92**, 146–151.

Oregon Department of Fisheries and Wildlife (ODFW) (1993) *Aquatic Inventory Project Stream Report, Deschutes River, Steelhead Falls to City of Bend*. Bend, OR: ODFW, 48 pp.

Papoutsoglou S.E. & Papaparaskeva-Papoutsoglou E. (1978) Effect of water temperature on growth rate and body composition of rainbow trout (*Salmo gairdneri*, R.) fry, fed on maximum ration in closed system. *Thalassographica* **2**, 83–97.

Preall R.J. & Rigler N.H. (1989) Comparison of actual and potential growth rates of brown trout (*Salmo trutta*) in natural streams based on bioenergetic models. *Canadian Journal of Fisheries and Aquatic Sciences* **46**, 1067–1076.

Reeves G.H., Everest F.H. & Hall J.D. (1987) Interactions between the redside shiner (*Richardsonius baltesatus*) and the steelhead trout (*Salmo gairdneri*) in western Oregon: the influence of water temperature. *Canadian Journal of Fisheries and Aquatic Sciences* **44**, 1603–1613.

Reeves G.H., Everest F.H. & Sedell J.R. (1993) Diversity of juvenile anadromous salmonid assemblages in Coastal Oregon Basins with different levels of timber harvest. *Transactions of the American Fisheries Society* **122**, 309–317.

Regier H.A., Holmes J.A. & Pauly D. (1990) Influence of temperature changes on aquatic ecosystems: An interpretation of empirical data. *Transactions of the American Fisheries Society* **119**, 374–389.

Regier H.A., Lin P., Ing K.K. & Wichert G.A. (1996) Likely responses to climate change of fish associations in the Laurentian Great Lakes BasIn concepts, methods and findings. *Boreal Environmental Research* **1**, 1–15.

Scott W.B. & Crossman E.J. (1973) Freshwater fishes of Canada. *Bulletin of the Fisheries Research Board Canada* **184**, 966 pp.

Scott D. & Shirvell C.S. (1987) A critique of the instream flow incremental methodology and observations on flow determinations in New Zealand. *Proceedings of the Third International Symposium on Regulated Streams, Edmonton, Alberta, August 4–8, 1985.* Alberta Enrivonment and Alberta Water Resources Commission, pp. 27–43.

Tennant D.L. (1976) Instream flow regimens for fish, wildlife, recreation and related environmental resources. *Fisheries* **1**, 6–10.

US Bureau of Reclamation, US Department of Interior (1993) *Surface Water Quality Study Report.* Denver, CO: Upper Deschutes River Basin Water Conservation Project, 54 pp.

Vigg S. & Burley C.D. (1991) Temperature-dependent maximum daily consumption of juvenile salmonids by Northern Squawfish (*Ptychocheilus oregonensis*) from the Columbia River. *Canadian Journal of Fisheries and Aquatic Sciences* **48**, 2491–2498.

Voelz N.J. & Ward J.V. (1990) Macroinvertebrate responses along a complex regulated stream environmental gradient. *Regulated Rivers: Research and Management* **5**, 365–374.

Walks D.J. (1997) Discharge and its consequences to physical habitat and trout populations in the Deschutes River of Central Oregon. MSc thesis, Oregon State University, Corvallis, Oregon, 79 pp.

Weatherley N.S., Campbell-Lendrum E.W. & Ormerod S.J. (1991) The growth of brown trout (*Salmo trutta*) in mild winters and summer droughts in upland Wales: model validation and preliminary predictions. *Freshwater Biology* **26**, 121–131.

Dynamics of a population of brown trout (*Salmo trutta*) and fluctuations in physical habitat conditions – experiments on a stream in the Pyrenees; first results

V. GOURAUD and C. SABATON

Electricité de France, Research and Development Division, Environment Branch – 6, quai Watier 78400 Chatou, France (e-mail: veronique.gouraud@edf.fr)

P. BARAN and P. LIM

Ecole Nationale Supérieure Agronomique de Toulouse, Avenue de l'Agrobiopôle BP107 Auzeville Tolosane, F-31326 Castanet-Tolosan Cedex, France

Abstract

The physical habitat is considered to be a potential limiting factor for fish populations. The object of this study was to test how physical habitat influences the dynamics of brown trout, *Salmo trutta* L., populations. Experiments were undertaken on a stream in the French Pyrenees with fluctuating physical habitat conditions. Parallel to the habitat study, the trout population was studied by means of three inventories per year at seven representative sites on the stream. Strong temporal and spatial variations were observed in biomass in the stream. The relatively high occupancy rates observed differed from one site to another. Changes in biomass at different life stages (fry, juveniles, adults) in the course of one year of monitoring were simulated according to observed growth, natural mortality and simultaneous fluctuations in carrying capacity. The juvenile biomass calculated over the course of 1996 was rarely limited by the habitat, which may explain the high survival rates found for this life stage. Adult biomass, on the other hand, was subject to habitat limitation in spring and in summer.

Keywords: Habitat, population dynamics, PHABSIM, *Salmo trutta,*

10.1   Introduction

In a natural environment, four types of biological processes regulate the abundance of organisms: recruitment, mortality, emigration and immigration. In each of these processes, there is strong interaction with the characteristics of the environment. Among the environmental factors, physical habitat plays a predominant role *vis-à-vis* salmonids, and especially the brown trout, *Salmo trutta* L. (Chapman 1966; Heggenes 1988; Baran *et al.* 1997). In general, dynamic and habitat-model approaches are rarely adopted in the same study (Binns & Eiserman 1979; Clark *et al.* 1980; Elliott 1985; Milner *et al.* 1985). Numerous authors separate spatial variations in fish stocks from temporal variations (e.g. Platts & Nelson 1988; Gouraud *et al.* 1998). Annual variations in fish abundance are often related to major fluctuations in recruitment (Elliott 1994), or to floods or low-water periods (Jowett & Richarson 1989; Underwood & Bennett 1992). Conversely, spatial variations are explained in terms of habitat features and especially physical habitat characteristics (Souchon *et al.* 1989; Baran *et al.* 1995). The development of numeric models such as IFIM (Bovee 1982) has enabled quantification of the potential of a physical habitat for the different stages in development of the brown trout. The WUA-abundance ratio is occasionally satisfactory during limiting periods corresponding to low carrying capacity (Orth 1987). It is clear, however, that the biomass of trout in a river at any given moment is dependent on many biotic or abiotic factors whose evolution over time determines the entire dynamics of the population. For salmonids, the link between fluctuations in stocks and carrying capacity can be explained by density-dependent mortality related to the territorial behaviour of the individuals; while extensively described for the fry stage (Mortensen 1977b; Elliott 1984), this phenomenon has been studied little or not at all for other life-stages. The hypothesis that density is limited by the size of the territory has been examined for salmonids (Symons 1971; Grant & Kramer 1990), but only rarely has this type of behavioural observation, often restricted to the scale of the individual, been applied to the dynamics of entire populations.

The objective of this study was to analyse the role of physical habitat, described in terms of WUA (Bovee 1982), in regulating density and biomass of trout juveniles and adults in a small stream in the central French Pyrenees, the Neste d'Oueil. In particular, an attempt was made to determine whether the phenomena of mortality or emigration observed among these size classes in the course of one year were related to an exceeding of the carrying capacity of a stream during limiting episodes.

10.2   Materials and methods

10.2.1   *Study site*

The Neste d'Oueil is a sub-tributary of the Garonne watershed in the Pyrenees. It originates at an altitude of 1450 m by the confluence of several streams. It joins with the One at 765 m in altitude, 9.2 km downstream. Its mean slope is 11.8% and the width varies from 4 to 6 m. The hydrological regime is of the pluvio-nival type with two low-water periods (winter and summer), and high water levels during the spring melt.

According to the geomorphological criteria defined by Amoros and Petts (1993), the primary segment of the Neste d'Oueil comprises three major types of reach. Successively, first a plateau reach downstream (3450 m in length – slight slope of <5% – broad alluvial floor), a mountainous reach (3320 m in length – moderate slope of 5–8% – relatively narrow alluvial floor) and a gorge reach (3200 m in length – very narrow alluvial floor – steep slope of >10%). Each reach is characterised by a succession of mesohabitats. A complete survey of the area enabled determination of the percentage of representativeness of the different groups of mesohabitats, as defined by Delacoste *et al.* (1995). On the basis of this morphodynamic survey, seven study sites were selected (Fig. 10.1). Their principal characteristics are given in Table 10.1.

10.2.2   *Fish populations*

At all seven study sites, fish populations were sampled in July and September 1996, February and July 1997, by electric fishing. Before sampling, each unit of the study site was isolated upstream and downstream by block nets (mesh size 8 mm). Fish

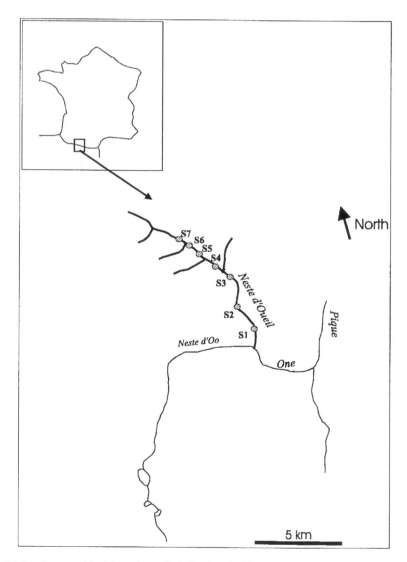

Figure 10.1   Geographical location of study sites in France

Table 10.1   Main features of the study sites on the Neste d'Oueil

Site	Slope (%)	Length (m)	Width (m)	Cascade (%)	Ladder (%)	Rapid (%)	Riffle (%)	Glide (%)	Pool (%)
Gorges (S1)	10	50.7	5.5	60	0	16	0	0	24
St-Paul (S2)	4.4	177.8	5.5	3	26	24	0	18	29
Lower Mayrègne (S3)	3.8	140.4	4.3	0	0	35	4	46	15
Upper Mayrègne (S4)	3.7	162.0	4.5	0	7	0	66	9	18
Caubous (S5)	3	54.9	4.9	0	21	27	29	23	0
Cirès (S6)	2	150.5	3.5	0	19	7	36	26	12
Bourg d'Oueil (S7)	5.3	112.8	2.9	0	12	0	63	25	0

numbers were counted for each mesohabitat and each age class. Fork length (mm) and weight (g) were measured. Scales were taken from 30 to 50 individuals at three sites to determine growth.

Trout biomass and density were calculated using the DeLury method (DeLury 1951). The efficiency of the method for fry density $(0+)$ was calculated by separating this age class from the others. At each site, the population structure was divided into three size classes (fry: <110 mm; juveniles: 110–160 mm; adults: >160 mm). Mean density and biomass of trout were compared between sites, inventories and meso-habitats. As a normal distribution of the variables was respected, an ANOVA was performed.

10.2.3   *Habitat study*

The quality of the physical habitat was estimated using the IFIM model (Bovee 1982). At each of the seven study sites, depth, velocity, and substrate composition were sampled at two different flow rates along predetermined transects. Weighted usable areas (WUAs) were calculated for the different life stages of brown trout (fry, juvenile, adult) using probability-of-use curves obtained on rivers in the Pyrenees (Belaud *et al.* 1989), subsequently modified to allow for the substrate parameter (Sabaton *et al.* 1995). A simplified hydraulic model (Sabaton *et al.* 1995) was used to enable interpolation of WUAs. This interpolation is based on measurements taken at the two flow rates studied only, and cannot be satisfactorily used for extremely different values.

10.2.4   *Physical habitat and trout biomass*

Spatial distribution of fish – concept of carrying capacity

WUAs in each mesohabitat inventoried were measured during the first two electric fishing surveys and extrapolated for the following two (when discharge was similar).

These WUAs were compared with the biomass fished to determine the different occupancy rates (in g m^{-2} of WUA) found at each site during the study. The maximum occupancy rates observed at each site served to estimate the corresponding carrying capacity per m^2 of WUA; the carrying capacity per m^2 of WUA is defined as the maximum biomass, for any given life stage, that the site can accommodate in one m^2 of WUA observed at any time in the year (expressed in g m^{-2} of WUA). It is then possible, for a given site and WUA value, to calculate the maximum biomass of the life-stage studied which can be accommodated at the site, with this biomass defining the total carrying capacity of the site.

Evolution of biomass and carrying capacity over time

Two liquid level recorders located upstream and downstream in the river monitored variations in discharge every 20 min. The simplified hydraulic model was used to reconstitute WUA chronologies for the different sites during the period, and subsequently the corresponding carrying capacity chronologies for the sites. Taking the numbers of individuals observed in July 1996, a monthly simulation of changes in biomass was performed, applying the growth rates actually observed to these numbers and testing different morality rate values. Mortality as conceived here includes two components: density-dependent and non-density-dependent. The non-density-dependent component refers to the mean natural mortality. Density dependent mortality was solely controlled by biomass and the habitat available to support it. Monthly biomass for a given life stage is calculated to allow for growth and natural mortality. When biomass exceeds the carrying capacity of the site (calculated on the basis of a monthly WUA chronology), additional mortality was accrued such that biomass was strictly limited to the carrying capacity.

The first step in the simulation applies the growth and natural mortality rates monthly to the biomass, as follows:

For $i \geq 1$ and $m \neq 1$

$$B_i\ (t,\ m) = S_i.B_i(t,\ m-1)\ W_i\ (t,\ m)/W_i\ (t,\ m-1)$$

where $B_i\ (t, m)$ is the biomass of age class i in month m of year t; $W_i\ (t, m)$ is the mean weight of age class i in month m of year t; and S_i is the rate of monthly natural survival (corresponding to the tested yearly survival rate) for age class i.

For $i \geq 1$ and $m = 1$(passage from year $t - 1$ to year t)

$$B_i(t,\ m) = S_i.B_i\ (t-1,\ 12)\ W_i\ (t,\ m)/W_i\ (t-1,\ 12)$$

For $i \geq 1$ and $m = 3$ (passage of each age class to the next higher age class)

$$B_i\ (t,\ m) = S_{i-1}.B_{i-1}\ (t-1,\ m-1)\ W_i\ (t,\ m)/W_{i-1}\ (t-1,\ m-1)$$

For each time step, the population is then divided into development stages (fry [al], juvenile [juv], adult [ad]) in accordance with the mean size of individuals in each age class. This gives the biomasses for each development stage: *Bad* (*t, m*), *Bjuv* (*t, m*), *Bal* (*t, m*). The second step compares these three biomasses with the carrying capacities of the site for the same month *m* –*CCad* (*t, m*), *CCjuv* (*t, m*), *CCal* (*t, m*) –, giving, for the adults, for example:

If *Bad* (*t, m*) ≤ *CCad* (*t, m*) : biomass not limited, therefore unchanged;

If *Bad* (*t, m*) > *CCad* (*t, m*) : *Bad* (*t, m*) = *CCad* (*t, m*).

The results of this simulation are compared with biomasses estimated during the fishing surveys, taking the different levels of natural mortality adopted (from 0 to 60% for juveniles and adults). For fry, the rate of natural mortality is set at 95% (Baglinière & Maisse 1991).

10.3   Results

10.3.1   *Densities and biomass of caught fish*

Total population

Trout density and biomass ranged from moderate to high, depending on the site (from 54–194) (Fig. 10.2). A significant difference was found in total biomass and density at the seven sites ($P < 0.05$); a significant difference was also observed among the four fishing surveys ($P < 0.05$), but only in terms of biomass. With respect to spatial distribution, a classic breakdown of the different size classes was found. Adults were considerably more abundant ($P < 0.05$) in pool and glide mesohabitats than in riffles or rapids. The different trout size classes were thus distributed in accordance with a physical habitat gradient.

Juveniles (110–160 mm)

Depending on the site and the season, the juveniles corresponded to individuals in age classes 1+ and/or 2+. The fluctuations observed in juvenile biomasses (Table 10.2) were regulated by various phenomena, including mortality, growth and passage to the adult stage. The increase in numbers between July and September 1996 at the Les Gorges, upper Mayrègne, Caubous and Cirès sites corresponded to immigration, which is reflected in the increase in juvenile biomasses over this period at Caubous and Les Gorges. At the other two sites, this was masked by the presence of 2+ individuals, still at the juvenile stage in July, and by their passage to the adult stage in September. Immigration also occurred at the Les Gorges and lower Mayrègne sites between September 1996 and February 1997.

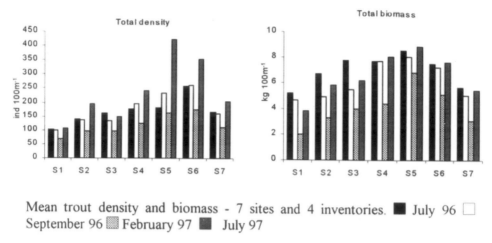

Mean trout density and biomass - 7 sites and 4 inventories. ■ July 96 □
September 96 ▓ February 97 ■ July 97

Figure 10.2   Mean trout density (individuals 100 m^{-1}) and biomass (g m^{-2}) – seven sites and
four inventories

Adults

Globally, a drop in biomass was observed from July 1996 to February 1997,
corresponding to mortality or emigration of adults. The increase observed between
February and July 1997 was due to the passage of 2+ individuals into the adult
population (Table 10.2). This passage to the adult stage occurs at different periods,
depending on the year and the site. Upstream – Bourg d'Oueil – individuals become
adult in October of their third year (when they are 2+), explaining the increase in
biomass observed during the inventory in February 1997. Downstream – Les Gorges
and Saint Paul – the growth rate was higher (due to higher temperatures) and the
passage occurred earlier (May–June), hence the increases observed in July 1997. At
intermediate sites, growth rates varied from year to year. In the upper Mayrègne,

Table 10.2   Biomass of juveniles and adults at each site (in g m^{-2}) – seven sites and four
inventories

Site	Juvenile				Adult			
Date	Jul 96	Sept 96	Feb 97	Jul 97	Jul 96	Sept 96	Feb 97	Jul 97
Gorges (S1)	9.6	11.1	11.5	9.7	41.1	35.0	7.1	32.8
St Paul (S2)	11.8	12.0	8.5	9.4	55.1	36.0	25.9	47.9
Lower Mayrègne (S3)	25.9	7.0	7.2	4.6	51.3	46.5	37.9	56.5
Upper Mayrègne (S4)	26.4	13.3	8.5	8.3	44.8	56.8	35.9	59.4
Caubous (S5)	8.1	13.4	11.4	32.5	77.6	64.4	66.3	60.1
Cirès (S6)	32.4	15.7	12.3	25.8	41.8	53.0	39.1	50.2
Bourg d'Oueil (S7)	24.5	31.6	8.3	12.2	25.6	18.3	25.2	35.7

lower Mayrègne and Cirès, the populations of juveniles observed in July 1996 comprised of two cohorts (1 + and 2 +); passage of 2 + individuals to the adult stage in September explains the differences found during the autumn 1996 survey. It should be noted that in July 1997, the juvenile stage consisted of only one 1 + cohort and that a similar increase must not have occurred in September 1997. Such increases were not found in the lower Mayrègne, suggesting a substantial disappearance of individuals.

10.3.2   *Mortality*

Survival of a given age class between fishing surveys was estimated by the ratio between the number of individuals found during the second survey and the number of individuals observed in the age class they formed in the first survey. Annual mortality thus estimated varied with the age class of the trout. It was low for the 1+ (7%) and then rose markedly for the other age classes (43%, 72% and 61% respectively for 2+, 3+ and 4+ individuals). In addition, a significant difference was found between seasons ($P < 0.05$). On average, the observed mortality of the different age classes was greater between September 1996 and February 1997 (autumn and winter: 38.2%) than during the other two seasons (from July 1996 to September 1996: 25.5%, and from February 1997 to July 1997: 21.6%).

10.3.3   *Growth*

Growth was relatively slow at all seven sites. There was a gradient from upstream to downstream, linked to the rise in water temperature. Fish do not reach the legal size for capture before their fourth year. Individuals lose weight during the winter. The mean weight loss varied with age: 11% for 1+, 9% for 2+, 5.2% for 3+, 8% for 4+, and 10.2% for 5+.

10.3.4   *Habitat*

At all sites, the adult and juvenile WUAs increased with discharge up to a maximum between 0.5 and 1 $m^3 s^{-1}$ depending on the site; they subsequently decreased with higher discharges. This decrease in WUA was greater for juveniles than for adults, juveniles being more sensitive to increased depth and velocity. Within the range of discharges studied, WUAs for adult stages at downstream sites (Les Gorges, Saint Paul, lower Mayrègne) were relatively homogeneous and notably larger than at upstream sites. For juveniles, it was the intermediate sites (upper Mayrègne, lower Mayrègne, Saint Paul) which had the best WUAs. At all sites over the entire study period, habitat quality on a monthly scale was better for juveniles than for adults.

10.3.5   *Biomass and habitat quality: determination of carrying capacity*

Globally for all sites, the trout biomass was highest in mesohabitats with best habitat quality. The biomass/WUA ratio (occupancy rate) was, however, not constant, and depended on the mesohabitat. The carrying capacities (m^{-2} of WUA) were estimated in each site by the maximum occupancy rates observed (Table 10.3).

10.3.6   *Evolution of simulated biomass and carrying capacity over time*

The results of monthly simulations of the change in biomass, taking as initial data the biomass observed in July 1996, are given in Figures 10.3 and 10.4. These figures also show the simultaneous changes in carrying capacity during the year simulated (July 1996 to July 1997).

Juveniles

At all sites except Caubous, the biomasses were subject to an insufficient carrying capacity at least once a year, resulting in limitation by the habitat (according to the hypotheses adopted in the simulation). These limitations occurred in the simulation when the juvenile stage was composed of two cohorts (1+ and 2+); this phenomenon was observed in reality at many sites between July and September, when 2+ individuals had not yet reached adult size. It should, however, be noted that the Saint Paul population was subject to limitation when composed of only one cohort. The very low estimations obtained for juveniles in July 1997 were most certainly due to a serious underestimation of fry in September 1996 (initial simulation values), which was attributed to the poor efficiency of the fishing surveys for small individuals. For this reason, the simulations performed for the fry stage between July 1996 and February 1997 are not presented here.

Adults

At almost all sites, the adult biomass was limited at least once between March and July (arrival of the 2+ and strong weight gain) for all hypotheses of natural mortality $\leq 40\%$. The end of the summer period (low-water period leading to a slight drop in WUA, coincidental with the arrival of 2+ individuals at some sites) may be

Table 10.3   Maximum occupancy rates observed at each site (in g m^{-2} of WUA)

	Gorges	St Paul	Lower May	Upper May	Caubous	Cirès	Bourg
Juvenile	12	8	16	17	31	28	39.5
Adult	51	58	63	79	125	102	87

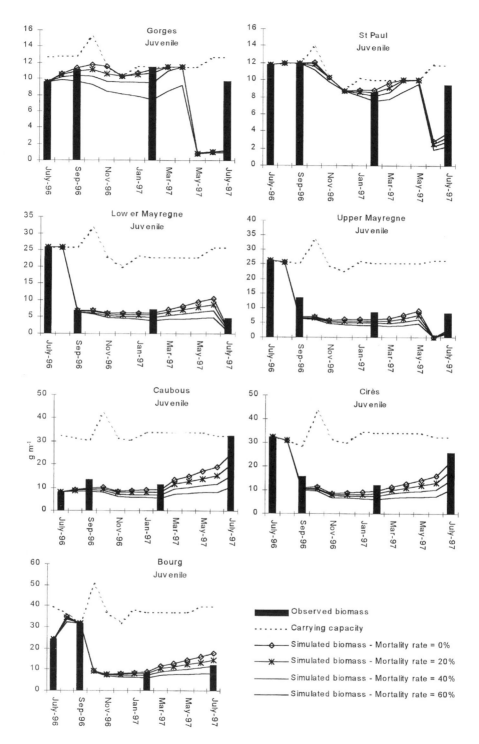

Figure 10.3 Simultaneous change in carrying capacity, observed biomass and simulated biomass – juveniles

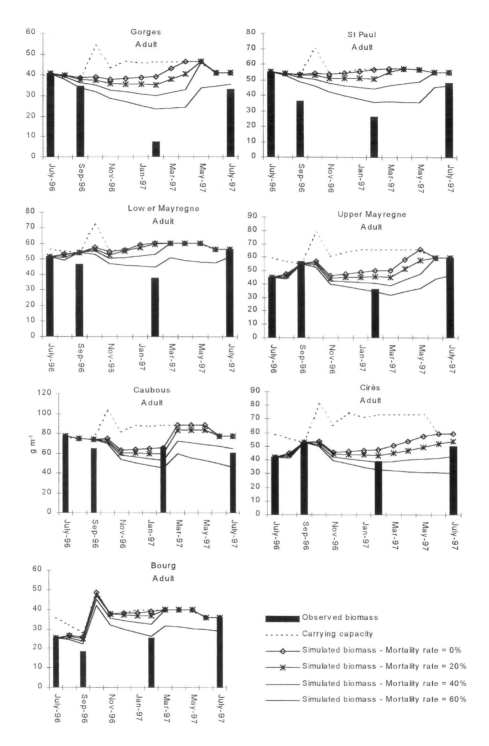

Figure 10.4 Simultaneous change in carrying capacity, observed biomass and simulated biomass – adults

limiting. Whatever the rate of natural mortality simulated, this limitation occurred at the lower Mayrègne, upper Mayrègne, Caubous and Cirès sites. The simulated biomasses were consistent with trends observed. This concordance is more or less satisfactory, depending on the rate of natural mortality adopted, with the best rate varying from 20 to 60%, depending on the site.

10.4   Discussion

10.4.1   *Variations in density and biomass*

Trout biomass and density showed both temporal and spatial variations in the Neste d'Oueil. Spatial variations have been widely documented as being closely linked with habitat conditions (Heggenes 1988; Haury *et al.* 1991). The concept of the 'carrying capacity' of an environment is frequently used in this connection (Platts & Nelson 1988; Baran *et al.* 1996). The significant differences in biomass observed from one site, and more especially one type of mesohabitat, to another, illustrate this relationship between populations and habitat quality. A classic distribution of abundances according to habitat type was found, with a strong preference on the part of adults for pools (Baran *et al.* 1997). This distribution is closely dependent on individual needs in terms of physical habitat (Baran *et al.* 1997). Many authors, however, also underscore the importance of year-to-year variations in abundance of brown trout populations (Crisp *et al.* 1974; Moyle & Vondracek 1985; Kelly-Quinn & Bracken 1988; Platts & Nelson 1988).

10.4.2   *Mortality*

Trout mortality differs with the site and the age class. On average, annual mortality of 1+ individuals was 7%. This was relatively low in comparison with other studies in which mortality varies from 55 to 83% (Egglishaw & Shackley 1977; Mortensen 1977b; Rasmussen 1986; Baglinière & Maisse 1991; Elliott 1994). It is important to note, however, that the studies mentioned were conducted on small streams (0.5–3 m in width) in which the carrying capacity was lower than in the Neste d'Oueil. This stream, varying in width from 4 to 7 m, provides a larger environment and, in particular, has deep areas more favourable to juveniles. Furthermore, the studies conducted by Egglishaw and Shackley (1977), Baglinière and Maisse (1991) and Elliott (1994) involved a type of watershed in which, in autumn and especially in spring, fish one or two summers old migrate from the streams to the main river. The mortality rates observed by these authors integrated both migration and natural mortality. Migration may occur from affluents into the Neste d'Oueil; this would explain the arrival of new 1+ individuals and the low estimated rates of mortality. This hypothesis could only be tested by marking the fish in tributary streams. For the other age classes, mortality increased in proportion to the age of the fish, as found in all studies of population dynamics. Note should be taken of the rapid increase in mortality after the age of 2 (42.4%).

10.4.3 *Role of the physical habitat in population dynamics*

Carrying capacity

The rates of use of the WUA are high in this study (on average, 81 g m^{-2} of WUA) in comparison with available data (65 g m^{-2}, Bovee 1982; 30 g m^{-2}, Souchon *et al.* 1989). These high occupancy rates can be linked to the considerable trophic resources in the river (mean invertebrate density > 12 103 ind. m^{-2}, mean invertebrate biomass > 14 g m^{-2}). Mortensen (1988), Rasmussen (1986) and Lobon-Cervia *et al.* (1986) pointed out the close concordance between high trout production and density and trophic richness in a river. However, this high trophic availability and the high occupancy rates may not be truly interdependent, but rather simultaneously related to other positive factors in the environment aside from the physical habitat (water quality, cover, shelter, etc.). On a mesohabitat scale, Baran (1995) showed the existence of a maximum level of habitat use (120 g m^{-2} of WUA) which corresponds to the carrying capacity of these habitat units. However, as in the sites studied here, on a scale larger than the mesohabitat, he found significant variations in occupancy which depend on other environmental factors. In particular, a decrease in the rate of use of the habitat was found in reaches with high gradient (in S1 and S2 with a slope of > 5%) (Table 10.3). These differences in relationship with the physical habitat depending on the gradient have already been described, with lower physical habitat occupancy rates, as in the present study, in steep sloping reaches (Kennedy & Strange 1982; Chisholm & Hubert 1986; Baran *et al.* 1995). The type and variety of mesohabitats at a given site or in one reach also influence the occupancy rates. Adult biomasses observed in deep areas increased with the percentage of riffles in the reach (Baran *et al.* 1997). The difference between the occupancy rates found upstream and downstream in this study was accentuated by a second phenomenon: 2+ individuals reach adult size (160 mm) later upstream than downstream. The juvenile stage then groups the 1+ as well as 2+ individuals; thus the juvenile biomasses observed during the July and September inventories are higher upstream than downstream, hence the higher occupancy rates upstream.

Regulation of biomass by carrying capacity

At all sites, the juvenile stage was rarely subjected to limitation by the habitat, which may partially explain the high survival rates found. For adults, however, the biomasses, simulated by growth and natural mortality only, may in certain periods exceed the carrying capacity of the environment, defined by the WUA. In summer 1996 and spring 1997, the physical habitat might play an important role in regulating adult biomasses, whereas in winter, this role does not appear to be significant. There would therefore appear to be two components to fish mortality: one habitat-dependent and one natural component. In trout populations, mechanisms of mortality or emigration are very often density dependent (Gibson 1988). These mechanisms described for fry (LeCren 1973; Mortensen 1977b; Elliott 1985) are directly related to the territorial behaviour of the brown trout (Grant & Kramer 1990). In the Neste d'Oueil, it seems that intra-species competition related to defense of a territory leads to fish mortality or emigration. However, the variations in occupancy rates suggest that the minimum territory varies greatly, depending on the site. Studies on the behaviour of salmonids are relatively consistent regarding the flexibility of territorial size (Dill *et al.* 1981; McNicol *et al.* 1985; Grant & Noakes 1988) depending on the abundance of food or pressure from predation. This flexibility might explain some of the variations in the rate of occupancy of the WUA in the Neste d'Oueil. For the spring and summer periods, physical habitat alone cannot explain the change in biomass. Natural mortality and emigration were also found not to depend on the habitat. Few data are available in the literature, because most studies on population dynamics use a global mortality, integrating the two components mentioned above. However, the role of fishing, which constitutes a mortality factor for trout over the legal size for capture (> 200 mm) may be important. In their simulation model, Clarke *et al.* (1980) used a 30% fishing mortality for a given age class. A survey of fishermen on the Neste d'Oueil in 1997 indicated that caught fish represented, on average, 25% of the biomass of trout > 200 mm (unpublished data).

For the winter period, it is important to note that the change in carrying capacity with the strong discharge in November 1997 was not simulated. It is probable that this hydrological episode led to deterioration in physical habitat conditions, and a corresponding response in trout populations. Jowett and Richardson (1989) indicated that adult trout biomasses in several New Zealand rivers drop markedly after a flood episode (between −26 and −57%). In their Habitat Quality Index (HQI) model, Binns and Eiserman (1979) noted that strong annually fluctuations in discharge were detrimental to trout biomass. The occasional deterioration in carrying capacity of the physical habitat may thus lead to a reduction in abundance of adults. Winter (September–February) also corresponds to the period of reproduction, which may induce non-habitat-dependent mortality. Also a decrease in individual weight (−9%) was found between the two fishing surveys, leading to a reduction in biomass which can be linked to harsh thermal conditions and to reproductive activity (Mortensen 1997a).

10.4 Conclusions – prospects

The role of the physical habitat in regulating trout density no longer needs to be demonstrated, but remains difficult to quantify. The future possibility of linking habitat changes to population dynamics opens up new horizons, both for estimating the short- and medium-term impact of destructive climatic episodes and for understanding the effects of a change in river management strategy.

The difficulty in estimating non-density-dependent trout mortality rates (old age, fishing, reproduction) and in determining the carrying capacity of a given reach makes it all the more difficult to highlight the density-dependent – habitat-related – element in total mortality. Nonetheless, comparisons between sites would already make it easier to explain the variations in occupancy rates observed in relation to different environmental parameters (potential for shelter, slopes, variety in mesohabitats, etc.), so as to better understand the concept of carrying capacity.

Moreover, given the strong variations observed in mortality, it would also be helpful to work on the scale of the river as a whole. This would eliminate local phenomena of emigration observed at individual sites. The response of fish to fluctuations in habitat is represented here by mortality, which limits biomass to the carrying capacity of each site. This approach will now be refined with the simulation of trout displacement through the entire river, corresponding to their search for available favourable habitats in the event of local excessive density. Specific experiments with marking and monitoring of trout will enable improved understanding of fish behaviour in the face of fluctuating carrying capacity, with a long-term view to integrating this behaviour into simulation tools.

References

Amoros C. & Petts G.E. (1993) *Hydrosystèmes fluviaux*. Paris: Masson, 295 pp.

Baglinière J.L. & Maisse G. (1991) Biologie de la truite commune (*Salmo trutta* L.) dans les rivières françaises. In J.L. Baglinière & G. Maisse (eds) *La Truite, Biologie et Ecologie,* Paris: INRA, pp. 25–46.

Baran P. (1995) Analyse de la variabilité des abondances de truites communes (*Salmo trutta* L.) dans les Pyrénées centrales Françaises. Influence des échelles d'hétérogénéité de l'habitat. Thèse de docteur en Sciences Agronomiques, Institut Nationale Polytechnique de Toulouse, 147 pp.

Baran P., Delacoste M., Poizat G., Lascaux J.M., Lek S. & Belaud A. (1995) Approche multi-échelles des relations entre les caractéristiques de l'habitat et les populations de truites (*Salmo trutta* L.) dans les Pyrénées centrales. *Bulletin Français de la Pêche et de la Pisciculture* **337/338/ 339,** 399–406.

Baran P., Lek S., Delacoste M. & Belaud A. (1996) Stochastic models that predict trout population density or biomass on a mesohabitat scale. *Hydrobiologia* **337,** 1–9.

Baran P., Delacoste M. & Lascaux J.M. (1997) Variability of mesohabitat used by brown trout populations in French central Pyrenees. *Transactions of the American Fisheries Society* **126,** 747–757.

Belaud A., Chaveroche P., Lim P. & Sabaton C. (1989) Probability-of-use curves applied to brown trout (*Salmo trutta* L.) in rivers of Southern France. *Regulated Rivers: Research and Management* **3,** 321–336.

Binns N.A. & Eiserman F.M. (1979) Quantification of fluvial trout habitat in Wyoming. *Transactions of the American Fisheries Society* **108,** 215–228.

Bovee K.D. (1978) Probability-of-use criteria for the family salmonidae, *Report FWS/OBS-78/07.* Fort Collins, CO: US Fish and Wildlife Service, 80 pp.

Bovee K.D. (1982) A guide to stream habitat analysis using instream flow incremental methodology, *Instream Flow Information Paper No. 12, FWS/OBS 82/86.* Fort Collins, CO: Cooperative Instream Flow Service Group, US Fish and Wildlife Service, 248 pp.

Chapman D.W. (1966) Food and space as regulators of salmonid populations in streams. *American Naturalist* **100**, 345–357.

Chisholm I.M. & Hubert W.A. (1986) Influence of stream gradient on standing stock of brook trout in the Snowy Range, Wyoming. *Northwest Science* **60**, 137–139.

Clark R.D., Alexander G.R. & Gowing H. (1980) Mathematical description of trout-stream fisheries. *Transaction of the American Fisheries Society* **109**, 587–601.

Crisp D.T., Mann R.H.K. & McCormack J.C. (1974) The population of fish Cow Green, Upper Teesdale, before impoundment. *Journal of Applied Ecology* **11**, 969–996.

Delacoste M., Baran P., Lek S., and Lascaux J.M. (1995). Classification et clé de détermination des faciès d'écoulement en rivière de montagne. *Bulletin Français de la Pêche et de la Pisciculture* **337–339**, 149–156.

DeLury D.B. (1951) On the planning of experiments for the estimation of fish populations. *Journal of the Fisheries Research Board of Canada* **18**, 281–307.

Dill L.M., Ydenberg R.C. & Fraser A.H.G. (1981) Food abundance and territory size in juvenile coho salmon (*Oncorhynchus kisutch*). *Canadian Journal of Zoology* **59**, 1801–1809.

Egglishaw H.J. & Shackley P.E. (1982) Influence of water depth on dispersion of juvenile salmonids *Salmo salar* L. and *trutta* L. in a Scottish stream. *Journal of Fish Biology* **21**, 141–155.

Elliott J.M. (1984) Growth, size, biomass and production of young migratory trout *Salmo trutta* in a Lake District stream. *Journal of Animal Ecology* **53**, 979–994.

Elliott J.M. (1985) Population regulation for different life-stages of migratory trout, *Salmo trutta* in a Lake District stream. *Journal of Animal Ecology* **54**, 617–638.

Elliott J.M. (1994) *Quantitative Ecology and Brown Trout*. Oxford: Oxford University Press. 286 pp.

Gibson R.J. (1988). Mechanisms regulating species composition, population structure, and production of stream salmonids; a review. *Polskie Archiwum Hydrobiologii* **35**, 469–495.

Gouraud V., Baglinière J.L., Sabaton C. & Ombredane D. (1998) Application d'un modèle de dynamique de population de truite commune (*Salmo trutta*) sur un bassin de Basse-Normandie – calage des fonctions biologiques et premières simulations. *Bulletin Français de la Pêche et de la Pisciculture* **350/351**, 675–691.

Grant J.W.A. & Kramer D.L. (1990) Territory size as a predictor of the upper limit to population density of juvenile salmonids in streams. *Canadian Journal of Fisheries and Aquatic Science* **47**, 1724–1737.

Grant J.W.A. & Noakes D.L.G. (1988) Aggressiveness and foraging mode of young-of-the-year brook charr, Salvelinus fontinalis (Pisces, Salmonidae). *Behavioral Ecology and Sociobiology* **22**, 435–445.

Haury J., Ombredane D. & Baglinière J.L. (1991) L'habitat de la truite commune (*Salmo trutta* L.) en eaux courantes. In J.L. Baglinière & G. Maisse (eds) *La Truite, Biologie et Ecologie*. Paris: INRA, pp. 25–46.

Heggenes J. (1988) Physical habitat selection by brown trout (*Salmo trutta*) in riverine systems. *Nordic Journal of Freshwater Research* **64**, 76–90.

Jowett I.G. & Richardson J. (1989) Effects of a severe flood on instream habitat and trout populations in seven New Zealand rivers. New Zealand *Journal of Marine and Freshwater Research* **23**, 11–17.

Kelly-Quinn M. & Bracken J.J. (1988) Brown trout, *Salmo trutta* L., production in an Irish coastal stream. *Aquaculture and Fisheries Management* **19**, 69–95.

Kennedy G.J.A. & Strange C.D. (1982) The dispersion of salmonids in upland streams in relation to depth and gradient. *Journal of Fish Biology* **20**, 579–591.

LeCren E.D. (1973) The population dynamics of young trout (*Salmo trutta* L.) in relation to density and territorial behavior. *Rapport et Procès-Verbaux des Réunions du Conseil Permanent International pour l'Exploration de la Mer* **164**, 241–246.

Lobon-Cervia J., DeSostoa A. & Montanes C. (1986) Fish production and its relation with the community structure in an aquifer-fed stream of Old Castile (Spain). *Polskie Archiwum Hydrobiologii* **33**, 333–343.

McNicol R.E., Scherer E. & Murkin E.J. (1985) Quantitative field investigations of feeding and territorial behaviour of young-of-the-year brook charr, *Salvelinus fontinalis*. *Environmental Biology of Fishes* **12**, 219–229.

Milner N.J., Hemsworth R.J. & Jones B.E. (1985) Habitat evaluations as a fisheries management tool. *Journal of Fish Biology* **27** (Suppl. A), 85–108.

Mortensen E. (1977a) Population, survival, growth and production of trout *Salmo trutta* in a small Danish stream. *Oikos* **28**, 9–15.

Mortensen E. (1977b) Density-dependent mortality of trout fry (*Salmo trutta* L.) and its relationship to the management of small streams. *Journal of Fish Biology* **11**, 613–617.

Mortensen E. (1988) The significance of temperature and food as factors affecting the growth of brown trout, *Salmo trutta* L., in four Danish streams. *Polskie Archiwum Hydrobiologii* **35**, 533–544.

Moyle P.B. & Vondracek B. (1985) Persistence and structure of fish assemblage in a small California stream. *Ecology* **66**, 1–13.

Orth D.J. (1987) Ecological considerations in the development and application of instream flow-habitat models. *Regulated Rivers: Research and Management* **2**, 171–181.

Platts W.S. & Nelson R.L. (1988) Fluctuations in trout populations and their implications for land-use evaluation. *North American Journal of Fisheries Management* **8**, 333–345.

Rasmussen G. (1986) The population dynamics of brown trout (*Salmo trutta* L.) in relation to year-class size. *Polskie Archiwum Hydrobiologii* **33**, 489–508.

Sabaton C., Valentin S. & Souchon Y. (1995) La méthode des micro-habitats: protocoles d'application, *Rapport EDF-HE-31/95/10*. Chatou: EDF, 30 pp.

Souchon Y., Trocherie F., Fragnoud E. & Lacombe C. (1989) Les modèles numériques des microhabitats des poissons: applications et nouveaux développements. *Revue des Sciences de l'Eau* **2**, 807–830.

Symons P.E.K. (1971) Behavorial adjustment of population density to available food by juvenile Atlantic salmon 1971. *Journal of Animal Ecology* **40**, 569–587.

Underwood T.J. & Bennett D.H. (1992) Effects of fluctuating flows on the population dynamics of rainbow trout in the Spokane River of Idaho. *Northwest Science* **66**, 261–268.

Chapter 11
Fish habitat associations, community structure, density and biomass in natural and channelised lowland streams in the catchment of the River Wensum, UK

N.T. PUNCHARD, M.R. PERROW and A.J.D. JOWITT

ECON Ecological Consultancy, Biological Sciences, University of East Anglia, Norwich NR4 7TJ, UK
(e-mail: m.perrow@uea.ac.uk)

Abstract

Fish community structure, and the density, biomass and habitat preferences of the principal fish species, were investigated over one seasonal cycle in two streams, the natural Whitewater and the channelised Upper Wensum, in the catchment of the River Wensum, Eastern England. Riffle/pool structure and a wooded riparian zone were identified as important features, promoting the high density and biomass of fish in the Whitewater. The associations of minnow, *Phoxinus phoxinus*, and bullhead, *Cottus gobio*, with pool habitat and associated variables such as woody debris, and three-spined stickleback, *Gasterostreus aculeatus*, and stone loach, *Barbatula barbatulus*, with silt and macrophytes, explained the major differences in community structure between the two rivers. The tendency for large trout (*Salmo trutta*) to dominate the population in the Whitewater and small trout to dominate in the Wensum was a reflection of differences in habitat preferences. The impact of channelisation and routine management upon the fisheries and conservation value of headwater streams is discussed.

Keywords: Woody debris, riparian zone, *Cottus gobio*, *Salmo trutta*, *Gasterosteus aculeatus*, *Phoxinus phoxinus*

11.1 Introduction

The upper reaches, or headwaters, of rivers may be of considerable fisheries value, being the principal nursery areas for some commercially-important salmonids (e.g. salmon, *Salmo salar* L. and sea trout, *Salmo trutta fario* L.), as well as forming the principal habitat of non-migratory brown trout, *Salmo trutta* L. In the UK, species such as bullhead, *Cottus gobio* L. and brook lamprey, *Lampetra planeri* (Bloch), may also be abundant in this zone. Providing water quality and quantity are not limiting, physical habitat structure is thought to be the major determinant of the abundance, biomass and diversity of fish (Gorman & Karr 1978; Milner *et al.* 1985). Important habitat features include the presence of riffle/pool structure providing diversity in flow and substrate type and the presence of a forested riparian zone (the natural state for many lowland streams; Rackham 1990). Coarse woody debris originating from the riparian zone has a crucial role in determining the hydrological, hydraulic, morphological and biological characteristics of the channel (Harmon *et al.* 1986; Gurnell *et al.* 1995).

Channelisation, incorporating widening, straightening, deepening and removal of obstructions, has a number of physical impacts. On a local scale, straightening increases the gradient, induces higher velocity and thus increases the potential for transportation of bed material. Riffle/pool structure, and thus depth and flow diversity, is replaced by rather uniform flow in riffle runs or glides. Downstream, there may be significant deposition of transported material, particularly if the channel has been over-widened, which may lead to a reduction in channel capacity (Brookes 1988). The objective of channelisation to improve the efficiency of drainage away from valuable land or property, may be compromised unless the channel is managed on a routine basis. This may include, for example, periodic dredging of deposited material. Moreover, as channel capacity may be reduced by roughness within the channel, routine maintenance typically involves the removal of submerged vegetation (weed-cutting) or coarse woody debris. All forms of routine management may have major impacts on flora and fauna (Crisp & Gledhill 1970; Kern-Hansen 1978; Garner *et al.* 1996). A further effect of channelisation, especially if the bed level of the river is lowered by dredging to improve drainage of the floodplain or riparian zone (perhaps for agricultural purposes), is to disconnect the channel from its floodplain and its associated functions. The latter includes the supply of nutrients, organic material, sediment and water to the channel. In the upper reaches the contribution from the riparian zone underpins the aquatic production of the channel (Perrow & Wightman 1993). Even where channel and riparian zone are connected, riparian trees are often managed or even removed to reduce the input of leaf litter and woody debris, which add to roughness of the channel. Reduction in tree cover also reduces the amount of food resources available to fish (Mason *et al.* 1984).

Consequently, there is an increasingly recognised need for riverine restoration/ rehabilitation (e.g. Cowx 1994; Cowx & Welcomme 1998). From a species perspective, in order to achieve this goal, detailed information on habitat requirements is essential. Moreover, if a favourable outcome (e.g. increased population density) is to be achieved in a predictable, cost-effective manner, the mechanisms behind any changes also need to be understood. For example, an increase in flow can lead to an increase in the area of suitable spawning gravels available for trout (Hermansen & Krog 1984).

The purpose of this chapter is to compare the community structure, density and biomass of fish in two connected streams in the same catchment, one of which has been modified whilst the other has more or less retained its natural form and function. The habitat relationships of the resident fish species were determined over a one-year cycle of seasons. The implications of the management of streams for land drainage and flood defence purposes are discussed.

11.2   Materials and methods

11.2.1   *Study sites*

The study sites were the River Wensum and one of its tributaries, the Whitewater stream in Norfolk in Eastern England. The Wensum is a chalk aquifer-fed river system supporting a number of characteristic, internationally, nationally or regionally notifiable fish, mammals, birds, plants and invertebrates (Perrow *et al.* 1996a). The upper reaches of the Wensum (sampled between national grid reference TF885239 and TF878253) have been modified considerably by straightening and widening (to 4 m) as it crosses valuable arable land, which requires efficient drainage. The depth of the stream is generally less than 50 cm. Deciduous and coniferous woodland dominates the riparian zone. There is little instream structure, apart from where gaps in the canopy have allowed littoral encroachment of vegetation, which restricts the width of the channel and increases flow diversity. Occasional fallen trees have resulted in natural debris dams leading to the creation of habitat features such as bars and pools. In contrast the Whitewater stream (sampled between national grid reference TG079204 and TG082202) has retained much of its original sinuous channel form and width (1.5–2 m), especially where it flows through damp woodland managed as a nature reserve. Riparian trees, particularly alder (*Alnus glutinosa*), extend their roots into the channel, which, combined with debris dams of fallen trees and finer woody material, enhance riffle/pool structure, resulting in diverse flow, depth (from 5 cm on riffles to 1 m in pools) and substrate type. Both rivers have good water quality (e.g. Class 1A with BOD <1.5 mg L^{-1}). Consequently, differences in invertebrate diversity (BMWP score typically >100 in the Whitewater compared to <75 in the Wensum) appear to reflect differences in habitat diversity.

11.2.2   *Fish sampling*

Sampling was undertaken in the summer (30 and 31 July 1996), late autumn (5 and 6 November 1996) and spring (2 and 7 May 1997). Each river was sampled over one day. Point-abundance sampling (PAS) using electric fishing (Copp & Peñáz 1988; Perrow *et al.* 1996b) was used.

On each sampling occasion (with the exception of the Upper Wensum on the first sampling occasion, when 93 points were sampled) 20 points were taken in each of five, 100-m sections (reaches) covering the full range of habitats within each river. Thus any comparison could be made both on a point and reach basis (see below). The points were systematically selected (5 m apart) with the electric fishing operator moving diagonally from bank to bank. The number of points in the littoral margin was determined by applying the relative proportion of the area occupied by the margin to that of the channel.

Electric fishing was conducted by wading. High frequency (600 Hz) pulsed DC (rectangular wave at 300 V with a variable duty cycle of 0–50%) electric fishing equipment (Electracatch WFC11–12 volt) was used, powered by a generator of 1.7 kVa. With a 40-cm diameter anode head, the effective sampling zone at each point was 1.3 m^{-2}, allowing quantitative estimates of density and biomass (see Copp & Peñáz 1988).

At each point sampled, fish captured were identified, measured to the nearest mm fork length and a representative sample of specimens of each species was weighed to the nearest gram, before being returned unharmed to the water.

At each point, several habitat variables within the sampling area were quantified: flow velocity (cm s^{-1}); water depth (cm); % cover of three broad categories of substratum- stones (>3 cm), gravel (<3 cm) or silt; % cover of woody debris; % cover of leaf litter; % cover of submerged tree roots and logs; and % cover of submerged, emergent, floating and overhanging macrophytes. All cover estimates were assessed visually.

11.2.3 *Data analysis*

Measures of the different habitat variables were compared between the rivers on the different sampling occasions using Kruskal-Wallis tests. The density (n m^{-2}) and biomass (g m^{-2}) of all fish combined, density of common small species including bullhead, minnow, *Phoxinus phoxinus* (L.), stone loach, *Barbatula barbatulus* (L.), and three-spined stickleback, *Gasterosteus aculeatus* L., and density (n m^{-2}) and biomass (g m^{-2}) of trout were compared in the same way.

Within each river on each occasion, the abundance of any particular variable, at points with and without a particular species of interest (except trout) was compared using Mann-Whitney-Wilcoxon U-tests. Although the functional groups and species of macrophytes were recorded, these were combined and tested under a general category.

As the number of trout captured was relatively low (6–32 individuals in each river on each occasion), analysis of habitat relationships was undertaken on a reach scale. Moreover, the habitat variables were modified in accordance with previous work on the habitat preferences of trout (e.g. Milner *et al.* 1985; Bowlby & Roff 1986). This led to the development of indices of both flow and depth variability, derived from the coefficient of variance (standard deviation divided by mean) for each; and the use of the mean % cover of gravel, mean % cover of woody material (debris and roots) and the mean % cover of all categories of macrophytes combined.

The relationships between the biomass (g) of trout and each variable in the five reaches of each river on each occasion, were explored using the Spearman rank correlation coefficient (r_s). Scatter plots, incorporating all 100 points, were then used to determine trends in the distribution of the biomass of the fish in relation to different variables. This was only undertaken for the summer, which was assumed to be the period in which critical thresholds (e.g. minimum flow and depth requirements) were most likely to be exceeded.

11.3 Results

11.3.1 *Fish community structure, abundance and biomass*

Eight species of fish (brown trout, eel *Anguilla anguilla* (L.), chub *Leuciscus cephalus* (L.), dace *Leuciscus leuciscus* (L.), minnow, stone loach, three-spined stickleback and bullhead) were recorded in the Whitewater. Chub, dace and minnow were not amongst the seven species captured in the Wensum, although roach, *Rutilus rutilus* (L.), and ten-spined stickleback, *Pungitius pungitius* (L.), were represented. The community structure of the two rivers was thus different, although small species dominated numerically in both. Minnow dominated in the Whitewater, whereas three-spined stickleback dominated in the Wensum (Fig. 11.1). Minnow and bullhead occurred at a significantly higher density in the Whitewater, whilst three-spined stickleback and stone loach were more abundant in the Wensum (Table 11.1).

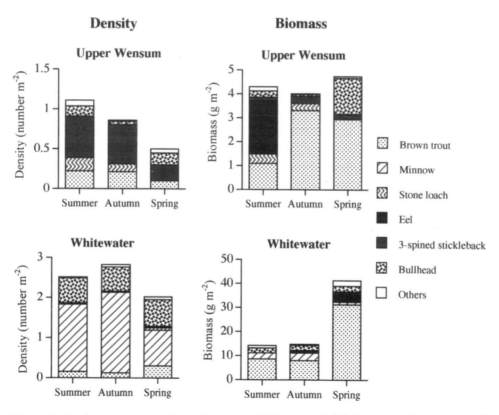

Figure 11.1 Species composition, density and biomass of fish in the study sites on all sampling occasions

Table 11.1 Comparison of habitat variables and fish species density and biomass between rivers on all sampling occcasions[a]

Variables	Summer 1996			Autumn 1996			Spring 1997		
	Z	P	Direction	Z	P	Direction	Z	P	Direction
Habitat									
Depth (cm)	−2.44	*	We > WW	−3.77	***	We > WW	−4.94	***	We > WW
Flow (cm s^{-1})	−0.51	n.s.		−2.18	*	We > WW	−4.51	***	We > WW
% silt	−3.00	**	We > WW	−6.80	***	We > WW	−3.92	***	We > WW
% gravel	−0.78	n.s.		−2.45	**	WW > We	−4.37	***	WW > We
% stone	−5.32	***	WW > We	−4.87	***	WW > We			
% leaf litter	−4.10	***	WW > We	−1.89	n.s.		−1.09	n.s.	
% woody debris	−6.57	***	WW > We	−3.33	***	WW > We	−1.86	n.s.	
% tree roots	−3.93	***	WW > We	−3.24	***	WW > We	−3.06	**	WW > We
% macrophytes	−4.59	***	We > WW	−6.05	***	We > WW	−6.01	***	We > WW
Fish									
Total fish n m^{-2}	−1.59	n.s.		−3.38	***	WW > We	−2.13	*	WW > We
Total fish g m^{-2}	−2.85	**	WW > We	−3.71	***	WW > We	−3.66	***	WW > We
Trout density n m^{-2}	−0.00	n.s.		−0.63	n.s.		−2.00	*	WW > We
Trout biomass g m^{-2}	−0.10	n.s.		−0.73	n.s.		−1.99	*	WW > We
Minnow n m^{-2}	−5.02	***	WW > We	−6.78	***	WW > We	−3.56	***	WW > We
Stone loach n m^{-2}	−3.97	***	We > WW	−3.32	***	We > WW	−1.28	n.s.	
Three-spined stickleback n m^{-2}	−4.81	***	We > WW	−6.12	***	We > WW	−6.38	***	We > WW
Bullhead n m^{-2}	−4.60	***	WW > We	−7.61	***	WW > We	−5.89	***	WW > We

[a]Mann-Whitney Z values and associated probabilities (*$P < 0.05$; **$P < 0.01$; ***$P < 0.001$) are shown.

The density and biomass of trout in the two rivers was similar in summer and autumn, but significantly higher ($P < 0.05$) in the Whitewater in spring.

The overall density of fish was higher in the Whitewater, being ≥ 2 m^{-2} (Fig. 11.2). The Wensum only exceeded 1 m^{-2} in summer and thereafter declined to 0.5 m^{-2}. Values were significantly lower ($P < 0.05$) in the Wensum in autumn and spring (Table 11.1). The biomass of fish was always significantly higher ($P < 0.01$) in the Whitewater, ranging from 15–40 g m^{-2}, compared to <5 g m^{-2} in the Wensum. Trout typically dominated the fish biomass in both rivers (Fig. 11.1).

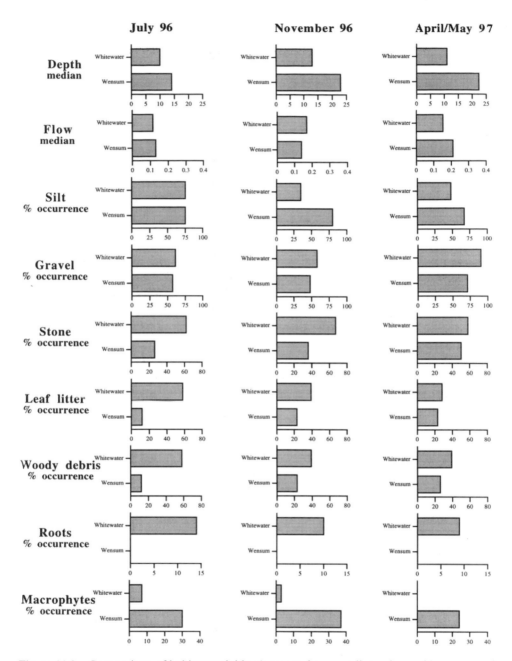

Figure 11.2 Comparison of habitat variables (expressed as a median value or % occurrence) between the study sites on all sampling occasions

11.3.2   *Comparison of habitat variables between rivers*

The Wensum had a greater median depth than the Whitewater. During the summer low flow period, the flow rate was similar in the two rivers. In autumn and spring, however, when flow (and thus water depth) within each river increased, the flow velocity in the Whitewater was significantly greater ($P < 0.05$) than that in the Wensum. By spring, this situation had reversed. Increased flow from autumn onwards appeared to have a scouring effect in the Whitewater, leading to a loss of silt (also leaf litter and woody debris – see below) and consequent exposure of gravel. This led to a significantly greater abundance of gravel and also stone in the Whitewater in both autumn and spring. Stone was also generally more abundant in the Whitewater. In contrast, the substrate of the Wensum was always characterised by a greater abundance of silt (Table 11.1).

Although the valleys of both rivers are wooded, riparian trees (alders) only interacted directly with the channel in the Whitewater, as illustrated by the significantly greater amount of tree root cover ($P < 0.001$) (Table 11.1). The presence of overhanging riparian trees also led to a significantly higher abundance ($P < 0.001$) of leaf litter and woody debris in the Whitewater, even though some of this appeared to be washed out or at least redistributed during higher flows. In the Wensum, the greater amount of light reaching the channel was manifested as a significantly greater ($P < 0.001$) abundance of macrophytes (both submerged and emergent). The abundance of these was reduced over the winter period, although they were maintained at a reasonable frequency of occurrence (Fig. 11.1).

11.3.3   *Habitat preferences of the principal fish species*

All of the common small species showed distinct habitat preferences, although there were some differences between seasons for some species (Table 11.2). In the Whitewater, minnows especially selected for pools, as indicated by positive associations with depth, woody debris and tree roots and negative associations with flow and stone. Although the basic pattern remained the same throughout the year, only depth and flow appeared to be consistently important (Table 11.2). A similar pattern of association was also noted in bullhead, with woody debris, leaf litter and tree roots preferred in summer and autumn, and greater depth (and thus silty habitat) preferred in spring. In the Upper Wensum, bullhead selected for woody debris at this time. The dominant species in the Wensum, three-spined stickleback and stone loach, tended to be positively associated with depth, silt and macrophytes and avoided higher flow velocities accordingly. Sticklebacks also avoided gravel and stone habitats (Table 11.2).

Trout showed few associations with any habitat variables in either river (Fig. 11.3), although in the Whitewater where the biomass was generally higher, positive associations between trout biomass and woody debris were recorded in summer ($r_s = 0.90$, $n = 5$, $P < 0.05$) and spring ($r_s = 0.90$, $n = 5$, $P < 0.05$). At that time,

Table 11.2 Habitat associations of particular fish species as shown by comparison of habitat variables at points where that species is present or absent[a]

	Summer 1996			Autumn 1996			Spring 1997		
	Z	P	+/−	Z	P	+/−	Z	P	+/−
Minnow – Whitewater									
Depth	−5.21	***	+	−5.56	***	+	−3.05	**	+
Flow (cm s^{-1})	−3.01	*	−	−4.12	***	−	−3.05	***	−
% silt	−4.88	***	+	−4.85	***	+			
% stone	4.64	***	−	−5.92	***	−			
% woody debris	−2.15	*	+						
% tree roots	−4.37	***	+						
Bullhead – Whitewater									
Depth	−1.13			−2.66			−2.88	**	+
% silt	−0.33			−1.99			−2.31	*	+
% gravel	−1.33			−0.07			−2.17	*	−
% leaf litter	−3.15	**		−2.60	**		−0.48		
% woody debris	−3.75	***		−6.00	***		−1.47		
% tree roots	−0.39			−2.27	*		−0.24		
Stone loach – Upper Wensum									
Depth	−2.44	*	+						
Flow (cm s^{-1})	−2.12	*	−				−2.04	*	−
% silt	−2.10	*	+						
% gravel	−1.96	*	−						
% leaf litter							−2.61	**	+
% macrophytes	−2.18	*	+	−2.21	*	+	−3.36	***	+
Three-spined stickleback – Upper Wensum									
Depth	−2.05	*	+	−2.81	**	+	−4.32	***	+
Flow (cm s^{-1})	−2.36	*	−	−2.20	*	−			
% silt	−2.07	*	+	−5.03	***	+	−2.83	**	+
% gravel	−1.97	*	−	−2.68	**	−			
% stone				−4.14	***	+	−3.05	**	−
% macrophytes	−2.65	**	+				−2.02	*	+
Bullhead – Upper Wensum									
% woody debris							−4.26	***	+

[a]Mann-Whitney Z values and associated probabilities where *$P < 0.05$, **$P < 0.01$ and ***$P < 0.001$ and the direction of the difference ($+/−$) are shown.

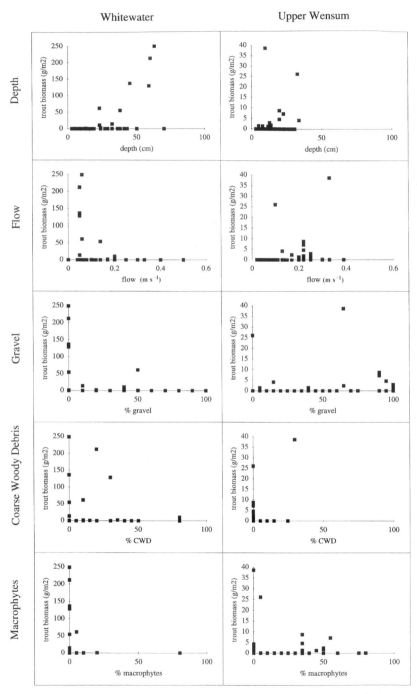

Figure 11.3   Scatter plots (*n* = 100) between trout biomass (g) and various habitat variables in both study sites in summer

a negative correlation was also observed for flow ($r_s = 1$, $n = 5$, $P = 0.001$). In the Wensum, trout were positively associated with macrophytes in autumn ($r_s = 0.90$, $n = 5$, $P < 0.05$).

11.4 Discussion

A wealth of literature is available on the effects of channelisation upon fish and fisheries (see Brookes 1988). Typical effects include a reduction in abundance, biomass and community diversity. However, the bulk of the literature, particularly from the USA, is concerned with commercially important salmonids. Moreover, the impact upon fish is often implied through a reduction in particular habitat variables that are generally thought to be important. Although the present study lacked controlled experimental manipulation or the monitoring of a before-and-after response, the use of PAS allowed many habitat variables to be quantified in relation to fish. This provided a sound statistical basis, which is often a major constraint of fisheries sampling. Further, few studies have been undertaken in the UK, especially in the headwaters of lowland streams, and several of the generally common and widely distributed species have been relatively poorly studied, even though they may be important in functional or conservation terms. Most importantly, the clear signal of strong preferences of the dominant species for particular habitat features characteristic of natural or modified streams, provided good evidence that differences in physical habitat structure were responsible for the observed differences in fish community structure, abundance and biomass between the study rivers.

The generally strong selection of bullhead for woody debris in the current study contrasts with that observed by Waterstraat (1992), Hoffmann (1996), and Roussel and Bardonnet (1997), who all suggested preferences for riffles. The differences in habitat selection may stem from differences in the nature of the streams studied. The upper reaches of lowland streams, which were the focus of this study, are also likely to have a naturally high component of woody debris where natural wooded riparian zones persist, but typically do not have the cobble and rock substrate of upland areas.

Minnow only occurred in the Whitewater, where it selected for several variables associated with pools, typically in large shoals (up to 118 individuals, 90 m^{-2}, were recorded at a point). Roussel and Bardonnet (1997) also observed this. It is likely that shoaling in structured environments may also be a response to predation risk, most likely from large trout. However, this does not explain the absence of minnow in the Wensum, where deep silty habitat was abundant and large trout were absent. As with bullhead, it is plausible that the pattern of habitat selection changes at dusk and riffles are important foraging habitats. Indeed, these may lead them to compete for resources with young trout (Hestagen *et al.* 1992). With limited invertebrate resources in the Wensum, it is possible that minnow is out-competed by young trout.

Sticklebacks and stone loach were associated with the silty habitats predominant in the Wensum. Although the surface flow recorded was as high, if not higher than that in the Whitewater, the flow near the bottom (as indicated by the deposition of silt) or amongst macrophytes is likely to be much lower. This would favour sticklebacks, which are not morphologically adapted for flowing environments. The presence of macrophytes is also likely to be critical as these provide nest sites (Cleveland 1984). Stone loach, with a laterally flattened body form, is adapted for flowing environments. However, it forages amongst the fine sediments deposited in slack water, using the olfactory senses of its barbels to detect prey, chiefly chironomid larvae (Hyslop 1982). The abundance of silty habitats and the lack of other potentially competing epibenthic species are suggested to be the reasons stone loach is only prevalent in the Wensum.

In general terms, small, short-lived species, which are likely to be dependent on more or less annual recruitment, are likely to respond rapidly to any habitat changes. Where conditions are favourable, recruitment may quickly lead to numerical dominance. Where conditions become unsuitable, populations of many species may be unsustainable, particularly if others are favoured. In the Wensum, there was some evidence of populations of species such as stickleback and stone loach, associated with depositing channels, disappearing over the winter months in the presence of higher winter flows. This effect may be typical in channelised streams, where flood flows may be achieved rapidly and be of high intensity. Functional floodplains in natural streams tend to smooth discharge into the river, which also has greater storage capacity with a longer length of channel (Brookes 1988). Greater habitat diversity and structure is also likely to lead to greater flow diversity with pools in particular providing refuges under flood conditions.

Larger, more mobile, longer-lived species, such as trout, may be expected to be less affected by habitat changes in the short term, unless these are particularly severe or unless fish are dependent on particular resources (i.e. habitat specialists). Many studies have illustrated strong relationships between trout and habitat variables such as woody debris, overhanging cover and pools (Gard 1961; Hunt 1976; Flebbe & Dolloff 1995). This is exacerbated by the territorial nature of trout, with an increase in the number of potential territories through provision of suitable habitat, leading to a corresponding increase in the density (biomass). This has led to the development of habitat suitability models for the species (e.g. Milner *et al.* 1985; Wesche 1985). However, habitat relationships may not always be clear-cut. For example, Bowlby & Roff (1986) showed that two habitat quality indices only accounted for 6.7 and 9.2% of the variance in trout biomass in Southern Ontario streams. In the current study, the lack of significant differences in trout density and biomass between the structurally different rivers (apart from in spring), the relative lack of habitat associations (only four significant results from a possible 30) and selection of different habitats on occasion and between rivers, all suggest that trout may not respond to habitat modification in a predictable manner. Sources of variation include diel and seasonal differences in habitat requirements, as well as differences between small and large fish (Roussel & Bardonnet 1997).

In the current study, the large trout in the Whitewater only occurred in water deeper than 25 cm with flow less than about 0.2 m^{-2}, a low proportion of gravel with no more than about 30% cover of woody debris (Fig. 11.3), whereas in the

Wensum, the small trout often occurred in flows exceeding 0.2 m^{-2}, often over shallow gravelly riffles with macrophytes, particularly where these provided cover at the edges of the stream. This may be a typical pattern where other types of cover are in short supply (Eklöv & Greenberg 1998). Differences in habitat selection by small and large fish may thus have compensated, in part, for differences in habitat structure between the two rivers. Small trout, like other small fish (see above) also appeared to be vulnerable to high flow, with a reduction from 32 in autumn to six individuals by spring.

Overall, riffle pool structure and the presence of a wooded riparian zone contributing woody debris to the channel appear to be particularly important elements, promoting diversity, abundance and biomass in natural streams. Clear habitat relationships, which are readily expressed in terms of abundance, may be observed for the principal small species. Extended sampling to include seasonal and diel variation (see Rousell & Bardonnet 1996) in habitat selection, supported by investigation of likely mechanisms and perhaps specific biological aspects (e.g. diet) where there is a paucity of information, should allow habitat quality indices to be developed. Such information may be used during schemes to improve the fisheries or conservation value of modified streams. Even in the absence of full habitat prescriptions, it is clear that significant benefits may be accrued by the reduction in, or cessation of, the practice of the removal of woody debris during routine maintenance.

Acknowledgements

We would like to thank Drs Charles Beardall and Robin Burrough of the Environment Agency for supporting the project. Dr Isabelle Côté helped with the field work/data collection and provided advice on statistical analysis. Martin Peacock and Mark Tomlinson assisted in the field, with the latter also helping in the preparation of the figures.

References

Bowlby J.N. & Roff J.C. (1986) Trout biomass and habitat relationships in southern Ontario streams. *Transactions of the American Fisheries Society* **115**, 503–514.

Brookes A. (1988) *Channelized Rivers: Perspectives for Environmental Management*. Chichester: Wiley, 326 pp.

Cleveland A. (1984) Nest site habitat preference and competition in *Gasterosteus aculeatus* and *G. wheatlandi*. *Copeia* **3**, 698–704.

Copp G.H. & Peñáz M. (1988) Ecology of fish spawning and nursery zones in the flood plain, using a new sampling approach. *Hydrobiologia* **169**, 209–224.

Cowx I.G. (1994) *Rehabilitation of Freshwater Fisheries*. Oxford: Fishing News Books, Blackwell Science, 485 pp.

Cowx I.G. & Welcomme R.L. (1998) *Rehabilitation of Rivers for Fish*. Oxford: Fishing News Books, Blackwell Science, 260 pp.

Crisp D.T. (1963) A preliminary survey of brown trout and bullheads in high altitude becks. *Salmon and Trout Magazine* **167,** 45–59.

Crisp D.T. & Gledhill T. (1970) A quantitative description of the recovery of the bottom fauna in a muddy reach of a mill stream in southern England after draining and dredging. *Archive für Hydrobiologie* **67,** 502–541.

Eklöv A.G. & Greenberg L.A. (1998) Effects of artificial instream cover on the density of 0 + brown trout. *Fisheries Management and Ecology* **5,** 45–53.

Flebbe P.A. & Dolloff C.A. (1995) Trout use of woody debris and habitat in Appalachian wilderness streams of North Carolina. *North American Journal of Fisheries Management* **15,** 579–590.

Gard R. (1961) Creation of trout habitat by constructing small dams. *Journal of Wildlife Management* **52,** 384–390.

Garner P., Bass J.A.B & Collett G.D. (1996) The effects of weed cutting upon the biota of a large regulated river. *Aquatic Conservation – Marine and Freshwater Ecosystems* **6,** 21–29.

Gorman O.T. & Karr J.R. (1978) Habitat structure and stream fish communities. *Ecology* **59,** 507–515.

Gurnell A.M., Gregory K.J. & Petts G.E. (1995) The role of coarse woody debris in forest aquatic habitats-implications for management. *Aquatic Conservation – Marine and Freshwater Ecosystems* **5,** 143–166.

Harmon M.E., Franklin J.F., Swanson F.J., Sollins P., Gregory S.V., Lattin J.D., Anderson N.H., Cline S.P., Aumen N.G., Sedell J.R., Lienkaemper, G.W., Cromack K. & Cummins K.W. (1986) Ecology of coarse woody debris in temperate ecosystems. *Advances in Ecological Research* **15,** 133–302.

Hermansen H. & Krog C. (1984) A review of brown trout (*Salmo trutta*) spawning beds, indicating methods for their re-instatement in Danish lowland rivers. In J.S. Alabaster (ed.) *Habitat Modification and Freshwater Fisheries* London: Butterworth, pp. 116–123.

Hestagen T., Hegge O. & Skurdal, J. (1992) Food choice and vertical distribution of European minnow *Phoxinus phoxinus* and native and stocked brown trout *Salmo trutta* in the littoral zone of a subalpine lake. *Nordic Journal of Freshwater Research* **67,** 72–76.

Hoffmann A. (1996) Impact of river maintenance and construction on habitat use of the bullhead *Cottus gobio. Fischoekologie* **9,** 49–61.

Hunt R.L. (1976) A long-term evaluation of trout habitat development and its relation to improving management-related research. *Transactions of the American Fisheries Society* **195,** 361–364.

Hyslop E.J. (1982) The feeding habits of stoneloach and bullhead. *Journal of Fish Biology* **21,** 187–196.

Kern-Hansen U. (1978) The drift of *Gammarus pulex* L. in relation to macrophyte cutting in four small Danish lowland streams. *Verhandlungen der internationalen Vereingung für theoretische und angewandte Limnologie* **20,** 1440–1445.

Mason C.F., MacDonald S. M. & Hussey A. (1984) Structure, management and conservation value of the riparian woody plant community. *Biological Conservation* **29,** 201–216.

Milner N.J., Hemsworth R.J. & Jones B.E. (1985) Habitat evaluation as a fisheries management tool. *Journal of Fish Biology* **27** (Suppl. A), 85–108.

Perrow M.R. & Wightman A.S. (1993) *The River Restoration Project, Phase 1: The Feasibility Study*. Oxford: River Restoration Project (RRP), 186 pp.

Perrow M.R., Jowitt A.J.D. & Hey R.D. (1996a) *River Rehabilitation Feasibility Study of the Wensum Tributaries*. Ipswich: National Rivers Authority, Anglian Region, 67 pp.

Perrow M.R., Jowitt A.J.D. & Zambrano González L. (1996b) Sampling fish communities in shallow lowland lakes: point-sample electric fishing versus electric fishing with stop-nets. *Fisheries Management and Ecology* **3,** 303–313.

Rackham O. (1990) *Trees and Woodland in the British Landscape*, revised edition. London: Phoenix, 234 pp.

Roussel J.M. & Bardonnet A. (1996) Differences in habitat use by day and night for brown trout (*Salmo trutta*) and sculpin (*Cottus gobio*) in a natural brook. Multivariate and multi-scale analyses. *Cybium* **20**, 43–53.

Roussel J.M. & Bardonnet A. (1997) Diel and seasonal patterns of habitat use in a salmonid brook: an approach to the functional role of the riffle-pool sequence. *Bulletin Français de la Pêche et de la Pisciculture* **346**, 573–588.

Waterstraat A. (1992) Investigation of the ecology of *Cottus gobio* L and other fish species from two lowland streams of Northern Germany. *Limnologica* **22**, 1347–149.

Wesche T.A. (1985). Stream channel modifications and reclamation structures to enhance fish habitat. In J.A. Gore (ed.) *The Restoration of Rivers and Streams: Theories and Experience.* Massachusetts: Butterworth, pp. 103–163.

Chapter 12
A methodology to evaluate physical habitat for reproduction of brown trout (*Salmo trutta* L.) and the relation with fry recruitment

M. DELACOSTE, P. BĄRAN and J.-M. LASCAUX

Laboratoire d'Ingénierie Agronomique, Equipe Environnement Aquatique et Aquaculture. ENSAT, 1 avenue de l'Agrobiopole, BP107, 31326 Auzeville Tolosan, France (e-mail: baran@ensat.fr)

Abstract

The relationship between spawning activity of brown trout, *Salmo trutta* L., and the characteristics of physical habitat was studied in the Aude stream in the Pyrenean mountains. Three methodologies measuring physical habitat characteristics were tested: (1) a transect method (IFIM) with calculation of weighted usable area (WUA); (2) a measurement of the area of suitable spawning gravel (SSG); and (3) a measurement of depth and velocity on all the areas of SSG calculated as WUA_{SSG}. WUA, SSG and WUA_{SSG} were significantly correlated with redd density ($P < 0.05$), but the best correlation was obtained with WUA_{SSG}. A significant annual variation in redd density ($P < 0.05$) was directly related to variation in physical habitat over the three years of study. Modifications of the areas of gravels or variations of hydraulic conditions were the most important parameters explaining the variability in redd density. Recruitment was correlated to the spawning activity ($P < 0.05$) but high flow conditions, carrying capacity for fry and water quality could greatly modified the survival of fry. The results are discussed in relation to the habitat preference of brown trout for spawning activity and its importance to the population dynamics.

12.1 Introduction

Numerous studies have described the habitat requirements and spawning behaviour of stream-dwelling salmonids (Jones & Ball 1954; Tautz & Groot 1975; Shirvell & Dungey 1983; Witzel & MacCrimmon 1983; Crisp & Carling 1989). Females select spawning sites on the basis of physical characteristics of the stream (Reiser & Wesche 1977; Ottaway *et al.* 1981; Shirvell & Dungey 1983; Delacoste *et al.* 1995a). Studies that include such habitat preference to predict redd densities are limited (Delacoste *et al.* 1993). Spawning habitat of stream-dwelling salmonids may, however, vary spatially and temporally across sections, streams or drainage basins, in relation to shifts in area of suitable spawning gravel (Delacoste *et al.* 1993, 1995a; Magee *et al.* 1996). Despite its importance for management of salmonid populations, the

---

Address for correspondence: P. Baran, Consantil Supérieur de la Pêche, 20 Rue Charrue, 21000 Dijon, France

quantification of carrying capacity of physical habitat for salmonid spawning is poorly documented, excepted with IFIM (Bovee 1982). Using this methodology, Delacoste *et al.* (1995a) showed a poor ability to predict redd densities. They suggested the basis of a methodology integrating the calculation of weighted usable area (WUA) of the PHABSIM model (Bovee 1982; Belaud *et al.* 1989) with the measurement of the areas of suitable spawning gravels.

The estimation of a carrying capacity for spawning may play an important role in determining the dynamics of salmonid populations, and also patterns of distribution of fry, juveniles and adults. Usually, salmonid recruitment has been examined in relation to the parent stock expressed as a number of eggs per m^2 or the spawner abundance (Elliott 1985b; Myers & Barrowman 1996), but rarely in relation to redd density or spawning habitat availability (Beard & Carline 1991; Bozek & Rahel 1991). Most of the habitat-based models predicting the abundance of trout do not integrate independent variables related to spawning habitat (Fausch *et al.* 1988). The basic assumption of these models is that recruitment is not limiting as suggested by McFacdden (1969) for resident trout.

The objectives of this study were: (1) to characterise the spatial and temporal variability of the spawning activity of brown trout, *Salmo trutta* L., in a Pyrenean stream; (2) to examine the factors influencing this variability; (3) to test the ability of three habitat descriptors (PHABSIM model, areas of suitable spawning gravel (SSG) and WUA on the areas of SSG) to predict redd density; and (4) to analyse the relationship between redd density and the recruitment of brown trout.

12.2 Methods

12.2.1 *Study sites*

The study was conducted in the headwaters of the Aude, a coastal stream in the east of the Pyrenees (Fig. 12.1). Stream flow is influenced by completion of Puyvalador reservoir (1450 m of elevation) in 1948. Water released from the reservoir is diverted into a pipeline for transport to a hydroelectric power station at 450 m elevation. Twenty-seven kilometres of stream are submitted to low flow conditions. In addition, three other small dams divert flow to hydroelectric power stations in these 27 km. Brown trout is usually the only fish species present, but sculpins *Cottus gobio* L. and gudgeon, *Gobio gobio* (L.), are also occasionally found. The morphological diversity of the study area was described using the channel morphodynamic unit (CMU) classification developed by Delacoste *et al.* (1995b) for Pyrenean streams. Twelve study sites were chosen between 1350 and 450 m of elevation to represent the morphological diversity of the different reaches (Table 12.1).

12.2.2 *Sampling*

In 1993, a preliminary study was conducted on five sections (Table 12.1, Fig. 12.1) without consideration of habitat quality. The survey on the 12 study sections was

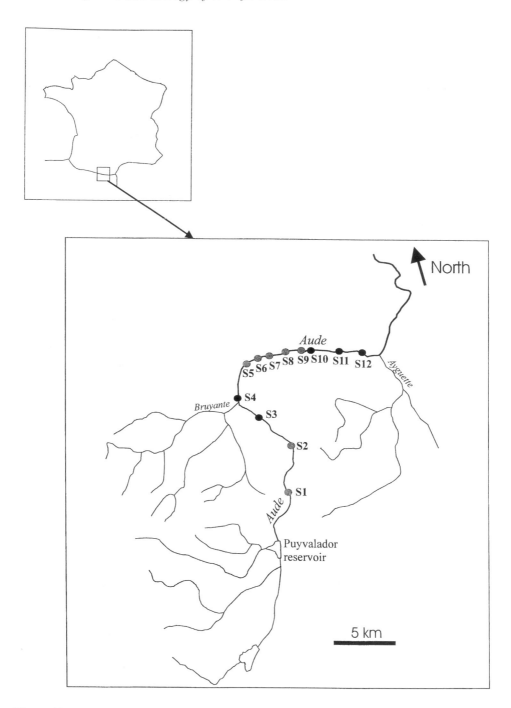

Figure 12.1 Area of study in the Aude River (see Table 12.1 for names of numbered locations)

Table 12.1 Main characteristics of the 12 sections studied in the Aude river

Sections	Elevation (m)	Gradient (%)	Mean flow in autumn ($m^3 s^{-1}$)	Width (m)
S1 Carcanet	1200	5.0	0.20	7.3
S2 Escouloubre	1005	4.2	0.31	6.2
S3 Sources chaudes	830	1.6	0.27	5.4
S4 Bruyante	780	4.8	0.62	8.9
S5 Aguzou 1	715	2.0	0.20	4.6
S6 Aguzou 2	708	2.4	0.15	5.2
S7 Pont Escouloubre	685	1.8	0.54	7.8
S8 Réserve de pêche	615	1.6	0.80	9.8
S9 Gesse 1	585	2.0	0.86	9.7
S10 Gesse 2	580	2.2	0.86	10.2
S11 Saint-Georges 1	500	1.7	0.21	10.1
S12 Saint-Georges 2	460	2.0	0.26	9.8

conducted between 1994 and 1997 inclusively. Each of the 12 study sites was divided into units representing different types of mesohabitat. Each mesohabitat was identified and classified among six main groups according to the classification established by Delacoste *et al.* (1995b). At the five sections assessed in 1993, fish populations were subsequently sampled for each year (1994–1997) at low flow in summer (July) by electric fishing using the removal method. In addition, a sample was taken in October 1994 to determine the age of first maturity and the sex ratio of fish in these five sections. Before sampling, the upstream and downstream limits of the study site were isolated by block nets (8 mm mesh size). Fish length (Fork length, mm) and weight (g) were measured. From November to January of each year, redds were counted on five occasions per site, and all redds were mapped at each sampling. The total number of redds was divided by the length of each section.

Physical habitat characteristics of the 12 study sites were sampled in 1994 and 1995 using the transect method. Depth, velocity (measured by electronic flow meter, March MacBirney) and substrate were described immediately after the spawning period according to PHABSIM analyses. The discharge value was identical during the measurement period and the spawning period. In earlier studies in Pyrenean streams, Delacoste *et al.* (1993, 1995a) showed the great importance of the area of suitable gravel (0.2–5 cm of diameter) for the spawning activity of brown trout. In 1994, 1995 and 1996, the area of suitable gravel (> 0.04 m^2) and also depth and velocity at different points of these areas on the 12 sections were measured. The number of points depended on the area (average: 1 point per 0.02 m^2).

12.2.3 *Analysis*

Trout biomass and density were calculated by the method of Seber and LeCren (1967) and separate calculations were performed for 0 group and for older fish. The fishing efficiency (E) was estimated as $E = (C_1/P) \times 100$, where C_1 is the number of trout in the first passage, and $\bar{P}$ is the estimated density. The mean sampling efficiencies for all sections and years were 73% ± 3.1% for 0 group and 81% ± 4.1% for older trout. The fry production per redd in each section was estimated by dividing the density of fry in July by the redd density. For the physical habitat, weighted usable area (WUA) (Bovee 1982), spawning suitable gravel (SSG) (Delacoste *et al.* 1993) and weighted usable area on SSG (WUA$_{SSG}$) were calculated.

Application of PHABSIM model with transect method

For each hydraulic cell (k) characterised by the different points of the transect, WUA was calculated as:

$$\text{WUA}_k = A_k \times f_v(V_k) \times f_v(D_k) \times f_v(S_k)$$

Application of PHABSIM model on the area of suitable spawning gravel

For each area of SSG, WUA was calculated as:

$$\text{WUA}_{SSGk} = A_{SSGk} x\, f_v(V_k) \times f_v(D_k)$$

where $f_k(V_k), f_k(D_k)$ and $f_k(S_k)$ are the suitability weighting factor for velocity, depth and substrate, respectively (only for PHABSIM) in cell k. A_k is the plan area of the cell k in PHABSIM model and A_{SSGk} is the area of suitable spawning gravels. Suitability weighting factors were obtained from the probability-of-use curves established on Pyrenean streams (Delacoste *et al.* 1995a).

For the stream section:

$$\text{WUA} = \sum_{k=1}^{n} \text{WUA}_k \quad \text{or} \quad \text{WUA}_{SSG} = \sum_{k=1}^{n} \text{WUA}_{SSGk}$$

WUA, WUA$_{SSG}$ and SSG were divided by the total length of each section.

Continuous density variables were non-normal (Lilliefors test, $P < 0.05$), thus correlation between density of redd and the quality of physical habitat were made using Spearman rank correlation coefficient ($P < 0.05$). Comparisons of mean redd densities between the different groups of channel morphodynamic units were made with the Mann-Whitney test ($P < 0.05$) and Kruskal-Wallis Analysis of Variance of ranks ($P < 0.05$).

12.3   Results

Spawning occurred in October, November and December with a peak in mid-November. Few active spawners were observed after the second week of December. The redd density ranged from 1 to 78 100 m^{-1} (Table 12.2).

Table 12.2 SSG, WUASSG, WUA and the density of redd for the 12 sections studied

Sections	WUA m² 100 m⁻¹			SSG m² 100 m⁻¹			WUASSG m² 100 m⁻¹			Redds no. 100 m⁻¹			
	1994	1995	1996	1994	1995	1996	1994	1995	1996	1993	1994	1995	1996
S1 Carcanet	0	0	0	7	8	8	2	3	2		15	21	19
S2 Escouloubre	0	0	0	7	7	7	1	1	1		9	10	9
S3 Sources chaudes	10	10	10	13	24	17	5	7	6	23	23	62	35
S4 Bruyante	0	0	0	13	18	21	4	6	1	17	12	26	4
S5 Aguzou 1	0	0	0	12	13	14	6	6	5		31	34	39
S6 Aguzou 2	3	3	3	7	7	8	3	3	2		18	17	22
S7 Pont Escouloubre	54	54	55	170	176	186	60	61	62		68	75	78
S8 Réserve de pêche	4	5	5	40	46	46	11	10	10		43	51	51
S9 Gesse 1	0	0	0	2	2	2	0	0	0		6	7	5
S10 Gesse 2	14	14	13	12	20	22	3	3	7	15	15	25	20
S11 Saint-Georges 1	5	5	0	9	31	36	1	2	0	11	11	34	1
S12 Saint-Georges 2	60	60	4	78	78	68	24	24	2	68	68	68	11

12.3.1   *Relationships between physical habitat and the density of redd*

There was a significant correlation ($P < 0.05$) between WUA, SSG or WUA$_{SSG}$ and redd density, but the best relation was with WUA$_{SSG}$ (Fig. 12.2). The transect method gave poor results for the estimation of physical habitat quality for spawning. At the channel morphodynamic unit scale (CMU), a significant correlation ($P < 0.05$) between density of redds and WUA, SSG or WUA$_{SSG}$ was also observed. The best correlation was obtained with WUA$_{SSG}$ (Fig. 12.3). The density of redd or the quality of physical habitat estimated by WUA, SSG or WUA$_{SSG}$ were significantly different ($P < 0.05$) between the six main groups of CMU (pool, glide, step, riffle, rapid and cascade). Glide and pool had the highest quantity of suitable physical habitat for spawning and also the highest density of redds (Table 12.3).

12.3.2   *Annual variability of the density of redds*

The density of redds in the 12 sections were significantly different between the three years of study ($P < 0.05$). At the same time, a significant variation of WUA, SSG and WUA$_{SSG}$ was observed. The inter-annual variability in redd densities was due to the yearly modifications of suitable spawning gravel (SSG) or hydraulic conditions.

The area of SSG on several sections varied significantly during the 3 years of study ($P < 0.05$). For example, in the section S4 (Bruyante), an increase of SSG was observed between 1994 and 1995 ($+38\%$) related to the sediment transport at high flow in 1995 and also an increase of the redd density ($+80\%$) (Fig. 12.2b). In section S3 (Sources Chaudes), the increase of SSG in 1995 was related to a restoration of spawning habitat conducted by the Federal Department of Fishing. Suitable spawning gravels were deposited on the stream bed. This increase of SSG and WUA$_{SSG}$ ($+82\%$ and $+48\%$, respectively) induced an increase of the density of redds ($+165\%$) (Fig. 12.2b).

In 1996, very low redd density was observed at sections S4 (Bruyante), S11 (St-Georges 1) and S12 (St-Georges 2) (Fig. 12.2b). These decreases (-84%, -97% and -84%, respectively) were related to modifications in flow conditions. These modifications were induced by relative high flow conditions at these three sections during autumn 1996 when hydropower generation ceased. These flow conditions (1–1.5 m^3 s^{-1}) compared to the normal low flow (0.19–0.5 m^3 s^{-1}) induced greater depth and velocity. The SSG was not different but the WUA$_{SSG}$ decreased markedly (between -84% and -91%). This example shows the importance of hydraulic conditions for the estimation of suitable spawning habitat.

12.3.3   *Relationships between densities of redd and the recruitment*

The relationship between the recruitment and redd density was studied on five sections. Densities of fry varied significantly between the three years of study

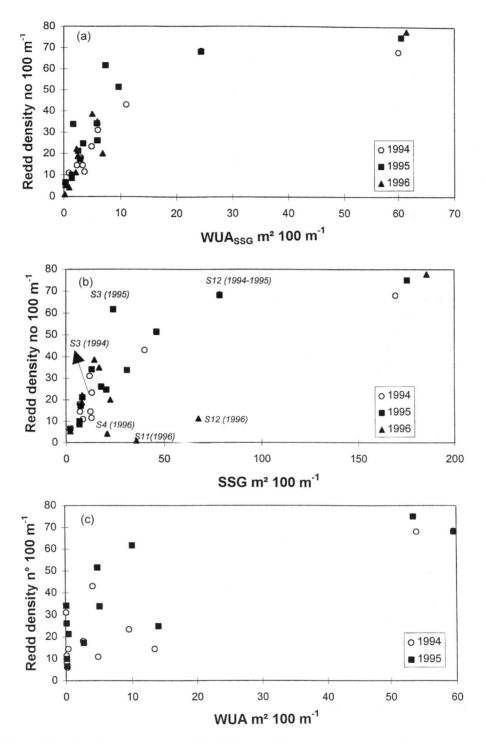

Figure 12.2   Relationship between SSG, WUA$_{SSG}$, WUA and the density of redd for the 12 sections

Table 12.3   Mean density of redd and standard deviation (in brackets) for the six main groups of channel morphodynamic units (CMUs) developed by Delacoste *et al.* (1995b)

Main CMU groups	SSG (m² 100 m⁻¹)		WUASSG (m² 100 m⁻¹)		Redds (no. 100 m⁻¹)	
Deep	15.6	(17.7)	3.2	(5.3)	25.8	(20.6)
Glide	54.1	(133.4)	15.4	(49.4)	68.5	(50.0)
Step	6.6	(9.4)	1.6	(2.5)	10.9	(6.7)
Riffle	54.6	(6.9)	13.9	(25.0)	43.4	(37.6)
Rapid	1.6	(1.7)	0.8	(1.0)	0.0	(0)
Cascade	0.0	(0)	0.0	(0)	0.0	(0)

($P < 0.05$) (Table 12.4). High densities were observed in 1996 and 1994 but there was a recruitment failure in 1997 and 1995 for the upstream sections. The densities of fry were significantly related with the densities of redd ($P < 0.05$) (Fig. 12.4). There was no relationship between the density of female normally present in the section during the year and redd density in autumn (Fig. 12.5). This observation confirmed that some spawners migrated from their resting habitat to a suitable spawning area in autumn.

If the number of fry per redd is considered, a significant variation was observed between years ($P < 0.05$). In 1994, a high mean number of fry per redd was found (6.3 fry redd⁻¹ ± 3.3) while in 1997, the mean number of fry per redd was very low (1.5 fry redd⁻¹ ± 0.8). The survival of fry was variable from year to year in relation to density-dependent mortality (1996), hydrology (1997) or water quality (1995).

In 1996, the restoration of spawning habitat in section S3 (Sources Chaudes) induced an strong increase of redd density (+165%) but a slow increase of the fry density (+32% in comparison with 1994). The number of fry per redd decreased (8.6 fry redd⁻¹ in 1994 and 4.3 fry redd⁻¹) indicating a low survival rate (Fig. 12.6) in 1996. Conversely, in section S4, the increase of fry in 1996 (+59%) was related to a significant increase in redds in 1995 (+50%) (Fig. 12.6). At this section, although the redd density increased, the density of fry stayed lower than the carrying capacity for this life-stage.

The effect of relative high flow conditions during the spawning period has been shown for sections S4, S11 and S12 in 1996. The low recruitment, especially the low number of fry per redd observed in 1997, suggests the effect of another factor. In December 1996, a severe flood occurred for two days (115 m³ s⁻¹ for a mean annual flow of 6.5 m³ s⁻¹) (EDF/DTG unpublished data). Many redds were damaged and a great number of areas of spawning gravels were washed out inducing a high mortality rate for fertilised eggs deposited in the gravel. This severe flood explained the very low density of fry in all the sections in 1997.

In 1995, a low density of fry and also a low number of fry per redd was observed. This was associated with the cleaning of Puyvalador reservoir in June 1995 which induce high concentrations of suspended sediments (1–2 g L⁻¹) over several weeks.

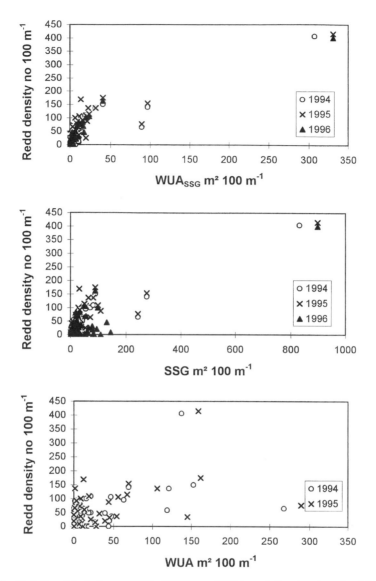

Figure 12.3 Relationship between SSG, WUA$_{SSG}$, WUA and the density of redd for the 88 channel morphodynamic units (CMU)

12.4 Discussion

12.4.1 *Densities of redd and physical habitat features: the importance of the substrate*

Spawning activity of brown trout in the Aude river showed a strong annual and spatial variability directly related to the variability of physical habitat features. At the mesohabitat scale, brown trout selected glides and riffles. Baran *et al.* (1997)

Table 12.4 The density of fry and the ratio fry per redd in the five sections

	Fry (no 100 m^{-1})				Fry per redd			
Sections	1994	1995	1996	1997	1994	1995	1996	1997
S3 Sources chaudes	201	38	265	27	9	2	4	1
S4 Bruyante	177	40	281	2	10	3	11	0
S10 Gesse 2	90	43	28	41	6	3	1	2
S11 Saint-Georges 1	51	32	61	2	5	3	2	2
S12 Saint-Georges 2	117	66	127	25	2	1	2	2

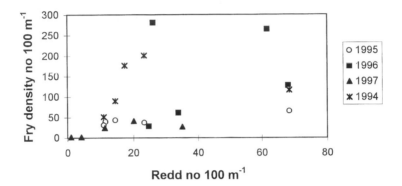

Figure 12.4 Relationship between the density of redd and the density of fry in five sections

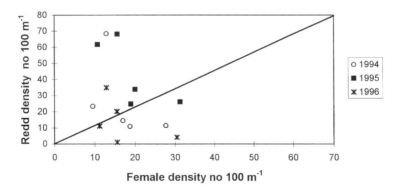

Figure 12.5 Relationship between the density of female and the density of redd in five sections

observed a similar distribution in the central part of Pyrenean streams. Previous studies indicated a strong relationship between physical habitat characteristics and the density of redds in Pyrenean streams (Delacoste *et al.* 1993; 1995a). The area of suitable spawning gravel is considered a key-factor for the density of redds (Magee *et*

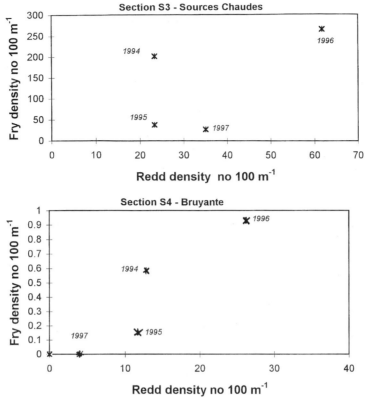

Figure 12.6 Relationship between the density of redd and the density of fry in sections S3 and S4

al. 1996). Brown trout selected spawning sites on the basis of physical characteristics of the stream, especially substrate composition (Shirvell & Dungey 1983; Witzel & MacCrimmon 1983; Grost *et al.* 1990). Females need a specific size-class of substrate between 0.2 and 5 cm diameter (Crisp & Carling 1989; Kondolf & Wolman 1993) depending on their ability to excavate this substrate (Reiser & Wesche 1977) for the redd. The availability of suitable microhabitats could thus directly influence the spawning activity of brown trout. Consequently, methodologies to estimate the carrying capacity of habitat for spawning must primarily integrate the composition of substrate.

12.4.2 *Methodologies to estimate carrying capacity of physical habitat for reproduction*

If the different methodologies measuring the physical habitat features for spawning are considered, a difference between the transect method and the suitable spawning gravel methods was found. WUA estimated by the PHABSIM model (Bovee 1982) gives poor estimates of carrying capacity of physical habitat for spawning. Shirvell (1989) and Delacoste *et al.* (1995a) indicated a low ability of the PHABSIM model to predict salmonid redd density. Comparing WUA, SSG and especially WUA_{SSG} allows the estimation of variability of the density of redd. The difference between the two types of approach seems to be related to the description of substrate. Indeed, for the transect method, the percentages of the different size-classes of substrate were noted at each point without any consideration of the type of distribution. For a same percentage of gravel, a large spread between boulders and cobbles or a succession of patches can be found. These two types of gravel distribution are not used equally for spawning by brown trout, because females need a minimum area of gravels (near 0.04 m^2) to dig its redd depending on the fish's length (Ottaway *et al.* 1981). When gravels were spread around boulders and cobbles, the trout cannot select the habitat for spawning, despite a suitable WUA. A patchy distribution of gravels is necessary for the reproduction of salmonids. Thus the calculation of WUA_{SSG} integrating the characteristic of substrate, and also the measurement of velocity and depth, are the best criteria for the estimation of the carrying capacity of physical habitat for this activity.

12.4.3   *Annual variability of redd density in relation to physical habitat modifications*

The survey showed a great variation of redd density for each of the four years of study. Similar inter-annual variation has been observed on many trout streams (e.g. Shrivell & Dungey 1983; Beaudou 1993). In the Aude river, this inter-annual fluctuation in spawning activity was directly related to physical habitat modifications, i.e. (1) modification of suitable spawning gravel (natural or artificial) or (2) shifts in hydraulic conditions during the spawning season.

When SSG and hydraulic conditions remained stable (for example section S1 and S2), redd densities did not change markedly (Coefficient of Variation (CV): 18.7% and 9.3%). Conversely, when physical habitat changed (for example, sections S3 and S4 in 1995), the redd densities changed drastically ($+165\%$, $+80\%$, respectively). For these sections, the availability of SSG naturally or artificially increased in 1995 in relation to sediment transport (sections S4) or habitat restoration (section S3). Higher flows in spring (maximum of mean daily flow: 6.9 m^3 s^{-1} in 1995 instead of 2–2.5 m^3 s^{-1} for the other years) induced some movements of sediment, particularly sand and gravel. Benda *et al.* (1992) indicated a relationship between the area of preferred spawning substrate for coho salmon, *Oncorhynchus kisutch* (Walbaum), and the stream power as an index of the stream's ability to transport sediment. This stream power was calculated using the discharge, the average gradient of the stream reach, a coefficient of fluid density and the gravitational acceleration. Stream hydraulics greatly influenced the transport of sediment particularly during high flow conditions (Shirazi & Seim 1979). Beschta and Platts (1986) considered discharge and the availability of sediment as the most important features influencing the channel morphology. This strong relationship between discharge, channel morphology, transport of sediment and consequently salmonid spawning habitat is not well documented, despite the implications on stream surveys and management.

When flow conditions were greatly modified during the spawning period (sections S4, S9 and S12 in 1996), the physical habitat changed when considering the WUA$_{SSG}$ The increase in velocity (mean bottom velocity: 0.23 m^3 s^{-1} in 1995 and 0.37 m^3 s^{-1} in 1996 for the section S12) induced a lower quantity of microhabitat suitable for spawning despite a similar area of SSG as in 1995, and most trout could not build their redds. In the case of high flow during spawning period, WUA$_{SSG}$ is an important indicator of the carrying capacity of physical habitat for the reproduction of brown trout.

12.4.4   *Relationship between recruitment and redd densities*

The relationship between the redd density and the fry density in July was significant, but very low, and characterised by a large annual variation. In the Aude river, if each year is considered separately, the higher recruitment was observed in the sections with higher densities of redd. Beard and Carline (1991) also indicated a significant and yearly correlation between fry and redd densities with an average of 3 fry redd^{-1} similar to that found in the Aude. This association implies a relatively limited dispersal of fry from their natal areas. This result is consistent with the findings of LeCren (1965), Elliott (1966) and Mortensen (1977a), who observed downstream dispersal of fry immediately after emergence in relation to the development of territorial behaviour among fry, but they also indicated that young trout became stationary after this early dispersal. Solomon and Templeton (1976) pointed out a limited dispersal of fry from redds for brown trout, as did Raddum and Fjellheim (1995) for Atlantic salmon fry, *Salmo salar* L. The annual variation between spawning activity and fry recruitment (variation of fry redd^{-1}) shows the importance of three parameters: carrying capacity for fry associated with density-dependent mortality, hydrological conditions and water quality. The shift in relationship between fry and redd density in section S3 supported the hypothesis of carrying capacity for fry and density-dependent mortality (LeCren 1973; Mortensen 1977b; Elliott 1985a). If the recruitment in this section is compared between 1994 and 1996, great increase of redd density in autumn 1995 (+165%) was found to induce a relative low increase of fry (+32%). The production of fry per redd decreased between the 2 years (1994: 8.6 fry redd^{-1}; 1996: 4.3 fry redd^{-1}). In 1996, the number of fry was sufficiently high to be regulated by density-dependent mortality. In a long-term investigation of the population dynamics of migratory trout, Elliott (1985b) showed that fry density could be related to egg density using the Ricker (1954) stock-recruitment model. For the Aude river, insufficient measures of redd density were taken to test this model. The survey also indicated annual variation in survival of fry most related to density-dependent mortality. In 1995, the recruitment of fry was very low, despite the redd density in autumn 1994 being similar to 1993. Cleaning of the

Puyvalador reservoir during June 1995 induced high concentrations of suspended sediments. Newcombe and MacDonald (1991) developed a stress index of suspended sediment for under-yearling salmon based on the concentration of sediments and the duration of exposure. Harmful effects for fish were placed in a hierarchy that spans the range from mild to severe (modification of growth to 100% mortality rate) in 14 graduated steps. It seems that in the case of the Aude river, the concentration of suspended sediment (> 2 g L^{-1}) induced a high mortality rate of fry, particularly in sections S3 and S4 downstream of the dam. The severe flood which occurred in December 1996 also induced a high mortality of fertilised eggs deposited in redds and subsequently a low density of fry in July 1997. Nehring and Miller (1987) observed that the hydrological conditions, and particularly the importance of floods during autumn and spring, can greatly influence the survival rate of trout fry in headwater streams. Thus recruitment in brown trout is by redd density, physical habitat features and water quality, but only a long-term investigation will indicate the real effect of fluctuations of the density of fry on the other age classes, particularly for a resident trout population, such as in the Aude river.

12.4.5 *Limits of the methodology to assess the carrying capacity for trout spawning*

Contrary to IFIM, the measurement of WUA_{SSG} cannot be used to predict instream flow needs for reproduction. By using models, hydraulic conditions could be simulated at different flow values for the areas of gravels measured at the basic flow, but the recruitment of new areas of gravels with an increase of flow could not be simulated in relation to a non-random distribution of gravel patches in mountain streams. Only measurements at different flow values could help to make a decision.

The study in the Aude river showed the importance of physical habitat characteristics for the spawning activity of brown trout. The WUA calculated on suitable spawning gravel areas could be used to analyse and predict the spatial and temporal variability of the spawning of brown trout. It could be integrated in long-term surveys of trout populations and in evaluations of management and stream rehabilitation projects.

Acknowledgements

We thank Alain Alric, Dominique Baril and the Fisheries Department of the Aude region for their assistance in the field. This study was funded by Electricite de France, Agence de l'Eau Rhône-Médittérannée-Corse and Direction Regionale de l'Environnement Languedoc–Roussillon as part of the survey programme of the high valley of Aude stream.

References

Baran P., Delacoste M. & Lascaux J.M. (1997) Variability of mesohabitat used by brown trout populations in French central Pyrenees. *Transactions of the American Fisheries Society* **126**, 747–757.

Beard D.T. Jr & Carline R.F. (1991) Influence of spawning and other habitat features on spatial variability of wild brown trout. *Transactions of the American Fisheries Society* **120**, 711–722.

Beaudou D. (1993) Impacts des déversements de truites domestiques dans les populations de truites communes (*Salmo trutta fario* L.). Etude dynamique et génétique. Cas du bassin de l'Orb (Hérault). PhD thesis, Université de Montpellier II, 308 pp.

Belaud A., Chaveroche P., Lim P. & Sabaton C. (1989) Probability-of-use curves applied to brown trout (*Salmo trutta* L.) in rivers of Southern France. *Regulated Rivers: Research and Management* **3**, 321–336.

Benda L., Beechie T.J., Wissmar R.C. & Johnson A. (1992) Morphology and evolution of salmonid habitats in a recently deglaciated river basin, Washington State, USA. *Canadian Journal of Fisheries and Aquatic Sciences* **49**, 1246–1256.

Beschta R.L. & Platts W.S. (1986) Morphological features of small streams: significance and function. *Water Resources Bulletin* **22**, 369–379.

Bovee K.D. (1982) A guide to stream habitat analysis using instream flow incremental methodology, *Instream Flow Information Paper No. 12, FWS/OBS 82/86*. Fort Collins, CO: Cooperative Instream Flow Service Group, US Fish and Wildlife Service, 248 pp.

Bozek M.A. & Rahel F.J. (1991) Assessing habitat requirements of young Colorado River cutthroat trout by use of macrohabitat and microhabitat analyses. *Transactions of the American Fisheries Society* **120**, 571–581.

Crisp D.T. & Carling P.A. (1989) Observations on sitting, dimension dimension and structure of salmonid redds. *Journal of Fish Biology* **34**, 119–134.

Delacoste M., Baran P., Dauba F. & Belaud A. (1993) Etude du macrohabitat de reproduction de la truite commune (*Salmo trutta* L.) dans une rivière Pyrénéenne, la Neste du Louron. Evaluation d'un potentiel de l'habitat physique de reproduction. *Bulletin Français de la Pêche et de la Pisciculture* **331**, 341–356.

Delacoste M., Baran P., Lascaux J.M., Segura G. & Belaud A. (1995a) Capacité de la méthode des microhabitats à prédire l'habitat de reproduction de la truite commune. *Bulletin Français de la Pêche et de la Pisciculture* **337/338/339**, 345–354.

Delacoste M., Baran P., Lek S. & Lascaux J.M. (1995b) Classification et clé de détermination des faciès d'écoulement en rivière de montagne. *Bulletin Français de la Pêche et de la Pisciculture* **337/338/339**, 149–156.

Elliott J.M. (1966) Downstream movement of trout fry in a Dartmoor stream. *Journal of the Fisheries Research Board of Canada* **23**, 157–169.

Elliott J.M. (1985a) The choice of a stock-recruitment model for migratory trout, *Salmo trutta*, in an English Lake District stream. *Archiv für Hydrobiologie* **104**, 145–168.

Elliott J.M. (1985b) Population regulation for different life-stages of migratory trout, *Salmo trutta* in a Lake District stream. *Journal of Animal Ecology* **54**, 617–638.

Fausch K.D., Hawkes C.L. & Parsons M.G. (1988) Models that predict the standing crop of stream fish from habitat variables. *US Forest Service General Technical Report PNW-GTR* **213**, 52 pp.

Grost R.T., Hubert W.A. & Wesche T.A. (1990) Redd site selection by brown trout in Douglas creek, Wyoming. *Journal of Freshwater Ecology* **5**, 365–371.

Jones J.W. & Ball J.N. (1954) The spawning behaviour of brown trout and salmon. *The British Journal of Animal Behaviour* **2**, 103–104.

Kondolf G.M. & Wolman M.G. (1993) The size of salmonid spawning gravels. *Water Resources Research* **29**, 2278–2285.

LeCren E.D. (1965) Some factor regulating the size of populations of freshwater fish. *Mitteilungen Internationale Vereinigung für theoretische und angewandte Limnologie* **13**, 88–105.

LeCren E.D. (1973) The population dynamics of young trout (*Salmo trutta* L.) in relation to density and territorial behaviour. *Rapport et Procès-Verbaux des Réunions du Conseil Permanent International pour l'Exploration de la Mer* **164**, 241–246.

Magee J.P., McMahon T.E. & Thurow R.F. (1996) Spatial variation in spawning habitat of cutthroat trout in a sediment-rich stream basin. *Transactions of the American Fisheries Society* **125**, 768–779.

McFadden J.T. (1969) Dynamics and regulations of Salmonid populations in streams. In T.G. Northcote (ed.) *Symposium on Salmon and Trout in Streams.* Vancouver: University of British Columbia, pp. 313–329.

Mortensen E. (1977a) Population, survival, growth and production of trout *Salmo trutta* in a small Danish stream. *Oikos* **28**, 9–15.

Mortensen E. (1977b) Density-dependent mortality of trout fry (*Salmo trutta* L.) and its relationship to the management of small streams. *Journal of Fish Biology* **11**, 613–617.

Myers R. A. & Barrowman N.J (1996) Is fish recruitment related to spawner abundance? *Fishery Bulletin* **94**, 707–724.

Nehring R.B. & Miller D.D. (1987) The influence of spring discharge levels on rainbow trout and brown trout recruitment and survival, Black Canyon of the Gunnison River, Colorado, as determined by IFIM/PHASIM models. *Proceedings of the Anglers Conference Western Association Fish and Wildlife Agencies, Salt Lake City, Utah*, pp. 87–96.

Newcombe C.P & MacDonald D.D. (1991) Effects of suspended sediments on aquatic ecosystems. *North American Journal of Fisheries Management* **11**, 72–82.

Ottaway E.M., Carling P.A., Clarke A. & Reader N.A. (1981) Observations on the structure of brown trout, *Salmo trutta* Linnaeus, redds. *Journal of Fish Biology* **19**, 593–607.

Raddum G.G. & Fjellheim A. (1995) Artificial deposition of eggs of Atlantic salmon (*Salmo salar* L.) in a regulated Norwegian river: hatching, dispersal and growth of fry. *Regulated Rivers: Research and Management* **10**, 169–180.

Reiser D.W. & Wesche T.A. (1977) Determination of physical and hydraulic preferences of brown and brook trout in the selection of spawning locations, *Water Research and Technology Completion Report C.7002.* Laramie: United States Department of the Interior, University of Wyoming, 100 pp.

Ricker, W.E. (1954) Stock recruitment. *Journal of the Fisheries Research Board of Canada* **11**, 559–623.

Seber G.A.F. & LeCren E.D. (1967) Estimating populations parameters from catches large relative to the population. *Journal of Animal Ecology* **36**, 631–643.

Shirazi M.A. & Seim K.S. (1979) Stream system evaluation with emphasis on spawning habitat for salmonids. *Water Resources Research* **17**, 592–594.

Shirvell C.S. (1989) Ability of PHABSIM to predict chinook salmon spawning habitat. *Regulated Rivers: Research and Management* **3**, 277–289.

Shirvell C.S. & Dungey R.G. (1983) Microhabitats chosen by brown trout for feeding and spawning in rivers. *Transactions of the American Fisheries Society* **112**, 355–357.

Solomon D.J., Templeton R.G. (1976) Movements of brown trout, *Salmo trutta* L., in a chalk stream. *Journal of Fish Biology* **9**, 411–423.

Tautz A.F. & Groot C. (1975) Spawning behaviour of chum salmon (*Oncorhynchus keta*) and rainbow trout (*Salmo gairdneri*). *Journal of the Fisheries Research Board of Canada* **32**, 633–642.

Witzel L.D. & McCrimmon H.R. (1983) Redd site selection by brook trout and brown trout in south eastern Ontario streams. *Transaction of the American Fisheries Society* **112**, 760–771.

Section IV
Anthropogenic impacts

Chapter 13
Wimbleball Pumped Storage Scheme: integration of water resource management, engineering design and operational control to compliment the needs of the salmonid fisheries in the River Exe

H.T. SAMBROOK
South West Water, Peninsula House, Rydon Lane, Exeter EX2 7HR, UK

I.G. COWX
University of Hull, International Fisheries Institute, Hull HU6 7RX, UK

Abstract

Wimbleball Reservoir is a strategic reservoir that regulates flows in the River Haddeo, a tributary in the headwaters of the River Exe. Increasing demand for water supply in East Devon resulted in the promotion and construction of Wimbleball Pumped Storage Scheme. The new scheme has increased the yield of the existing reservoir and reduced the refill period from 2–3 years to a single year. The component parts and operating philosophy are relatively simple, with abstraction from the Exe at times of high river flows and releases of water during low flow periods. Throughout the planning and implementation of the scheme, water resource management, engineering design and operating rules have been adapted and refined to compliment the needs of the river and its fisheries. Particular emphasis was placed on the indigenous salmon and trout stocks in the upper catchment areas and the potential impacts of the modified flow regimes on the various life stages. Concerns relating to the sustainability of the Haddeo as a spawning and nursery habitat for juvenile salmonids were assessed. In addition, the effects of abstractions and augmentation releases on adult salmon behaviour were assessed using radio-telemetry. Various intake and river weir designs were considered to ensure minimum impact on fisheries, especially the upstream and downstream migration of salmon. All environmental studies and assessments identified flow, and associated parameters, as the most significant factor. As a result, integration of flow control via the abstraction and releases, to compliment the known needs of the fisheries and the river, ensured that any potential impacts were designed out, reduced to an acceptable level or that suitable mitigation measures were established.

Key words: River Exe, salmon, trout, migration, water resource management

13.1 Introduction

Wimbleball Reservoir was the first of three strategic reservoirs constructed in the South West of England over a 14-year period (1976–1989). Wimbleball has a storage of 21 320 ML, which, together with Colliford (28 540 ML) and Roadford (34 500 ML) provide 75% of the total storage available for reliable public water supplies (PWS) in Devon and Cornwall. This new generation of reservoir was required to meet increasing demands across the region. By operating in conjunction with established river and small reservoir sources, support was given to large water supply zones, rather than just to single communities. Water demands continue to increase in the South West region due to population increases (the second fastest growth area in England), increase in per capita consumption, marked growth in the tourist industry and the potential onset of climatic change. The ever-increasing risk of climatic change is predicted to result in an increase in annual rainfall, with higher levels in winter and decreased rainfall in summer. In addition, the increasing temperatures will mean that evapotranspiration will be higher, with a resultant reduction in summer run-off. Such conditions will impact raw water supplies and a warmer climate will inevitably increase consumer demand.

Potential solutions to future increases in demand within the Wimbleball Supply Zones of East Devon have been considered since 1989. These options formed the basis of the evaluation study associated with Wimbleball Pumped Storage Scheme (WPSS) and included:

(1) Demand management – leakage control, metering policy, conservation and education have been implemented.
(2) Otter Valley Groundwater – optimisation and rehabilitation of existing borehole sources to increase yield. Any development could be constrained by water quality and environmental issues.
(3) Axe Valley Reservoir – a feasibility study was undertaken to assess a new pumped storage reservoir in East Devon. This proposal was rejected due to insufficient yield, geological problems, environmental sensitivity and operational constraints.
(4) Desalination – shelved due to environmental and financial constraints.
(5) Wimbleball PSS – development of a pumped storage scheme to support an existing strategic reservoir.

Throughout its strategy, South West Water acknowledged the importance of demand management, but recognised that future reliable supplies in East Devon will be dependant on a twin-track approach of demand management coupled with new resource development. Following extensive feasibility studies (1991), a decision was made to promote, assess and implement the Wimbleball PSS. This chapter presents a snapshot of important issues and selected abstracts of relevant and applied benefit to both fisheries and water resource management.

13.2   Wimbleball Reservoir and Exe Water Resources

Wimbleball Reservoir is an existing strategic reservoir that regulates flows in the River Haddeo, a headwater tributary of the River Exe (Fig. 13.1). Construction of

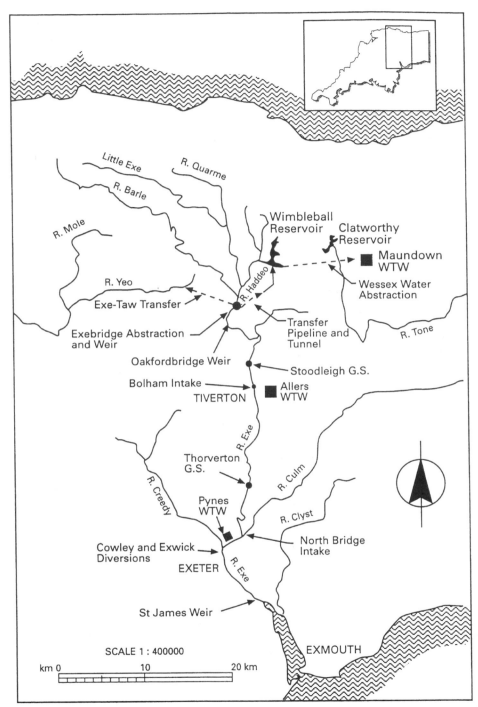

Figure 13.1 Map of the River Exe catchment showing the main features of the water resource schemes

the reservoir was completed in 1978. The original water resource scheme comprised river abstractions, an intercatchment transfer, river augmentation and direct supply abstraction (Fig. 13.1). There are two licensed public water supply (PWS) abstractions on the main stem Exe: at Bolham (maximum licensed quantity 32.0 MLd^{-1}) and Northbridge (maximum licensed quantity 66.46 MLd^{-1}). The abstractions from the unsupported river flows are controlled by a prescribed flow (pf) threshold at Thorverton gauging station (pf: 273 MLd^{-1}). Both licences comprise two parts, a Licence of Right and a formal licence condition. As natural river flows recede, releases are made from Wimbleball Reservoir to augment the base flow in the Exe and hence support the PWS abstraction. The maximum augmentation release is 99.9 MLd^{-1}. In addition, releases can be made to support the intercatchment transfer between the rivers Exe and Taw for abstraction at the head-of-tide, Newbridge. This supported abstraction is to a maximum of 23 MLd^{-1}. At all times a compensation flow of 9.1 MLd^{-1} is maintained to the River Haddeo. Finally, there is a direct supply abstraction of 31.82 MLd^{-1} (maximum licensed quantity) from Wimbleball to Maundown Water Treatment Works (Wessex Water).

Wimbleball is situated within Exmoor National Park and is surrounded by a mixture of upland agricultural land and woodlands. The reservoir is overlooked by Haddon Hill on the eastern flanks, which is a Site of Special Scientific Interest [SSSI], and within a short radius are many County Wildlife Sites. Water quality is high throughout these upper river reaches and in the reservoir. The Barle and Little Exe continue to support the best salmon, *Salmo salar* L., and trout, *Salmo trutta* L., spawning and nursery areas in the Exe catchment. The Haddeo and Pulham also support these salmonid species. Overall, the rivers in the upper Exe catchment have high fisheries, ecological and conservation status.

13.3   Wimbleball Pumped Storage Scheme

Until February 1997, the existing Wimbleball scheme operated as described. The ecology and environment of the river and reservoir have been impacted and possibly adapted to a range of hydrological changes over a 20-year period. The introduction of the pumped storage scheme gives added yield and security to water resources in the catchment. Such proposals could potentially have a serious environmental impact unless appropriate assessments and actions are taken to reduce the risks.

The components associated with the pumped storage scheme are relatively basic, comprising a suitably located abstraction point and a pipeline to deliver the water to the reservoir. The refill of Wimbleball is no longer solely reliant on natural inflows; now refill can be guaranteed in a single year, i.e. compared to the original scenario of 2–3 years. The abstraction at Exebridge effectively increases the catchment collection area by a factor of 10 times (29.1–283.3 km^2), and increases the daily mean flow (dmf) from 0.748 to 9.706 m^3 s^{-1}.

In engineering terms, the main elements are relatively simple, but each element proved to be environmentally sensitive and challenging such that appropriate modifications were required. The engineering elements (Fig. 13.2) include:

- Wimbleball Reservoir – existing;
- intake and flow measurement weir;
- pumping station;
- water main and tunnel;
- outfall structure.

Fisheries issues associated with the intake, weir design and operational procedures were among the main environmental concerns.

The timetable for Wimbleball PSS extends over an 18-year period from the first feasibility studies in the mid-1980s to the post-scheme monitoring in 2002. The main milestones are:

- mid-1980s: various feasibility studies undertaken after the drought events of 1984 and 1989;
- late 1991: intensive environmental studies and licence application;
- March 1993: abstraction licence granted;
- Autumn 1996: construction commenced;
- February 1997: operation commenced;
- on-going: post-scheme monitoring to 2002.

13.4   Environmental Assessment

The remainder of this chapter relates to the approach adopted for the Environmental Assessment, with the focus on specific salmon and trout issues that were instrumental in influencing engineering design, operational procedures and mitigation measures.

The general approach to the Environmental Assessment and Scheme promotion included:

- scheme outline and investigations;
- existing situation: fisheries and environmental status;
- scheme evaluation and impact assessment;
- instigation of additional investigations;
- undertaking of extensive consultation;
- development of operating rules through the integration of legislative and regulatory constraints, environmental and political issues, engineering and operational matters, etc.;
- iteration and refinement of the scheme detail and consolidation of the operating rules;
- achieving an environmentally acceptable strategy that attains the balance between the environment, scheme yield, sustainability and operating costs;

- design of appropriate mitigation measures;
- implementation: construction and operation;
- compilation of an Operating Agreement and post-scheme monitoring.

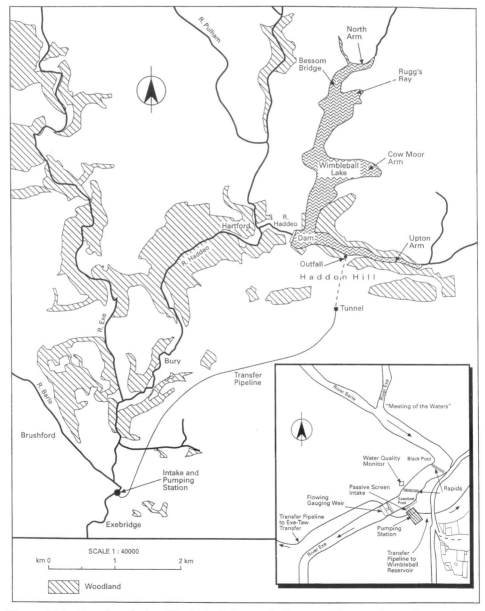

Figure 13.2 Details of the Wimbleball Pumped Storage Scheme abstraction point at Exebridge

An extensive Environmental Assessment was undertaken covering both aquatic and terrestrial concerns associated with each aspect of the scheme. Most work was required on the aquatic environments, with a complex array of interactions considered. Potential issues were identified and priority given to studies at the outset. Investigations were undertaken, targeting the highest priority issues that were scheme-impact related. Throughout, it was essential to remain focused on specific impacts of the scheme and not to attempt to undertake peripheral and often academic issues. Annual patterns of flow changes (abstraction and augmentation) were assessed, together with the effects on the river and the environment. Such studies included flow-related parameters (velocity, volume, depth and wetted area), sedimentation, substrate composition, channel morphology, water quality, macrophytes, macroinvertebrates, fisheries, birds and otters. This information ensured adequate baseline data against which impacts could be assessed, priorities refined and guidelines produced to give environmental protection. Throughout the assessment, flow was identified as the critical factor. Hence the approach was to build the known flow needs of fish and the river into the control procedures for the scheme. Such an approach ensured that the impacts were either designed out, reduced to an acceptable level or mitigated against. Wherever possible, environmental gain was encouraged and certainly not ignored.

13.5   Reservoir scheme impacts on fish

Although the rivers Exe and Haddeo are already impacted by the operation of Wimbleball Reservoir, the new pumped storage scheme could impinge on the fish and fisheries of the Exe catchment in several ways. These include:

- modification of the flow regime in the rivers Haddeo and Exe;
- modification of the physico-chemical characteristics, particularly temperatures of both rivers;
- modification of the reservoir environment;
- impact on salmon migration and performance of the fisheries;
- potential for entrainment and impingement of juvenile fish at the abstraction point;
- disturbance of the river environment during the construction phase.

All these potential impacts were addressed and key issues are selected for discussion, with a particular bias towards salmon.

The potential impact of the scheme operation on river flow could influence both the adult and juvenile stages of the salmon in fresh water. Altering the flow regime and related parameters could influence the various life-history components throughout the year:

- stimulus to adult migration – estuary-to-river and throughout the river;
- distribution of adults;
- choice of spawning grounds;

- survival of eggs and alevins in gravel;
- fry emergence and distribution;
- habitat requirements for fry and parr life stages and smolt production;
- parr displacement;
- downstream migration of smolts.

In addition, the possible effects on the performance of the salmon rod fishery and estuary net fishery were assessed, together with the brown trout in the rivers, the reservoir fishery (rainbow trout, *Oncorhynchus mykiss* (Walbaum)) and the coarse fishery in the lower and mid reaches. (roach, *Rutilus rutilus* (L.), dace, *Leuciscus leuciscus* (L.), bream, *Abramis brama* (L.), grayling, *Thymallus thymallus* (L.))

13.6   Status of the fish populations and fisheries of the River Exe

The River Exe supports both good game and coarse fisheries, and in the past was considered one of the finest spring salmon rivers in England. The salmonid fisheries are associated with the estuary (commercial seine netting for salmon; primarily grilse in June–August) and the river, with the majority of salmon and resident brown trout caught upstream of Bickleigh. Historical catch statistics enable the comparison of the performance of both the rod and net fisheries on an annual basis (see Fig. 13.3).

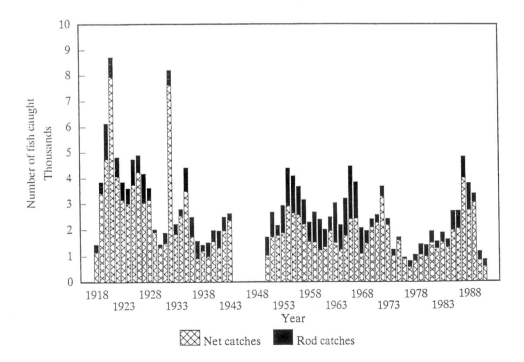

Figure 13.3   Annual net and rod catch statistics for salmon in the River Exe (1918–1991)

A combination of temporal and spatial trends are useful in assessing the initial effects of any water resource proposal. In addition, knowledge of the date of each rod-caught salmon enables flow to be allocated to compliment other information regarding the location of the catch, size, method, etc. This in turn allows analysis of the rod fishery and behaviour of salmon along the river in association with flow and time using frequency distribution and regression analysis. Figure 13.4 shows the daily distribution of salmon catches along five sections of the River Exe in 1988. The importance of spates to the salmon is evident and analysis enabled flow thresholds for migration and fishery performance to be determined. Such data provide similar general findings as that derived from the tracking studies (Fig. 13.5), but lack the precision required for threshold derivation. The performance of the rod fishery is flow dependent and extremes of flow, high (>3000 MLd^{-1}) and low (<100 MLd^{-1}) are associated with poor catches, while medium flows (range: 250–1000 MLd^{-1}) support relatively good catches. The distribution of radio-tagged salmon in 1992 (Fig. 13.5) provided the precise information on flows required to stimulate upstream migration of salmon. In particular it allowed evaluation of the minimum flows necessary for adult salmon to bypass the numerous obstructions on the river. It is information from these fish that was used to set thresholds at strategic locations, particularly at critical times to the upstream migration of adult salmon.

Upstream migration of adult salmon is not only flow dependent but is affected to differing degrees by 14 weir obstructions along the main river. The radio tracking studies identified that the most problematical of these are at St James Weir, Exwick/ Cowley weirs and Oakfordbridge Weir (Fig. 13.1). Detailed analysis of the radio tracking studies using a new flow-frequency analysis method identified specific flow thresholds below which free migration of salmon is impeded. (Solomon *et al.* 1999). Figures 13.6 and 13.7 show examples of the different flows used by salmon arriving and ascending St James Weir and Oakfordbridge Weir, respectively. In both cases, the weir itself is an impediment to movement of salmon at certain flows. At St James Weir, the threshold flow is about 5.8 m^3 s^{-1}, at which the peak difference is about 13%, i.e. 13% of the total run of tagged salmon was delayed by this weir due to flow falling below 5.8 m^3 s^{-1}, which represents the majority of fish arriving at the weir at such flows. Similarly at Oakfordbridge Weir, comparison of the fish-flow lines for arrival and departure indicate a delay of 24% of the total run at flows below 8.9 m^3 s^{-1}. About 15% of the run arrived at flows below 6.8 m^3 s^{-1} but no fish ascended the weir at such flows. Of all salmon arriving at flows below 10 m^3 s^{-1}, over half were delayed until flows exceeded that value.

The main salmon spawning grounds are on the main river and its tributaries upstream of Exebridge. The River Barle, and to lesser extent the Little Exe, are the main spawning reaches, contributing more than 70% of the total redd counts in any one year. The main river downstream of the Barle confluence and the Haddeo catchment are relatively poor spawning areas but are important nursery areas. Trout populations in the Exe catchment are primarily comprised of resident brown trout with small numbers of migratory sea trout. Trout spawning and nursery grounds are located in the upper reaches of most tributaries (Table 13.1). By comparison to the Barle, the Haddeo supports lower densities of salmon (Table 13.1).

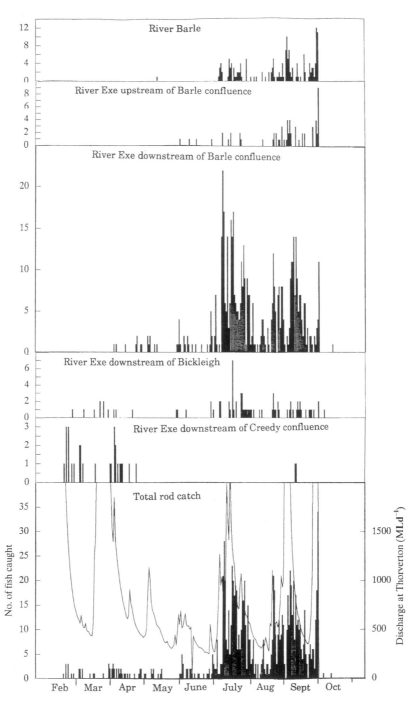

Figure 13.4 Distribution of reported daily catches in different reaches of the River Exe and daily mean flow at Thorverton during 1988

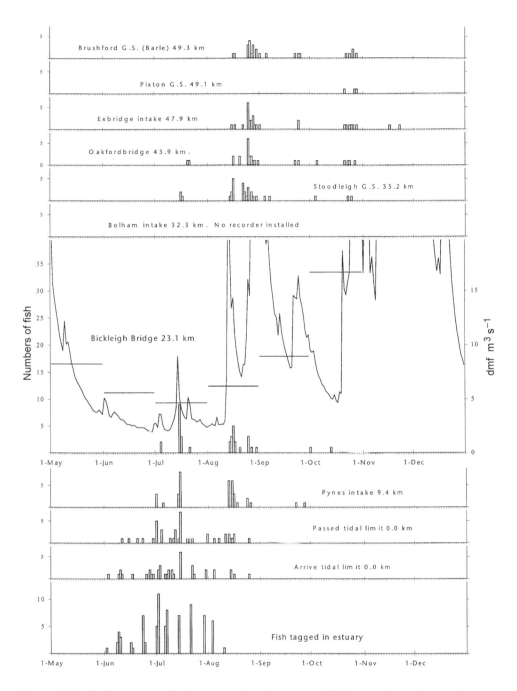

Figure 13.5   Daily records of salmon tagged and passage of fish passed recording stations in 1992. Also shown is the daily mean flow for Thorverton gauging station and the long-term monthly mean (horizontal line). The distances shown are from the tidal limit at St James Weir

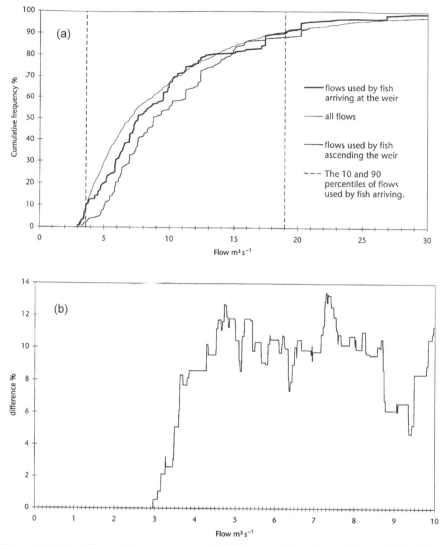

Figure 13.6   (a) Cumulative frequency of daily mean flow: dmf values (all flows) and dmf values used by salmon arriving at and ascending St James Weir. (b) Differences in cumulative frequency of flows used by salmon arriving at and ascending St James Weir

The River Exe supports good stocks of coarse fish, especially roach, dace, bream, grayling and gudgeon, *Gobio gobio* (L.) (Cowx 1983, 1988). In addition, eel, *Anguilla anguilla* (L.), bullhead, *Cottus gobio* L., stoneloach, *Barbatula barbatulus* (L.), minnow, *Phoxinus phoxinus* (L.) and lamprey species are distributed throughout the river and its tributaries.

Impoundment of Wimbleball began in 1977 and the indigenous brown trout of the inundated stream formed a small natural stock for the new fishery. The wild stock is still maintained by recruitment from the reservoir inflow streams, albeit at low levels.

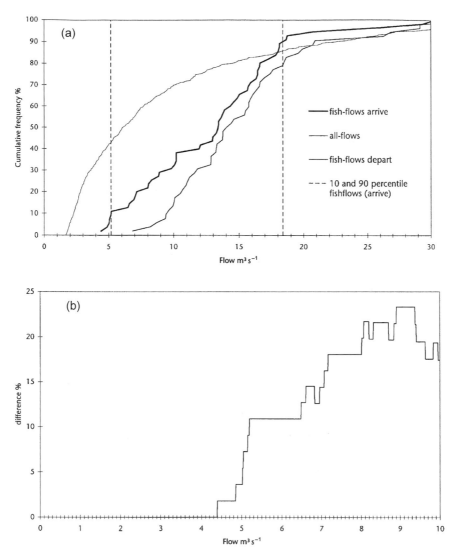

Figure 13.7   (a) Cumulative frequency of daily mean flow (all flows) and dmf values used by salmon arriving at and ascending Oakfordbridge Weir. (b) Differences in cumulative frequency of flows used by salmon arriving at and ascending Oakfordbridge Weir

As a result, Wimbleball has been developed as a stocked rainbow trout fishery with stock introduced on a regular basis at a catchable size. Analysis of the historic catch statistics reveals that extensive draw-down of the reservoir has an adverse effect upon fishing activity, but not upon catches. The pumped storage scheme will increase draw-down (frequency and extent), particularly in dry summers, followed by a faster recharge and guaranteed refill in the spring. Overall, there is unlikely to be an effect on catch rate, but there will be a loss of the aesthetic appeal of the location that may deter anglers from fishing.

Table 13.1   Mean densities (*n* ind. 100 m^{-2}) of juvenile salmon and trout in selected sub-catchments of the upper River Exe in 1992

Sub-catchment	Number of sites	Salmon fry 0+	Salmon parr >0+	Trout fry 0+	Trout parr >0+
Little Exe	16	14.03	8.73	19.70	14.03
Quarme	3	5.99	3.59	26.03	24.53
Exe	4	1.28	2.39	0.18	3.30
Haddeo	6	2.72	5.78	2.68	10.69
Pulham	4	2.49	4.24	98.53	17.46
Barle	9	61.87	19.36	3.26	9.28
Brockley River	1	0.00	8.18	1.10	24.86
Iron Mill Stream	2	3.53	4.04	1.59	15.15
Batherm	7	2.74	0.49	7.33	8.86

13.7   Impacts on flow

Extensive studies identified that flow was the main factor in determining impacts of the scheme operation. The studies had to differentiate between the existing regulated flow regime and the associated impacts from those of the new regime of abstraction and additional enhanced flows. Hydrographs were produced for wet, dry and average years for the different operating regimes of demand to the maximum authorised licensed quantities (e.g. Fig. 13.8). These hydrographs were generated for various locations along the river length, together with other information on depth, velocity, and wetted area. Such data formed the basis of assessing impacts, their significance and hence their priority. This approach ensured that all priority issues were adequately addressed. The main features of the extensive model scenarios and filed studies were as follows.

13.7.1   *Flow impacts: releases to River Haddeo (existing situation to worse case scenario)*

- Haddeo flows are already affected by existing operations, with compensation flow and enhanced flows as and when needed. Superimposed on this base pattern is the natural flow modulation from the River Pulham. These flows will be altered further by increased releases.
- Annual and seasonal variations result from weather and climate conditions that directly impact on demand patterns. Also, as demands grow, then the release volume will increase, becoming more consistent and steady for longer periods. Augmented flows in average years will increase markedly in comparison to both wet and dry years; the existing flow regime could be doubled. The reliance on

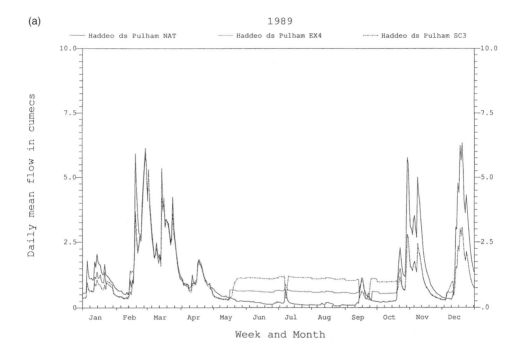

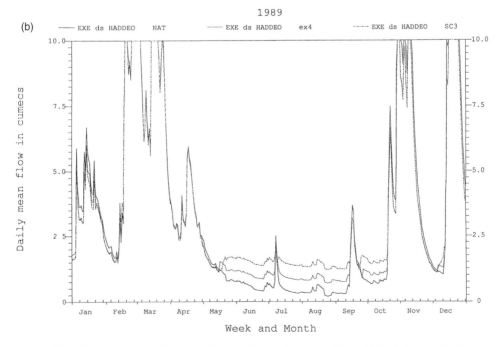

Figure 13.8 Hydrographs showing the predicted change of Wimbleball Pumped Storage Scheme operation on flows under two different operating rules: (a) on the River Haddeo downstream of the River Pulham; (b) the River Exe downstream of the River Haddeo confluence

stored water in dry or drought years means that there will be little change to the existing flow regime. Similarly in wet years, abstractions are sustained more by unsupported flows with only few releases.

- Throughout the summer months (May–September), the natural low flows will be eliminated and daily flows enhanced throughout the Haddeo and Exe.
- Greatest impacts on the flow changes will be local to the reservoir in the upper Haddeo. These impacts are reduced along the lower Haddeo and throughout the Exe as a result of the increased channel dimensions and ingress of tributaries along the downstream reach.
- Comparison of parameters for existing and predicted scenarios show that the greatest impact in the upper Haddeo would result in an increase in volume (46 MLd^{-1}), velocity (0.18 m s^{-1}), depth (9 cm), and wetted width (0.42 m). Inevitably, the impacts on the main stem Exe are markedly less, reflecting the greater river width within the confines of the banks.
- Under a worse case of predicted conditions there would be an eight-fold increase in flow downstream of the Pulham–Haddeo confluence, with a two-fold increase in the Exe, downstream of the Haddeo confluence. This comparison represents the worse case of predicted maximum release with the historic summer base flows, i.e. pre-reservoir impoundment.
- The control curve for the reservoir will reduce reservoir spillage (incidence and quantity) in the winter–spring period. Such spillage, together with spates, are important to adult salmon. Without spill to increase the baseflow and duration of winter flow, the spate events in Haddeo are reliant on the natural flow from the River Pulham. Too often the upper Haddeo only receives the compensation flow for much of the winter.

13.7.2   *Flow impact at Exebridge due to the abstraction*

Fisheries concerns restrict the abstraction to winter flows only, November–March. The impact on flows local to the abstraction point were modelled and the main features of the worse case abstraction of 150 MLd^{-1} were:

- greatest impacts occurring at times of maximum take as flows approach prescribed flow conditions;
- reduction in depth of 2–5 cm;
- reduction in water velocity of 0.05–0.15 m s^{-1};
- reduction in wetted width downstream is negligible.

Overall, any detectable flow-related parameter changes are reduced downstream along the gaining river reach. Other operating conditions relating to the control curves, water quality thresholds and operating rules affect the frequency of operation and hence the period over which these worse case scenarios occur. Such conditions reduce the effect of the abstraction during the winter months.

Impacts on fisheries are flow-dependent and all evidence suggests no significant effects downstream of Exebridge throughout the year. Hence there will be no impacts on either the coarse fishery, net fishery in the estuary, or the salmon rod fishery. Any concerns relate to salmon migration at spawning time and at obstructions along the river.

13.8   Outcome of the fisheries studies

The Environmental Assessment identified certain issues and concerns that were considered sufficiently important to warrant attention. The following represent examples that exhibit the range of issues in which fisheries had an influence in the engineering design, formulation of operating rules and promotion of mitigation measures.

13.8.1   *Operating rules: River Haddeo*

Under steady state conditions, the scheme is unlikely to cause any major detrimental effects to the fish species in the Haddeo. The predicted changes to volume, velocity, etc. will result in minor effects local to the reservoir only, as all parameters will remain within acceptable thresholds. However, the greatest risk to fry and parr was considered to be the pattern of any release, especially the rate of initiation and curtailment of releases. All future release patterns or changes to existing releases will be controlled to mimic natural spate characteristics, wherever possible. Valve control will require step-wise progression to implement a rapidly rising limb and a slow recession characteristic of a hydrograph. Significant and rapid changes in releases represent a major threat to the river and result in damaging velocities associated with the hydraulic wave front. Such events have been designed out by the establishment of procedures in an Operating Agreement. In conjunction with these rules, a preference will be made to the use of the uppermost draw-off, hence reducing risk to the river due to changes in water temperature and water quality. This approach has benefited from the existing practices that were developed at Roadford (Sambrook & Gilkes 1994). Although manual control has to be adopted at Wimbleball compared to PLC control at the newer Roadford Reservoir, the philosophy remains the same. Whenever possible, flow changes are made to benefit the fisheries and ecology of the receiving rivers. In addition, further risks to the river are reduced by adopting a procedure to minimise the need for frequent changes in flow and avoid pulsed patterns, hence retaining longer periods of steady block releases in response to demand.

The significance of reservoir spill was identified earlier, in terms of quantity, timing and duration, and noted as being important to the distribution of spawning salmon in the Haddeo. Refinements in reservoir operation will inevitably reduce the frequency and quantity of spill, together with a marked shift to later spill, hence avoiding the main migration period for salmon onto the spawning grounds. Evidence from redd counts and juvenile fish surveys show that spill at the correct time (and volume) is critical for improving upstream penetration and distribution of adults, influencing the number of resulting redds and generated densities of fry and parr in subsequent years. Without reservoir spill, the salmon in the Haddeo are solely dependent on spates emanating from the smaller Pulham catchment. Such spates are small and short-lived, hence reducing the effectiveness to attract spawning fish from the main stem Exe. Under such conditions, the salmon are confined to the lowermost reaches of the Haddeo only. The combination of Wimbleball spill and spates from the Pulham in October and November increase the chance of attaining optimum distribution of adults and resulting juvenile recruitment. Such a combination occurs infrequently and so special releases from Wimbleball have been promoted to benefit salmon stocks in the Haddeo. These releases would be timed to coincide with the availability of adult salmon and controlled to compliment spate events in the adjacent catchments. This option is one of three being considered by the Environment Agency for the use of the fisheries water bank (see Section 13.8.4).

13.8.2   *Operating rules: Exebridge abstraction point*

During the preparation of the licence application, various operating rules were promoted to avoid impact and minimise risk to the fisheries and river overall. The combination of measures neither compromised scheme yield nor imposed impossible and onerous operating practices. Fisheries issues were foremost in shaping the proposals and eventually all the issues were endorsed by the regulator and incorporated in the Abstraction Licence Conditions. The main features were as follows.

(1) Period of operation: Abstraction would be restricted to the period, November–March, inclusive. This period purposely avoids the salmon rod season and the known periods of migration of adult salmon (upstream spawning migration: October–November) and smolts (downstream spring migration: April–May). In most years, the great majority of the migration passed Exebridge is completed by November and operation of all abstractions at full take would represent minimal impact (Fig. 13.4). However, in some years, low flows during the autumn would mean that much of the migration would not be completed by November and early scheme operation could have an impact. As a result, reservoir operational control curves for Wimbleball were designed to minimise November pumping. Overall, the impact of Wimbleball refill abstraction on salmon migration will be negligible, except in years when a prolonged dry period persists into November. Even then, there is scope for impact and risk to be reduced. The radio tracking studies identified critical thresholds for salmon passage at strategic locations. One of these sites is Oakfordbridge Weir, which is a deterrent to free migration (Fig. 13.7). Easing fish passage at this location would reduce the critical threshold by some 2 m^3 s^{-1}. Under such conditions, abstraction must not reduce

river flow below the threshold flow during November. An alternative would be the use of part or all of the fisheries water bank (13.8.4) to protect critical flows in November. If the allocated annual quantity of water was exhausted, it is possible that access to part of the following year's quota could be negotiated.

(2) Prescribed flow conditions: Studies show that a prescribed flow of 100 MLd^{-1} (Q95) would be sufficient to optimise scheme yield and minimise environmental impact. This condition would operate in conjunction with a 50% take rule when applied to a minimum take of 10 MLd^{-1} and maximum take of 150 MLd^{-1}. Such a supplementary rule maintains the hydrograph profile, but it also introduces a further precautionary measure at times of maximum abstraction that is equivalent to a markedly higher prescribed flow of 250 MLd^{-1}.

(3) Authorised licensed quantity: The pump configurations and engineering constraints at Exebridge restrict the range of daily abstraction quantities to 10–150 MLd^{-1}. In addition, an instantaneous flow rate of less than or equal to 1.74 m^3 s^{-1} was stated. The abstraction was also restricted by an annual authorised quantity.

(4) Decision processes: A daily decision-making process was introduced to integrate the prescribed flow controls for all the public water supply abstractions along the Exe based on Thorverton GS and the management of releases from Wimbleball Reservoir.

(5) Spate sparing: While spates are important for salmon migration, they are also an essential part of the channel-forming processes of rivers which ultimately influence the conservation and environmental worth. The effectiveness of spate sparing is minimal in the November–March period but it will give added protection following dry autumns in the November–December periods. As opposed to summer spate sparing, no specific procedures are required. The essential need for water quality control and intake protection for the reservoir ensures avoidance of abstraction from spates, indirectly benefiting the salmon and river. Extensive studies have characterised the water chemistry associated with the winter spates and individual thresholds have been identified to reduce the risk of a water quality challenge to the reservoir. Throughout, turbidity was identified as the main surrogate for suspended solids, nutrients, metals, organics and pollution events. Turbidity, together with pH, dissolved oxygen, conductivity and ammonia form the primary protection determinands. The field studies identified the alarm thresholds that control the operation of the abstraction, without comprising yield or jeopardising water quality. These conditions avoid abstraction from the peaks of winter spates that are known to be important to environmental processes.

(6) Flow measurement: Adequate flow monitoring and measurement are normal requirements of abstraction licence conditions. Such conditions relate to both abstraction meters but also gauging structures in rivers, to ensure that the prescribed flow conditions are enforced continuously during the abstraction period. As a result, a new gauging weir was designed that enabled accurate flow measurement without compromising fisheries and other river users.

13.8.3   *Fisheries influence of engineering design*

The selection of the abstraction point at Exebridge represented a balance between cost, engineering design and scheme yield. Potentially, the location of the site downstream of the Exe–Barle confluence was a major fisheries concern. Exebridge is the 'gateway' to the two main spawning and nursery rivers (Barle and Little Exe) in the Exe catchment. Any intake design or operating regime that did not adequately address the fisheries issues alone could be responsible for a decline of salmon in the catchment. The capitalised value of the Exe salmon rod fishery is estimated at between £5 and £11 million.

The period of abstraction avoids the peak period of salmon smolt migration, which in the South West of England usually occurs in April and May each year. Irrespective, protection of the intake was required to prevent damage to both salmon fry and parr that are known to undertake downstream dispersion in the spring and autumn. In addition, the design must be sufficient to minimise impact to the other indigenous fish species in the Exe by preventing entrainment and impingement. Various designs were considered to fulfil the engineering and operational criteria, but the most effective intake design to protect fish was considered to be the Johnson Passive Screen (Cowx & Welcomme 1998). Such intakes have a proven performance on the Exe and the technology employs flow and velocity control to minimise impingement and entrainment of debris and aquatic life at surface water intakes. The screens have 3-mm slots with a maximum approach velocity of <0.15 m s^{-1}.

Contrary to the usual siting of these screens in the river, this screen array is located in an isolated bankside chamber that was constructed as an extension to a natural pool. The intake is located on the left bank at the extreme edge of the backwater to this pool. A deepwater channel exists on the right bank that supports the majority of low–medium flows. This natural feature adds to fish protection by attracting fish away from the intake. The screen chamber has a large surface area intake of 12.6 m long and 2.0 m deep to ensure a reduced approach velocity. There are coarse trash screens to protect the intake with 8 mm bar and 10 mm gap. The outer screen dimensions alone are sufficient to prevent ingress of the larger fish that inhabit the pool, i.e. adult salmon and trout, and eels.

A new weir was required at Exebridge to fulfil part of the monitoring and measurement condition of the licence. The original design was a standard compound crump weir, which could be adapted to support a resistivity fish counter. While such a design would have satisfied the criteria for fish passage, a major objection was made by the British Canoe Union, who represent canoeists that use the river in the winter months, outside the fishing season. Even with modifications to the standard design, the resulting standing wave remained a potential concern to the safety of the canoeists. This design was abandoned and an alternative weir design promoted. The final design was a modified flat-V weir, which complied with International Standards (no: ISO 4377:1990(E)). Once installed and operated, such a structure was considered to have minimal impact to fish and river users. At the outset it was recognised that any potential risk to the river would be associated with the construction phase. As a result, adequate precautions were taken, based on a detailed method statement specifically designed to minimise impacts to fish and the environment which were linked to best practice and pollution control measures. Fisheries and environmental criteria formed the basis of the final design and method statements. The low profile and low velocity features offered no impediment to free migration. Construction was undertaken in the winter months, outside the fishing season and avoiding the peak migration period for spawning salmon passing Exebridge. The high winter flows caused engineering problems throughout the construction period, including two-stage construction, higher and reinforced bunds, and greater pumping of seepage water to discharge via land irrigation. Such efforts and benefits were considered to outweigh the risks and extra engineering costs.

The raised bed profile (0.15–0.28 m) associated with the weir acts as a minor impoundment with the resulting backwater extending upstream into Sawdust Pool, the natural pool supporting the abstraction point. This resulted in a gain in average depth of 8–15 cm in the reach and a slight reduction in velocity of 0.01–0.04 $m^3 s^{-1}$. Water velocity over the centre of the weir increased from 0.46 to 0.88 $m^3 s^{-1}$ in the flow range 140–400 MLd^{-1}. Such velocity increases occur over approximately 1 m of the downstream face before the merging of the hydraulic jump and the tailwater levels.

13.8.4 *Mitigation*

As discussed in Section 13.6, the physical presence of St James Weir at the head of tide and the existing flow arrangements over the weir represent a significant obstruction to salmon migration at times of low flows. The current abstraction regime exacerbates this effect to a minor extent on unsupported flows in the range 3–10 $m^3 s^{-1}$. The effect of abstraction is greater on fish ascending St James Weir than on fish arriving off the tide to the base of the weir. Those salmon that have already arrived at the foot of the weir are stopped from ascending and upstream migration delayed until conditions improve. Any effects of the abstraction on fish migration could be removed completely, along with much of the delaying effects of natural low flows by the provision of an effective fish pass at St James Weir. Based on the information gained from the radio tracking studies, South West Water is currently funding the provision of improved fish passage facilities at St James Weir as part of the mitigation measures.

The pre-scheme commitment by South West Water to a major radio-tracking study of salmon movement throughout the River Exe enabled considerably more information to be gathered on migration patterns and flow needs at obstructions along the river throughout the year. In addition to St James Weir, two further obstacles to salmon migration were identified at Exwick/Cowley weirs and Oakfordbridge Weir. These weirs are not affected by abstractions and are sites where improvements are required. The studies will aid this process through the use of sound scientific data. Such added benefit to the salmon fisheries is derived from the extension of the water resource-focused programme at little additional cost (Solomon *et al.* 1999).

During the promotion of the original Wimbleball Reservoir Scheme, a Fisheries Water Bank was established as a mitigation measure. A volume of 900 ML was allocated and made available in each refill period, i.e. 2–3 years. The advent of the pumped storage scheme offers the potential for an annual refill and so provision was made for the 900 ML water bank to be made available each year. While South West Water is required to implement the use of the water bank, the Environment Agency has responsibility for the overall best use of the volume. In the early years of experimentation, three possible options are being considered.

(1) Releases to stimulate and sustain adult salmon migration in the lower and middle reaches of the Exe. The volume of water available is small relative to peak flows and its use is more of a supplement to natural spates by delaying flow recession. Provided that the release timing is correct, then there could be some benefit to the lower river, but any perceived gain to migration in the estuary is considered to be minimal.
(2) More local use to enhance flows in the River Haddeo by building on the experiences gained at Roadford Reservoir and the River Wolf (Sambrook & Gilkes 1994). This option is discussed in Section 13.8.1 and has been promoted many times by South West Water. The aim is to simulate the needs of the various freshwater life stages of salmon and trout through the control of regulated flows. This will involve simulating spill (timing and volume) and balancing flows in the Haddeo in response to natural flow changes from the unregulated River Pulham.
(3) Protection of critical spate events and flow thresholds that are known to be important to salmon migration in the upper Exe at spawning time. This is particularly relevant in those years when November pumping would be needed following a dry autumn and resulting delay to adult migration into the headwaters (as discussed in Section 13.8.2).

At the original Wimbleball Public Inquiry in 1977, a commitment was given to offset the effects of the scheme operation on salmon fisheries. The initial mitigation programme was designed on a hatchery-based programme and the stocking of 1500 salmon smolts each year, using the indigenous genetic stock. In recent years the effectiveness of such stocking programmes has been questioned (Cowx 1994). As a result, a mitigation package was agreed and implemented in 1997. The aim is to cease the operation of all licensed commercial salmon and sea trout exploitation in the Exe estuary. South West Water instigated the 'buy-back' arrangement and makes financial compensation for each of the netsmen not to fish for salmon and sea trout in the first 14 weeks of the authorised netting season. The benefit of such action is to provide a target escapement of 100 adult salmon into the river, with the added potential gain in egg deposition of 300 000 eggs. Such an approach was supported by the Environment Agency and gives further protection to a unique race of spring salmon in the Exe, which is historically one of the best spring salmon rivers in England.

13.9   Conclusions

The main objective for any well-prepared and implemented major water resource project is to balance fisheries/environmental needs, scheme yield, sustainability and operating costs. This chapter highlights many of the issues that must be addressed if this is to be achieved and sets principles for the way forward. Certain key principles evolved following the experiences gained on the Wimbleball Pumped Storage Scheme that can be summarised as follows.

- It is essential that all the stakeholders are involved at an early stage and the underlying concerns relating to the development strategy are discussed as widely as possible. It is only when the principal issues are brought out into the open that they can be resolved. Clarity and understanding of ideas are essential at this stage.
- An initial assessment should be carried out to identify and prioritise problems that are likely to occur. It is important to differentiate between potential and actual concerns, in addition to the need to formulate and offer practical and technical solutions to mitigate any problems. Where possible, the potential benefits from the development scheme should be enhanced. This can only be achieved by integrating realistic fisheries and environmental perspectives into the engineering design and operating philosophy.
- A number of options for the proposed development scheme should be offered and the best possible scenario identified through consultation with all user groups. This can avoid a lot of conflict. Each option should take into account socio-economic arguments, risk assessment and the wider political perspective. A flexible approach must be adopted throughout the project up to the point where the scheme is finalised, and the final balanced perspective appreciated.
- In identifying and quantifying the issues, it is important to present sound scientific and applied perspectives based on the best available data in a manner that is easily understood by engineers, operators, regulators and the public alike. Always be professional, as well as fair and aware of all the issues.
- There is also a need to improve post-scheme assessment and to incorporate appropriate funding into the project design. Be realistic, prove what is good, set control/operating systems objectively, and disseminate information gained so that others can benefit.

Acknowledgements

The authors acknowledge the encouragement and permission by South West Water to publish this chapter. The views expressed are those of the authors and should not be considered those of South West Water.

References

Cowx I.G. (1983) The biology of bream, *Abramis brama* (L.), and its natural hybrid with roach, *Rutilus rutilus* (L.), in the River Exe. *Journal of Fish Biology* **22**, 631–646.

Cowx I.G. (1988) Distribution and variation in the growth of roach, *Rutilus rutilus* (L.), and dace, *Leuciscus leuciscus* (L.), in a river catchment in south-west England. *Journal of Fish Biology* **33**, 59–72.

Cowx I.G. (1994) Stocking strategies. *Fisheries Management and Ecology* **1**, 15–31.

Cowx I.G. & Welcomme R.L. (1998) *Rehabilitation of Rivers for Fish*. Oxford: Fishing News Books, Blackwell Science, 260 pp.

Sambrook H.T. & Gilkes P. (1994) Roadford Reservoir: Enhanced flows, fisheries and hydroelectric power generation. *Proceedings of the 8th British Dam Conference 1994*, pp. 119–133.

Solomon D.J., Sambrook H.T. & Broad K.J. (1999) Salmon migration and river flow: Results of tracking radio tagged salmon in six rivers in south-west England, *Environment Agency R & D Publication No 4*. Bristol: Environment Agency, 110 pp.

Chapter 14

Impacts of hydraulic engineering on the dynamics and production potential of floodplain fish populations in Bangladesh: implications for management

A.S. HALLS

Renewable Resources Assessment Group, Imperial College of Science, Technology and Medicine, 8 Prince's Gardens, London SW7 1NA, UK (e-mail: a.halls@ic.ac.uk)

D.D. HOGGARTH and K. DEBNATH

Marine Resources Assessment Group, 47 Prince's Gate, London SW7 2QA, UK

Abstract

Fish production in floodplain river systems is largely dependent upon the timing, extent and duration of the flood pulse, all of which can be severely modified by hydraulic engineering. This chapter brings together the results of a number of separate studies conducted at the Pabna Irrigation and Rural Development Project (PIRDP) flood control, drainage and irrigation (FCDI) scheme during 1995 and 1996, to examine the impact of hydraulic engineering on fish production in Bangladesh.

Catch per unit area (CPUA) was found to be 38–51% lower inside the scheme during both sampling years, although fishing effort could not account for these differences. In spite of major modifications to the hydrological regime within the scheme, comparisons of the population dynamics of six representative members of the floodplain fish community indicated that the production potential of fish within the scheme was unaffected. However, a tagging programme, multivariate comparison of species assemblages and monitoring catches of fish at sluice gates revealed that lower rates of recruitment from external sources were responsible for lower yields within the scheme. Simple, cost-effective mitigating management measures for improving sustainable yield and maintaining biodiversity are discussed.

Keywords: Hydraulic engineering, floodplain, river fisheries management, Bangladesh.

14.1 Introduction

Hydraulic engineering in Bangladesh is characterised by earth embankments or levees, which form poldered areas or flood control, drainage and irrigation (FCDI) schemes upon the floodplain. These schemes provide an important defence against extreme flooding, and a semi-controlled hydrological environment for growing high yielding varieties (HYV) of crops, particularly rice. Manipulation of water levels within FCDI schemes by means of sluice gates provides very reliable high cropping intensities. Sluice gates are managed by the Bangladesh Water Development Board

(BWDB), as advised by committees of farmers and landowners, to optimise water levels for agricultural production. In the north-west region of the country, local farmers report that crop production inside FCDI schemes is at least double that outside. With 85% of the population of the country living at the agrarian subsistence level, and with half the national product derived from agriculture (Pramanik 1994), FCDI schemes appear vital for the welfare of the people of Bangladesh, and to the economy as a whole. Currently, flood control schemes cover approximately 30% of the total land area of the country, and with increasing demand for greater agricultural output and economic stability, more schemes are planned for the future (Rahman *et al.* 1994).

Against this backdrop, fish provides more than 80% of the total animal protein consumed in the country and is the second most important export commodity after jute. Fishing also provides livelihoods and supplementary incomes for millions of people in Bangladesh (Rahman *et al.* 1994). Manipulating water levels within FCDI schemes to optimise agricultural output has the potential to significantly modify the extent, duration and timing of floodplain inundation, the principal driving force responsible for the productivity of floodplain fisheries (Junk *et al.* 1989). Levees also have the potential to obstruct lateral migrations and the passive drift of various life history stages of fish onto the floodplain (see Halls (1998) for review).

The impact of FCDI schemes on fish productivity (measured as catch per unit area, or CPUA) in Bangladesh has been previously examined as part of the World Bank/ODA funded Flood Action Plan Project 17 (FAP 17, 1995), based largely upon comparisons of catch rates inside and outside FCDI schemes throughout the country, but with little attempt to gain an understanding of the biological processes underlying them. Firm conclusions on the impacts of FCDI schemes were prevented by non-significant or contradictory results, and the effects of breached embankments and partially functioning schemes (Halls 1998).

This chapter re-examines, and attempts to explain, the impact of FCDI schemes on fish productivity in Bangladesh from a focussed investigation of one fully functioning scheme.

Recruitment upon the floodplain is generated by sedentary spawning stocks or takes the form of fish migrating or passively drifting from external sources, e.g. the main river or adjacent floodplains. It was hypothesised that the latter external sources of recruitment may be most impacted by FCDI scheme embankments acting as physical barriers to migrations,[1] whilst the production potential of recruits depends largely upon the relative flooding intensities (water levels) and the resources available for production.

Armed with this information and the results of two other surveys on recruitment potential from internal and external sources (see below), this chapter identifies appropriate mitigating management measures designed to improve sustainable floodplain fish yields and maintain species biodiversity.

---

[1]Halls (1998) showed through simulation modelling that lower water levels inside the PIRDP have only a marginal influence on the recruitment generated by sedentary spawning stocks.

14.2 Materials and methods

14.2.1 *Study site*

The investigation centred around the south-east corner of the Pabna Irrigation and Rural Development Project (PIRDP) located in the north-west region of Bangladesh, at the confluence of the Padma and Jamuna Rivers (Fig. 14.1). Two adjacent sampling regions were selected for study, one inside and one outside the embankment that protects the PIRDP lands from flooding. The sampling regions were believed to be similar, containing a mixture of habitats including secondary rivers, floodplains and natural depressions (beels). Flow of water between the two regions is controlled by one large and two small sluice gates (Fig. 14.1) spaced along the embankment.

Data were collected over a 2-year period between January 1995 and December 1996. Water height above mean sea level was measured daily inside and outside the PIRDP at the main Talimnagar sluice gate (Fig. 14.1).

14.2.2 *Impact on productivity (CPUA)*

Fish catches and fishing effort were estimated separately in the two sampling regions using a village-based frame survey, followed by twice-monthly interviewing of 40 randomly-selected respondent fishermen (MRAG 1997). To allow for the mobility of fishermen, respondents were selected both in the two main sampling regions, and in an adjacent region (Fig. 14.1). Data from the respondents in adjacent areas were used to estimate the additional catches in the main inside and outside regions taken by fishermen living nearby. Subsistence fishermen were not included in the surveys. Each respondent provided data on the catch and fishing effort of each type of fishing gear used in each sampling region during the preceding 1–5 days. Fishing effort was measured as gear-unit-hours, with times reported as active fishing hours or soak hours depending on gear type. Separate market surveys were undertaken by project staff to estimate the species compositions of fish catches by gear type, region and month, to enable subdivision of the estimated catch weights between species. Mean monthly catch rates were raised to the total catches in each region using the estimated numbers of fishermen from the frame survey.

14.2.3 *Impact on production potential of recruits*

Relative production potential of recruits was examined for six representative members of the floodplain species assemblage, hereafter referred to as 'key species': climbing perch, *Anabas testudineus* (Bloch); Indian major carp, *Catla catla* (Hamilton); snakehead, *Channa striatus* (Bloch); marbled goby, *Glossogobius giuris*

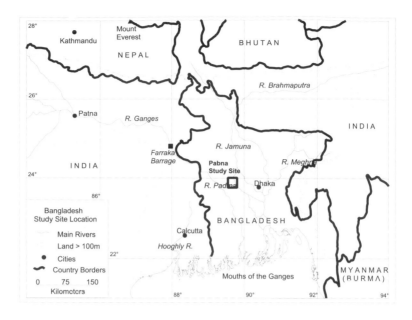

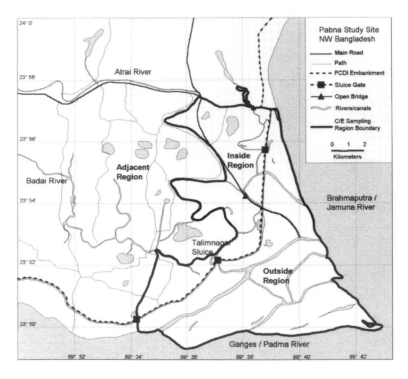

Figure 14.1   The catchment position of the study site located at the confluence of the Padma and Jamuna Rivers in North West Bangladesh (above) and the sampling regions straddling the PIRDP flood control embankment (below)

(Hamilton); spot barb, *Puntius sophore* (Hamilton) and a silurid catfish, *Wallago attu* (Bloch).

Growth and mortality rates, and patterns of recruitment were estimated from bimonthly sampled length frequency data. Attempts were made to model the growth of each species by fitting the seasonal version of the von Bertalanffy growth function (VBGF) (Pitcher & Macdonald 1973) to the length frequency data with length frequency distribution analysis (LFDA) software (Holden *et al.* 1995). However, the VBGF was inadequate for describing growth because: (a) inter-annual variations in growth were significant in most cases; (b) inter-annual variations in the time of recruitment (spawning time) were also common; and (c) the majority of distributions contained only a single mode (the 0+ cohort). Relative growth performance was instead examined by comparing observed mean length of the 0+ cohort in November ($\bar{L}_{Nov}$) when most of the year's growth had been achieved.

Total annual instantaneous mortality rate, Z, was estimated using the simple exponential mortality relationship (Gulland 1983). Estimates were made from unaggregated length frequency samples collected from low selectivity gear types which exhibited two clear modes assumed to represent the numbers of new recruits of age t (N_t) and the relative numbers of survivors from the previous years recruitment (N_{t+1}) (Gulland & Rosenberg 1992). By separating the two modes, Z was estimated as:

$$Z = -\ln(N_t + 1/N_t) \qquad (1)$$

Relative fish condition and fecundity of each key species inside and outside the PIRDP was examined by testing for significant ($P < 0.05$) differences in their $\log_e$ transformed length–weight, and length–fecundity relationships respectively using analysis of covariance (ANCOVA) (Brown & Rothery 1993).

The diet of each species in the two sampling locations was compared qualitatively using the mean percentage contribution of each food category over the 24-month sampling period. Feeding intensity and seasonality were compared using time series plots of mean monthly stomach fullness estimates (Halls 1998) with 95% confidence intervals.

The length at maturity for each key species and sex inside and outside the PIRDP was estimated as the length, Lm_{50}, at which half the sampled individuals were sexually mature. A logistic function was fitted to each data set using a non-linear least squares approach (SYSTAT 1990), providing 95% confidence intervals around Lm_{50}. Finally, the relative spawning times of the key species were compared from time series plots of monthly gonadosomatic indices (GSI) with 95% confidence intervals. Further details of all the data collection and analytical methods are given in MRAG (1997) and Halls *et al.* (1999).

14.2.4 *Impact on external sources of recruitment*

The impact of the PIRDP on external sources of recruitment was investigated with two studies: (a) a mark–recapture programme to examine the impact of the PIRDP on the migratory behaviour of the six key species; and (b) an examination of the aggregate impact of several Bangladeshi FCDI schemes, including the PIRDP on their fish species assemblages.

The mark-recapture programme

Fish were purchased live from fishermen and tagged with small, individually numbered, plastic tags attached through the dorsal musculature. Tagged fish were released both inside and outside the FCDI embankment, and in as many different locations around the study site as possible. A total of 4618 fish were released between December 1994 and August 1996 (see Halls *et al.* (1998) for further details).

Species assemblages

The impact of FCDI schemes on species assemblages was examined using a multivariate non-parametric multidimensional scaling (MDS) routine (Clarke 1993; Clarke & Warwick 1994) applied to species abundance data (small-meshed seine net catch per unit effort during 1993) sampled by FAP17 (1995) at 74 sites located inside and adjacent to seven FCDI schemes in the following four districts of Bangladesh: Pabna, north-west (NW), Tangail, north-central (NC), Sylhet, north-east (NE) and Faridpur, south-west (SW). These FAP17 study sites covered a broad range of habitat types including main and secondary rivers, canals and floodplains/beels (natural depressions or lakes upon the floodplain). Permutation (analysis of similarity; ANOSIM) tests were used to test for significant differences in species assemblages among different habitat types and inside and outside the FCDI schemes (see Halls *et al.* (1998) for further details).

14.2.5   *Recruitment potential*

The majority of fish populations at the study were virtually annual with individuals reaching sexual maturity and spawning within their first year (see below). Recruitment or year class potential must therefore be closely linked with rates of survival during the dry season prior to the spawning period at the start of the following year's flood (April–June). Recruitment potential generated from within the PIRDP was therefore examined by estimating the survival rates of exploited spawning stocks residing in dry season water bodies from November 1995 to May 1996 using depletion modelling, in support of the estimates derived from the length frequency analysis.

Recruitment potential from external sources was examined by monitoring catches taken with interceptory fishing gears operating within a 100 m radius of sluice gates during a 5-month period (July–November) of the 1996 flood season. These gears, which include lift nets, (veshal or khora jal), bag nets (suti jal), and jump nets (urani), are deliberately positioned to take advantage of the movements of fish towards or through the sluice gates, and their orientation reflects the directional behaviour of the fish migrations (Hoggarth *et al.* 1999).

14.3 Results

14.3.1 *Impact on hydrology*

Water levels at the study site are determined by the flooding patterns of the Jamuna and Padma Rivers and by local rainfall. During the rainy season, waters rise approximately 6 m (20 ft) to reach a peak flood by July or August. Floodplain inundation may continue for around 3 months before waters begin to fall in September or October. The floodplains at the study site are at an altitude of approximately 8 m (25 ft). During the 2-year sampling period, floodplains outside the PIRDP were covered by up to 2.6 m (8.6 ft) of water in the flood season (Fig. 14.2). Due to the embankment, flood depths inside the PIRDP reached a maximum of 2 m (6.4 ft). The PIRDP reduced, delayed and smoothed the flood curve inside the scheme compared to that outside, but flood waters still inundated almost all the land within the project site, except raised roads and housing. Delayed flooding inside the PIRDP shortened the period of floodplain inundation compared to outside by approximately 20 days. The combined effect of the delayed inundation and reduced water height significantly reduced the magnitude of flooding measured by the integral flood index (Welcomme (1979); here measured as foot-days of inundation over the average floodplain depth of 25 ft) inside the PIRDP in both sampling years. In 1995/1996, the flood index was 32% lower inside than outside, and 25% lower in 1996/1997 (Halls *et al.* 1999).

14.3.2 *Impact on productivity (CPUA)*

Total annual fish productivity, measured as catch per unit area (CPUA), was consistently greater (60–104%) in the outside region than the inside region, and higher in 1996 than in 1995 (Table 14.1). Fishing effort was, however, estimated to be only 53% and was 18% greater outside the PIRDP in 1995 and 1996 respectively. Such fishing effort differences are unlikely to be sufficient to explain the 104% and 60% observed differences in CPUA in the two regions, given that both regions are very heavily exploited (see later), and at a position where total yields in these multispecies fisheries are relatively unresponsive to exploitation levels (Hoggarth & Kirkwood 1996).

14.3.3 *Impact on production potential of recruits*

Time series length frequency data (e.g. *W. attu*, Fig. 14.3), revealed that all the sampled populations were dominated by a single mode (0+ cohort) for most of the year, with increasing mean length, typical of intensively-exploited short-lived species (Gulland & Rosenberg 1992). Recruitment occurred both inside and outside the PIRDP in July of each year, corresponding to the time of initial floodplain

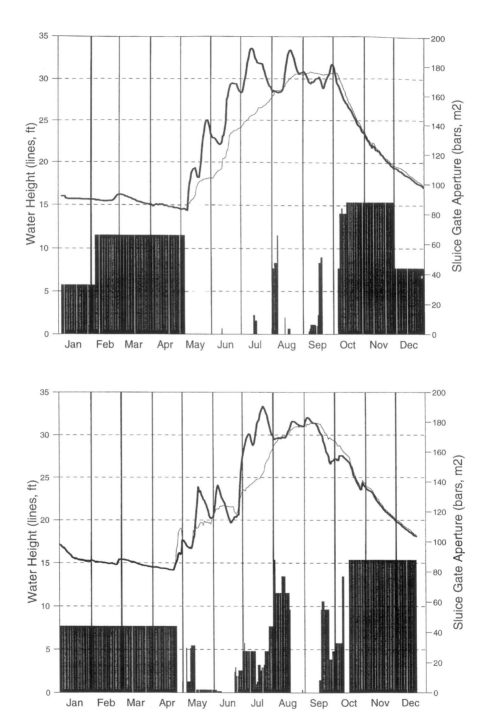

Figure 14.2 Daily sluice gate apertures (bars) and water heights measured inside (thick lines) and outside (thin lines) Talimnagar sluice gate at the PIRDP study site in 1995 (above) and 1996 (below)

Table 14.1 Total annual productivity (kg ha^{-1} yr^{-1}) estimates inside and outside the PIRDP

	Estimated productivity kg ha^{-1} yr^{-1}	
Region	1995	1996
Inside	51	81
Outside	104	130

inundation, before, or just after, the previous cohort virtually disappeared. In general, growth was highly seasonal, with the majority of the year's growth achieved between July and November, corresponding with the period of floodplain inundation. Very little growth occurred during the dry season (November–April).

Growth rates of the key species, measured by the mean length of the 0 + cohort in November ($\bar{L}_{Nov}$), were, with the exception of *C. catla,* significantly greater ($P < 0.001$) (up to 20%) inside the PIRDP, or not significantly different inside and outside the scheme, as in the case for *A. testudineus*. Five of the six key species also exhibited significantly greater ($P < 0.001$) growth rates (up to 23%) during 1995/1996 compared to 1996/1997. Higher growth rates inside the PIRDP and during the first year of sampling were found to be strongly correlated with lower fish densities inside the PIRDP and longer deeper flooding respectively (Fig. 14.2) (Halls *et al.* 1999).

Fish condition was significantly greater ($P < 0.05$) inside the PIRDP than outside for *C. catla*, *P. sophore*, *W. attu* and *G. giuris*, but not significantly different ($P > 0.05$) for *A. testudineus* and *C. striatus*. Fecundity was significantly greater ($P < 0.05$) inside the scheme for *A. testudineus* but not significantly different ($P > 0.05$) for the remaining species tested (Halls *et al.* 1999).

No significant differences in stomach fullness, seasonal pattern of feeding, diet composition, length at maturity of each sex and spawning time were detected inside and outside the PIRDP for any of the key species. Combining the length at maturity estimates with the observed patterns of growth imply that, with the exception of *C. catla*, all the key species reached sexual maturity by the end of their first year.

A total of 64 estimates of total mortality, Z, were calculated from 402 individual length frequency distributions examined. The 64 estimates were divided approximately equally across each bimonthly sampling period and hydrological year. Those distributions for which no estimate could be made either contained too few fish to detect reliable modes or simply comprised a single mode.

The estimates for Z for individual gear and month/year combinations ranged between approximately $Z = 2$ yr^{-1} and 6 yr^{-1}, with only four estimates below 2 yr^{-1} and only one estimate above 6 yr^{-1}. No significant ($P > 0.05$) differences in $\bar{Z}$ were found for the key species inside and outside the PIRDP (Table 14.2). Estimates of Z covering both sampling locations were therefore pooled, to give an overall mean value for each species. Total mortality rates were highest for *C. catla* ($Z = 4.7$ yr^{-1}),

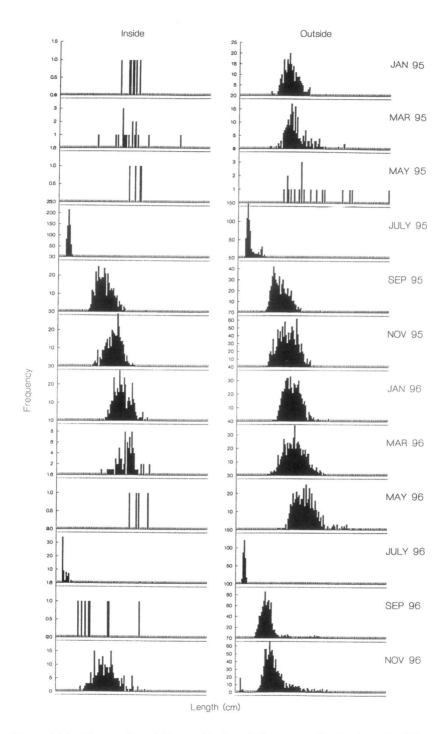

Figure 14.3 Time series of bi-monthly length frequency distributions for *W. attu* sampled from inside and outside the PIRDP from non-selective gear types. Scale range 0–127 cm in 1 cm intervals

Table 14.2 Comparison of mean total annual instantaneous mortality rate Z for each species, inside and outside the PIRDP. Where no significant ($P > 0.05$) difference in mortality exists, estimates of Z from inside and outside the PIRDP are pooled to give an overall $\bar{Z}$ with 95% confidence intervals. Equivalent mean annual survival rates $\bar{S}$ (%) for the overall $\bar{Z}$ are given in columns 8–10[a]

Species	$\bar{Z}$ IN	OUT	P	$\bar{Z}$ (yr^{-1})	IN and OUT $\bar{Z}_{lower}$	$\bar{Z}_{upper}$	$\bar{S}$ (%)	$\bar{S}_{lower}$ (%)	$\bar{S}_{upper}$ (%)
A.testudineus	3.4 (5)	4.5 (1)	NA	3.5 (6)	2.7	4.3	3	1.3	6.7
C. catla	4.0 (4)	5.7 (3)	0.11	4.7 (7)	3.4	6	0.9	0.2	3.3
C. striatus	3.6 (4)	4.1 (6)	0.62	3.9 (10)	3	4.7	2	0.9	4.9
P.sophore[b]	2.5 (10)	3.7 (12)	0.06	3.2 (22)	2.5	3.9	4	2	8.2
W.attu	3.5 (3)	3.7 (11)	0.78	3.7 (14)	3.1	4.2	2.5	1.5	8.2

[a]Sample size in parentheses. Confidence intervals are calculated as $\bar{Z} \pm t_{\alpha(2)\nu}$ SE. NA – no test possible.
[b]Excludes 1995, passive trap, IN ($Z = -0.03$) and May 1995 Seine net, IN ($Z = 0.4$).

implying a survival rate of just below 1%. The remaining species showed marginally lower mortality rates between 3.2 yr^{-1} for *P. sophore* and 3.9 yr^{-1} for *C. striatus*, equivalent to between 4 and 2% survival, respectively (Table 14.2).

14.3.4 *Impact on external sources of recruitment*

The mark–recapture programme

The strongest migrations (the percentages of fish moving from their release locations) were seen for all of the key species in the flood period. Although most fish were recaptured within only 1 or 2 km of their release locations in this heavily exploited fishery, all of the key fish species showed some ability to migrate significant distances. Between 2% (*C. striatus*) and 16% (*G. giuris*) of the tagged fish of each species were recaptured more than 2 km from their release locations. The maximum (straight-line) migration observed was 8.55 km for a *W. attu*. Longer migrations may have been achieved by tagged fish which migrated outside the study site, and were recaptured by fishermen unaware of the project.

Of the 1389 tagged fish recaptured, 14 *C. catla*, six *C. striatus* and 15 *W. attu* were released and then recaptured on opposite sides of the PIRDP embankment. Of these fish, 23 migrated into the PIRDP while 12 migrated outwards. Although the numbers of fish are small, these data confirm the ability of these species to negotiate the sluice gates in both directions. The other three key species, *A. testudineus, G. giuris* and *P. sophore* were not observed to cross the embankment, although they were present in the catches at the sluice gates (see below). Based on the release and recapture locations, it is likely that both *C. catla* and *C. striatus* passed through both the main 94.6 m^2 aperture Talimnagar sluice gate on the 50-m wide Badai River, and also through the smaller 8.1 m^2 Baulikhola sluice gate on the narrow Natuabari Canal (Fig. 14.1). Migrations across the embankment by *W. attu* were only seen in the vicinity of the Talimnagar gate (Hoggarth *et al.* 1999).

Species assemblages

Species assemblages sampled from sites outside FCDI schemes exhibited similarities according to both habitat type and geographical location; thus, differences among species assemblages inside and outside the FCDI schemes were tested separately for each habitat/region combination. Sufficient replicates were only available to test the null hypothesis (H_0: there are no differences in species assemblages at sites inside and outside the FCDI schemes) for the floodplain/beel habitat in the NE and NW regions. In both cases, the null hypothesis was rejected ($P < 5\%$). Examination of the remaining ordinations revealed statistically untestable, but visually convincing evidence that species assemblages from canal and secondary river habitat were also different inside and outside FCDI schemes in those regions which could be considered (Halls *et al.* 1998).

For floodplain/beel habitats in the north-west region, including the PIRDP scheme, indicator species analysis (Clarke & Warwick 1994) revealed that 25 species were absent or less abundant inside FCDI schemes compared to outside (Fig. 14.4). The majority of these species are large predators or conspicuous members of the highly-prized migratory 'whitefish' category (Welcomme 1985) including silurid catfish: *Pseudeutropius atherinoides* (Bloch), *Clupisoma garua* (Hamilton), *Mystus bleekeri* (Day), *Ailia coila* (Hamilton), *Silonia silondia* (Hamilton), *Heteropneustes fossilis* (Bloch), *Gagata youssoufi* (Rahman), *Ompok pabda* (Hamilton), *Mystus vittatus* (Bloch) and *W. attu*; large cyprinids (including the Indian major carps): *Labeo rohita* (Hamilton), *Labeo bata* (Hamilton), *Labeo calbasu* (Hamilton), *Cirrhinus reba* (Hamilton), *Cirrhinus mrigala* (Hamilton) and *C. catla*; clupeids: *Hilsa ilisha* (Hamilton), *Gudusia chapra* (Hamilton) and *Corica soborna* (Hamilton); the loach *Nemachilus botia* (Hamilton), *G. giuris*, and the mullet *Rhinomugil corsula* (Hamilton).

In the absence of these species, assemblages inside FCDI schemes are dominated by much smaller resident 'black' and 'greyfish' species (Welcomme 1985; Reiger *et al.* 1989) including the spiny eel *Mastacembelus pancalus* (Hamilton), the garfish *Xenentodon cancila* (Hamilton), the leuciscine minnow *Salmostoma phulo* (Hamilton) and the glass perches *Chanda nama* (Hamilton) and *Chanda baculis* (Hamilton) (Fig. 14.4). As well as also being less species rich, assemblages inhabiting floodplain/beel habitat inside FCDI schemes have a 25% lower unit value (mean price per kg) based upon average market prices quoted by FAP17 (1995) for the same sampling year. Very similar differences were observed for the other statistically significant comparison of floodplain/beel habitat in the north-east (Halls *et al.* 1998).

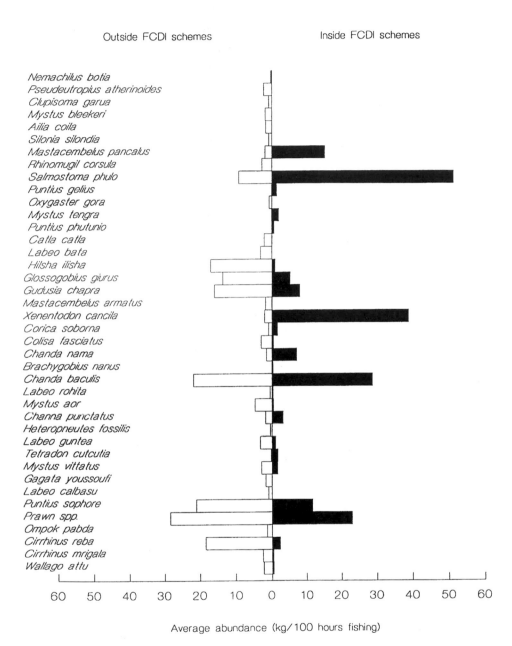

Figure 14.4 Average abundance (seine net CPUE (kg 100 h^{-1}) of species sampled from inside (solid bars) and outside (open bars) FCDI schemes in the north-west region. Species are arranged from top to bottom in descending order of their contribution to the average assemblage dissimilarity between the two groups of sites. Only those species contributing to 75% of the cumulative average dissimilarity are shown

14.3.5   *Recruitment potential*

Internal sources of recruitment – dry season survival

Dry season survival rates of the resident spawning stocks were estimated to be 0.54–1.9% inside the FCDI scheme and 0.6% outside, which are largely consistent with the estimates derived from the length frequency distribution analysis (Table 14.2). The numbers of fish species in the catches declined as the season progressed, presumably as the least common species became fished out, or as the non-air-breathing species died in the deoxygenated conditions. Large carps *C. catla*, *C. reba*, and especially *L. rohita*, were caught in both inside and outside water bodies, particularly in the perennial beels and rivers in the early dry season; such species were then noticeably less common in the late dry season, although some *L. rohita* and *C. reba* were still caught (MRAG 1997). Late dry season catches were dominated by the two predators: the snakehead *Channa marulius* (Hamilton) inside the FCDI scheme, and by *W. attu* outside.

In the majority of dry season water bodies, fish communities were richer outside the FCDI scheme than inside. Only the beel fish communities comprised more species inside the scheme, reflecting the diverse beel habitats available in the inside region. Fish species compositions were noticeably different inside and outside the FCDI scheme. Smaller species, such as the climbing perch *Anabas testudineus* (Bloch), *C. punctatus* and *P. sophore*, were generally more common inside than outside (MRAG 1997). The large catfish *W. attu* was especially common outside the FCDI in all water bodies. (Hoggarth *et al.* 1999).

External sources of recruitment – migrations through sluice gates

In the presence of the existing fishing operations and catches, an estimated 5000 kg of small fish migrate into, and hence recruit to, the PIRDP inside region during the early flood phase, mostly through the Talimnagar gate. It was also estimated that a further 300–2500 kg of fish, mostly major carps, may then migrate into the PIRDP during the early ebb active immigration phase, although it is less certain that such fish are actually able to get through the gates during the strong flows at this time.

These results confirm that fish are able to recruit to the PIRDP fishery through the sluice gates even with their attendant fishermen. The significance of this external source of recruitment is reflected in the total productivity (CPUA) inside the PIRDP being 50% lower inside the PIRDP in 1995, when the sluice gates at Talimnagar remained closed for almost the entire flood season due to strong flooding and high local rainfall, but only 37% lower in 1996 when the sluice gates were opened more frequently due to lower rainfall and weaker flooding (Table 14.1; Fig. 14.2).

14.4   Discussion

It is widely recognised (see review in Welcomme 1985; Junk *et al.* 1989; Halls 1998) that floodplain fish production is largely dependent on the timing, extent and duration of inundation or the flood pulse. In spite of significant modifications to the flood pulse within the PIRDP FCDI, the results of this assessment indicate that the growth performance of several of the species examined was higher within the scheme, correlated with lower fish density. Moreover, with survival rates and reproductive behaviour not significantly different, it was concluded that the production (elaboration of fish tissue) potential of recruits within the scheme was at least as high inside as outside, and therefore unaffected by existing hydrological modifications.

The dynamics of the six key species examined imply that fish populations at the study site were virtually annual, sustained only by their ability to mature and spawn by the end of their first year. In these fish stocks, recruitment failure due to excessive fishing mortality or failure to reach sexual maturity due to poor growth could have terminal implications. The extent to which the flood pulse may be modified in schemes like the PIRDP before growth and reproductive dynamics are significantly affected remains uncertain, but similar studies in other river systems (e.g. Wilton 1985; Winston *et al.* 1991; Liu & Yu 1992) suggest that some threshold must exist.

Given that the production potential of recruits was found to be at least as high inside the PIRDP scheme as outside, it was concluded that the 38–51% reduction in observed productivity within the scheme reflects the impact of the flood embankments on recruitment from external sources, in the form of fish (and their life history stages) migrating or passively drifting from the main river or adjacent floodplains.

This conclusion is supported by: (a) the tagging programme and the multivariate analysis of species assemblages, which confirmed that some fish are able to migrate into the PIRDP via sluice gates – conspicuous members of the migratory 'whitefish' category were absent or had much lower abundance inside the scheme compared with outside; (b) the substantial quantities of fish caught whilst migrating through the PIRDP sluice gates; (c) the impact of the PIRDP on productivity (CPUA), which was substantially lower in 1996 than 1995 (38% lower and 51%, respectively) when the main Talimnagar sluice gate was opened more frequently due to the lack of local rainfall; and (d) the numbers of species caught during the dry season, which declined, with progressively more sedentary 'blackfish' species forming the bulk of catches.

The continued existence of modified Bangladeshi fisheries indicates that floodplain fish stocks are extremely well adapted to surviving harsh conditions and high exploitation rates. Although fish stocks may not be about to collapse completely, there are clearly opportunities for increasing recruitment generated by sedentary populations residing within (and outside) FCDI schemes by improving survival rates. Such improved survival rates could be achieved in virtually any of the deeper, perennial water body types (which become discrete during the dry season) by limiting the final dewatering catches in the shallower water bodies, or other final fishing activities in the deepest non-dewaterable water bodies. Community-based management mechanisms by which such restraint might be achieved are currently being investigated. The results of simulation modelling also suggest that small improvements in survival rates, achieved through closing the fishery for 1–2 months towards the end of the dry season and start of the new flood, can produce considerable increases in both overall yield and yield-per-recruit for relatively small sacrifices in catch (Halls 1998). Simulations also indicate that survival rates and yield can be improved by retaining as much water as possible during the dry season, consistent with the findings of Welcomme and Hagborg (1977).

External sources of recruitment provide a significant contribution to floodplain productivity (CPUA). Indeed, the impoverished species assemblages remaining within the PIRDP towards the end of the dry season implies that external sources of recruitment are necessary for maintaining biodiversity. Potential exists to improve yield and maintain biodiversity within the PIRDP and other FCDI scheme fisheries through simple sluice gate management practices, designed to take advantage of the strong migratory tendencies for fish to enter such schemes, and improving sluice gate design to maximise the passive drift of fry to modified floodplains, without the need for costly fish passes or further research work. As a first priority, a fishermen's representative on the sluice gate management committees could give advice on synchronising sluice gate openings with peak fry densities in outside waters to maximise recruitment whilst minimising the threat to agricultural production. Some limitations on the use of interceptory fishing methods at sluice gates may also ensure that immigrating fish are able to disperse on to the floodplain to have some chance to grow before capture.

Acknowledgements

This research was funded by the UK Department For International Development (DFID), under Project R5953 'Fisheries Dynamics of Modified Floodplains in S. Asia' of the Fisheries Management Science Programme. Fieldwork was undertaken by Messrs Emdad Hossain, Shahabuddin Sheikh and Ranjan Kumar Dam, and field data were entered into a computer database by Mr Balai Debnath.

References

Brown D. & Rothery P. (1993) *Models in Biology: Mathematics, Statistics and Computing.* Chichester: Wiley, 688 pp.

Clarke K.R. (1993) Non-parametric multivariate analyses of changes in community structure. *Australian Journal of Ecology* **18**, 117–143.

Clarke K.R. & Warwick R.M. (1994) Change in marine communities. An approach to statistical analysis. Swindon: Natural Environment Research Council, 144 pp.

Flood Action Plan Project (FAP) 17 (1995) *Final Report – Main Volume.* London: Overseas Development Administration, 185 pp.

Gulland J.A. (1983) *Fish Stock Assessment. A Manual of Basic Methods.* Chichester: Wiley, 223 pp.

Gulland J.A. & Rosenberg A.A. (1992) A review of length-based approaches to assessing fish stocks. *FAO Fisheries Technical Paper* **323**. Rome: FAO, 100 pp.

Halls A.S. (1998) An assessment of the impact of hydraulic engineering on floodplain fisheries and species assemblages in Bangladesh. PhD thesis, University of London, 526 pp.

Halls, A.S., Hoggarth D.D. & Debnath, D. (1998) Impact of flood control schemes on river fish migrations and species assemblages in Bangladesh. *Journal of Fish Biology* **53** (Suppl. A), 358–380.

Halls A.S., Hoggarth D.D. & Debnath D. (1999) Impacts of hydraulic engineering on the dynamics and production potential of floodplain fish populations in Bangladesh. *Fisheries Management and Ecology* **6**, 261–285.

Hoggarth D.D. & Kirkwood G.P. (1996) Technical interactions in floodplain fisheries of south and south-east Asia. In I.G. Cowx (ed.) *Stock Assessment in Inland Fisheries.* Oxford: Fishing News Books, Blackwell Science, pp. 280–292.

Hoggarth D.D., Halls A.S., Dam R.K. & Debnath K. (1999) Recruitment sources for fish stocks inside a floodplain river impoundment in Bangladesh. *Fisheries Management and Ecology* **6**, 287–310.

Holden S., Kirkwood G.P. & Bravington M.V. (1995) *LFDA Version 4.01 User Manual.* London: MRAG, 43 pp.

Junk W.J., Bayley P.B. & Sparks R.E. (1989) The flood pulse concept in river–floodplain systems. In D.P. Dodge (ed.) Proceedings of the International Large Rivers Symposium. *Canadian Special Publications of Fisheries and Aquatic Science* **106**, 110–127.

Liu J. K. & Yu Z. (1992) Water quality changes and effects on fish populations in the Hanjiang River, China following hydroelectric dam construction. *Regulated Rivers: Research and Management* **7**, 359–368.

MRAG (1997) *Fisheries Dynamics of Modified Floodplains in Southern Asia. Final Technical Report.* London: MRAG, 275 pp.

Pitcher T.J. & MacDonald P.D.M. (1973) Two models for seasonal growth in fishes. *Journal of Applied Ecology* **10**, 597–606.

Pramanik M.A.H. (1994) Natural Disasters. In A.A. Rahman, S. Huq, R. Haider & E.G. Jansen (eds) *Environment and Development in Bangladesh.* Dhaka: Dhaka University Press, pp. 224–235.

Rahman A.A., Huq S., Haider R. & Jansen E.G. (eds) (1994) *Environment and Development in Bangladesh.* Dhaka: Dhaka University Press, 524 pp.

Regier H.A., Welcomme R.L., Steedman R.J. & Henderson H.F. (1989) Rehabilitation of degraded river ecosystems. In D.P. Dodge (ed.) Proceedings of the International Large Rivers Symposium. *Canadian Special Publications of Fisheries and Aquatic Science* **106**, 86–97.

SYSTAT (1990) *The System for Statistics.* Evanston, IL, SYSTAT, Inc., 1990, 677 pp.

Welcomme R.L. (1979) *The Fisheries Ecology of Floodplain Rivers.* London: Longman, 317 pp.

Welcomme R.L. (1985) River fisheries. *FAO Fisheries Technical Paper* **262**, 330 pp.

Welcomme R.L. & Hagborg D. (1977) Towards a model of a floodplain fish population and its fishery. *Environmental Biology of Fishes* **2**, 7–24.

Wilton M.L. (1985) Water drawdown and its effect on lake trout (*Salvelinus namaycush*). *Ontario Fisheries Technical Report Series* **20**, 12 pp.

Winston M.R., Taylor C.M. & Pigg J. (1991) Upstream extirpation of four minnow species due to damming of a prairie stream. *Transactions of the American Fisheries Society* **120**, 98–105.

Chapter 15
Management of fisheries in rivers used for potable abstractions in drought conditions: public health versus ecology

S. AXFORD

Environment Agency, Coverdale House, Aviator Court, Amy Johnson Way, Clifton Moor, York YO3 4UZ, UK

Abstract

Problems caused by the severe drought in Yorkshire, North East England in 1995/1996 are described. A number of measures were successfully adopted to prevent, reduce and mitigate acute adverse effects of Drought Orders taken out to ensure potable supplies. Monitoring of effects on ecology was instituted, but difficulties arose with discerning the effects of Drought Orders from natural drought and normal variations. Even if changes can be detected, there is a need to define whether they should be considered as environmental damage.

15.1   Introduction

Hindsight enables the most effective management strategies to be determined quite readily, but, during a drought, its duration and intensity are unknown. Droughts are natural events, periodic occurrence of which may be essential for maintenance of an ecosystem in a particular form. Potable water abstractions from rivers can accentuate drought conditions, but may be supported by inputs from reservoirs or groundwater; thus, some reaches may have greater flows than would occur under natural drought conditions. Nevertheless, at some point the need for water for potable abstraction will conflict with the need for water to remain in rivers to support fisheries. As Uff (1996) indicated, it is a major drawback to the use of rivers for water supply that they are most vulnerable to environmental impact at the time when the need for abstraction is greatest.

Licences to abstract water from rivers in England and Wales are issued by the Environment Agency and usually contain conditions on maximum rates of abstraction and minimum river flows at which these rates can take place. These are arrived at after due consideration of the likely environmental effects. However, under drought conditions, applications may be made for Drought Orders or Emergency Drought Orders that will override these licences. Such an application brings no necessity to consider environmental matters; the only issue arising under the statute is whether there is a serious deficiency of supplies and, for an Emergency

Drought Order, whether this is likely to impair the economic or social well-being of persons in the area. From this it is clear that considerations of public health may override those of possible environmental damage, although conditions may be placed on the operation of a Drought Order to protect the environment.

At the time of the severe drought in west Yorkshire in 1995/1996, rapid response had to be made to applications for measures to maintain potable water supplies that had potentially severe effects on fisheries and river ecology. This chapter details the measures taken to monitor, protect and mitigate effects on fisheries and river ecology, and provides indications of ecological effects detected part way through the monitoring programme that was set up. Interpretation and attribution of ecological effects has proved difficult. Some of the lessons learned about effects of drought on fisheries and river ecology, formulation of effective monitoring and surveillance programmes, and management of water resources, environmental protection and public relations are outlined.

15.2   The 1995/1996 drought

In 1995/1996, there was a severe drought in the west of Yorkshire, with rainfall well below the long-term averages in most months between May 1995 and November 1996 (Fig. 15.1). Reservoir stocks became severely depleted, and there was an increasing reliance on river abstraction and reductions in reservoir compensation flows to rivers that were not abstracted, so as to maintain reservoir potable supplies.

The 1995/1996 drought started in April 1995, when rainfall fell below the long-term average (LTA), but there was not much concern until June, when there had still been very little rainfall. The first application for a Drought Order was made on 26 July 1995 for the River Wharfe, and this was followed by a series of other Drought Orders and restrictions as the drought progressed (Table 15.1). Water demand

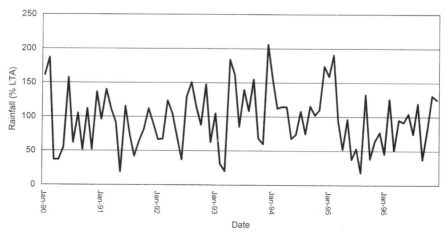

Figure 15.1   Rainfall data in the Yorkshire Dales as percentage of long-term average

Table 15.1 Drought Orders and restrictions of use 1995/1996

Type of Order	Number of Orders, 1995/1996
Abstractions from rivers, boreholes and still waters	22
Reservoir compensation discharges	12
Restrictions of use and cuts in domestic supplies	11

remained high despite restrictions. By August 1995, trial standpipes were erected in the streets of Bradford, in preparation for cuts in domestic supplies. Lack of potable water supply brought the possibility of major health hazards. Rota cut-offs of supply were threatened for West Yorkshire in 1995, which might have deprived people of their normal means of maintaining personal hygiene and cleanliness, and might have brought the risk of contamination of water supplies through periodic drops in pressure. Road tankering of water from more abundant water stocks to the north of the region started in September, thereby relieving some of the pressure on the water environment in West Yorkshire.

Rainfall continued to be below the LTA in each month, except September, until February 1996 (Fig. 15.1), following which the drought resumed until November 1996. The drought was worst in the west of Yorkshire; one rainfall station had only 46 % of the LTA between April 1995 and January 1996, with a calculated return period of 1 in more than 1000 years. The validity of expressing such events in terms of return periods must, however, be called into question by changing weather patterns, as evidenced by global warming and the increased frequency and severity of the El Niño phenomenon.

A region-wide hose pipe ban was imposed by Yorkshire Water Services (YWS) on 29 April 1996 and was lifted on 1 November 1996. The company ceased operating under drought conditions on 1 January 1997.

15.3 Management dilemmas

As the drought progressed, there was increasing management tension relating to compensation and river flows necessary to prevent severe environmental damage but maintain public water supplies. In a straight choice between people or fish, it was clear that the fish would suffer first, so it was a case of eking out water supplies to minimise environmental damage until the rain came to replenish stocks. This created a serious dilemma on rivers where flows were maintained by compensation discharges from reservoirs. Reducing the compensation discharge by, say, 50%, could maintain some water in both the river and reservoir for perhaps an extra 50 days before extinction of the fisheries, without too much harm in the intervening period, or reducing the flow by 75% could allow survival for an extra 150 days, but with much greater potential harm in the intervening period. If it did not rain until after this additional period it would not matter, except to question whether flows

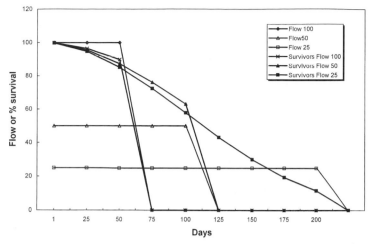

Figure 15.2 Predicted extension to availability of flow in rivers (survivors flow) under various options for adjusting compensation flow (100%, 50% or 25% of existing compensation flow)

should have been reduced still further, but if it rained after about 100 days, the additional harm under the lowest flows would have been unnecessary (Fig. 15.2).

It was this very question of the degree of harm related to different flow and abstraction scenarios and its permanence, which led to specification of a post-drought monitoring programme. This was designed to assist definition of the changes likely to occur at low flows, additional effects related to abstraction or augmentation, and the rate of recovery, so that management decisions in any future drought could have a firmer basis. The 1995/1996 drought could be regarded as a massive experiment on the resilience of the ecology of Yorkshire's rivers and their fisheries.

15.4 Management decisions

The National Rivers Authority (NRA, now incorporated into the Environment Agency) and YWS were to some extent prepared for the drought, as there had been a series of years with abnormally dry summers, which opened regular discussions about likely Drought Orders and other restrictions that might have been needed. Applications for Drought Orders and other restrictions were made in many years through the 1980s and early 1990s (Table 15.2). However, no detailed environmental monitoring was started in relation to these, as a wet spell occurred before or shortly after the implementation date in each year up to 1995.

Usually, there was little notice to the National River Authority (NRA) of the intention of YWS to apply for a Drought Order, the official consultation period being just 7 days. In this period the NRA had to decide whether or not to object to the granting of an Order. It was normally too late to insist that YWS should do more to reduce demand, and public health considerations would probably cause the Order to be granted, despite objections on environmental grounds. Thus the NRA was

Table 15.2 Frequency and location of applications for drought orders (DO) and hose pipe bans (HPB)

1983 – HPB – west
1984 – HPB + DO – west and north
1989 – HPB + DO – region wide
1990 – HPB + DO – region wide
1991 – DO – north, east, west
1992 – HPB + DO – east
1994 – HPB + DO – west
1995 – HPB + DO + EDO – most areas

usually confined to consideration of what protection and mitigation measures might be incorporated as conditions for not objecting. No evidence was available of minimum flow or water level requirements for any of the waters involved. The time of year and unknown duration of any Drought Order caused additional problems, since most animals and plants will tolerate poor conditions, provided that they do not coincide with sensitive periods of the life cycle or continue for too long. Consequently, discussions were held at various times on the forms of monitoring that might accompany an Order.

Problems likely to require remedial or mitigation measures for fisheries could be categorised as acute, requiring urgent and evident actions, and chronic, usually being displayed at a later date and having less evident management actions. Management measures to protect river fisheries in Yorkshire were rapidly formulated in 1995 in the face of Drought Orders. These took the form of: immediate measures to reduce damage, where its source was known; provision for actions triggered by indications of impending damaging conditions and mitigation measures where some damage was unpreventable; and a comprehensive monitoring and assessment programme to determine effects that might require mitigation, abstraction licence revision or consideration in any future Drought Order application. These are given in more detail below. The types of conditions attached to the Orders are shown in Table 15.3.

15.4.1 *Acute problems*

The immediate problems for fisheries were to remove risks to stocks caused by such factors as:

- fish stranding;
- confinement of fish and constraints on movements;
- increased impingement and entrainment at intakes;
- poor water quality.

Table 15.3   Drought Order conditions on the River Wharfe

WHARFE DROUGHT ORDER
1. The installation of oxygenation equipment to maintain dissolved oxygen concentration in the river above 60% saturation.
2. Monitoring of water quality in relation to fish habitats during the operation of the Order and after its cessation. This should include turbidity, pH, temperature, suspended solids, dissolved oxygen, and a range of chemical parameters. Patterns of flow also to be monitored.
3. Survey work on the invertebrate communities within and upstream of the areas affected by the operation of the Order. This should include studies over at least 3 years when no order is in force to determine the recovery of species populations if detrimental effects have occurred.
4. Monitoring of fish entrainment, especially fry, at Lobwood intake.
5. Monitoring of native white-clawed crayfish in the River Dibb, to cover impact of releases upon population numbers, breeding success, population recruitment, individual growth rates and maturation of this acid-sensitive species. Study to extend beyond Order period to determine population recovery.
6. Study long-term effects of possible acidification upon plant communities, as typified by water crowfoot

WHARFE – EMERGENCY POWERS
1. Maintain sewage works discharge quality at current performance meeting tighter temporary consent limits.
2. Continued provision of dissolved oxygen facilities on standby.
3. Reimbursement of costs of fish rescue.
4. Restocking of fish as identified by NRA. Brood stock for trout and grayling rearing to be taken from the Wharfe.
5. Compensation to fishery owners and operators for (a) loss of fish stock (b) loss of angling.
6. Intensive survey/monitoring of fish populations and physical habitat for next 3 years, particularly trout – Grassington to Tadcaster.
7. Restoration of spawning areas damaged by low flows. Removal of obstructions blocking fish access to tributaries.
8. Monitor population and passage of eels (particularly in relation to otter prey) over next 3 years.
9. Collate all existing information on river birds.
10. Compensation flow from Lindley Wood Reservoir to be maintained at statutory flow and present quality.

ARTHINGTON DROUGHT ORDER
1. By autumn 1996, assess the need for grayling and dace restocking downstream of Arthington and restock if necessary.
2. Identify sites and contribute to the creation of fish havens on this and lower stretches of the Wharfe.
3. Gather information on predators of fish e.g. piscivorous birds and otters, to assess likely effects on fisheries.
4. Collect information from fishery owners between Arthington and Tadcaster on changes in the distribution of fish and changes in angling activities. Prepare and publish a report.
5. Install suitable screens on Arthington intake. Install sonic or bubble screen in interim. Monitor entrainment losses.
6. Contribute to measures to install or improve fish passes on weirs.
7. Institute a weekly programme of water quality monitoring in the stretch from Arthington to Tadcaster for the period of the Drought Order.
8. Revised consent for Wetherby Sewage Treatment Works to apply for the period of this Drought Order, as well as the earlier Orders. Ensure that all sewage treatment works discharging to the River Wharfe are maintained at optimum performance.

The NRA therefore drew up plans for inspections to identify problems as they started to arise, and preferably in advance of this, and priority action plans for:

- fish rescues;
- maintenance of water quality (e.g. by having aeration equipment on standby and requiring more stringent effluent standards).

Action plans were also drawn up for:

- allowing fish access to suitable habitat (e.g. removal of obstructions to fish movement from mouths of tributaries used as spawning areas);
- protection of fish from predation and disease (e.g. voluntary bans on the use of keepnets that might subject fish to additional stress and disease susceptibility).

15.4.2 *Chronic problems*

These were identified as:

- change in factors affecting recruitment;
- change in factors affecting mortality;
- change in factors affecting movement.

All of these required detailed monitoring and assessment in order to identify the extent of variations expected in normal conditions, natural drought conditions and additional effects of changes in water regime imposed by the Drought Orders. Therefore, in formulating the assessment programmes, use was made of controls wherever possible to separate natural drought effects from those caused by abstraction, changes were to be assessed once the drought was over, and other possible influences were to be monitored so that their effects could be discounted.

15.5 Assessments

During the drought and in the first two years afterwards, assessments covered abiotic and biotic parameters.

Physico-chemical parameters monitored were: flows; temperature; dissolved oxygen; biochemical oxygen demand; ammonia; plant nutrients; reductions in water velocities and wetted widths; obstruction of feeder streams; exposure of gravel spawning areas and increased siltation; and river habitat. Aquatic ecological characteristics monitored included: macrophytes; benthic invertebrates; fish populations; angling; birds; and mammals.

On the rivers Wharfe and Ure, a programme of regular inspections and photographs at selected sites was implemented by the NRA to enable the appearance of the river to be linked to flows. A series of liaison meetings was held with anglers and environmental groups, who were asked to provide observations on any matters relating to flows. Regular liaison meetings continued between the NRA and YWS to review progress on the assessments.

15.6   Key issues/impacts

The ecological consequences of the 1995–1996 drought and associated abstractions sparked much public interest. Many of the predictable environmental responses to drought and low flows were observed across a wide range of Yorkshire's river systems during 1995 and 1996. In the relatively unpolluted rivers of North Yorkshire, consistent patterns of change were observed in the invertebrate and fish communities in accordance with the view expressed by Everard (1996), i.e. drought can have a beneficial as well as deleterious effect on freshwater ecosystems. In general, biodiversity and productivity were maintained and even enhanced, and several species, including many coarse fish, had better than average recruitment and growth, as well as expanding their range. In the same way that it may be difficult to distinguish adverse effects of abstractions from adverse effects of drought, beneficial effects may also be hidden.

Summaries of the findings have been given in an Environment Agency report (Environment Agency 1997). Changes could be classified as either detrimental or beneficial responses, at least at face value.

15.6.1   *Detrimental responses in rivers*

- Decreases in chemical and biological water quality, some downstream of effluent discharges (sewage treatment works, industry), due to lack of dilution.
- Decreases in dissolved oxygen and increases in pH.
- Lack of dilution for pollution incidents.
- Abundant algal growth, both filamentous species and epilithic diatoms, due to higher temperatures and nutrient levels, and longer retention time.
- Decline in some flow-loving invertebrates, such as Plecoptera, due to low flows and higher temperatures.
- More entrainment of fish and crayfish at intakes and impingement losses on intake screens.
- Some movement of flow-loving fish into remaining stretches of faster water.
- Some downstream displacement of fish.
- Blockage to side streams, preventing access for spawning salmonids.
- Increased exposure and siltation of spawning areas.
- Changes in angling activity, both in timing and distribution, due to weather conditions and movement/aggregation of target fish.

15.6.2   *Beneficial responses*

Most detrimental impacts noted were localised, but it is too early to say with certainty that some longer river stretches were not affected.

- Improvements in chemical water quality in some areas due to less frequent operation of storm overflows, and operation of sewage treatment works by YWS to stricter consent standards during the drought.
- Increased macroinvertebrate diversity and productivity in grayling, *Thymallus thymallus* (L.) and coarse fish zones of good quality rivers.
- Increased growth rates and fry survival of coarse fish, especially those favoured by slow flows.
- Support for future angling with strong year-classes of several fish species.

15.7   Surveillance and monitoring

Short-term impacts of reduced flows were readily detected by surveillance during the drought, and proactive mitigation measures and trigger levels for actions were successfully used. No attempts were made to ascribe the degrees of contribution to short-term impacts that were caused by the operation of the Drought Orders above those caused by natural drought.

Long-term monitoring to date has shown, in many instances, only subtle effects, if any, and it is difficult to discern whether these were due to flow or other causes. The additional effects of changes in flow due to operation of the Drought Orders have mostly been almost impossible to attribute with certainty.

15.8   Ecological effects and implications for monitoring

In areas free from effluent influences, reductions in compensation releases or increased abstraction, it was evident that the natural drought had an impact on the aquatic environment in Yorkshire, including drying out of stretches of some streams on limestone. Drying of the streambed is not an ecological disaster for many stream systems, as it can be a regular feature of their dynamics (e.g. winterbournes on the chalk of southern England), or is intermittent and coincident with drought periods. In these situations, the fauna and flora are adapted to such occurrences, change according to the prevailing conditions and rapidly recolonise as flow recommences.

It is clear that flow regime can have major effects on the ecology of streams. Poff and Ward (1989) analysed streamflow patterns across the USA in relation to biological communities. Highly variable flow regimes had dominant effects, but, in more stable flow environments, biotic interactions such as competition or predation were more important. Streams with stable flows tended to support larger, more specialised fish and invertebrates, and more long-lived species. This would suggest that the highly variable flow regimes of Pennine rivers, characterised by occasional summer as well as winter spates, would lead to physically controlled ecosystems. Presumably, the animals and plants would show a more rapid and resilient response to unusually high or low flows than in the more stable, lowland rivers, where droughts and floods could have dramatic and long-lasting impacts, especially if they occur in summer months.

The streams below reservoirs in the Pennines have unnatural and relatively stable flow regimes. The aquatic communities have not been exposed to periodic drying and may have developed accordingly. Thus, theoretically, responses to reduced compensation flows may be more severe than in natural streams.

Cause and effect linkages between physico-chemical characteristics of drought affected rivers and biota remain poorly understood, even though some general responses are known. It is obvious that simplistic correlations with periods of low flows alone will not suffice. Flow variability and range, duration of threshold flows, timing of events and the nature of the 'starting point' characteristics of the ecological community being monitored, are all features that need to be taken into account when trying to distinguish effects of low flows.

Great variability has been a feature of much of the ecological data analysed to date. Despite this, some significant changes linked to the drought and low flows have become apparent. Distinguishing between short- and long-term effects on fish populations has helped to clarify and weight the responses to drought that were recorded. A particularly difficult point was to decide when change in excess of normal variability should be regarded as environmental damage. Variability is a reflection of the unstable physico-chemical regime in spate rivers and may be protection against environmental damage in the short term.

Even extreme variation cannot be considered to be environmental damage if it can be demonstrated that a return of normal flows leads to a brisk reversal. Disturbance to the general pattern can be important if it is maintained, and if ecological successional changes occur which are irreversible or only very slowly reversible. Paradoxically, although the flow stability during the drought in the upland spate rivers of Yorkshire led to the formation of biological communities which were more diverse and productive than normal, greater proportions of these taxa would be more susceptible to high flows events than would otherwise be the case. Thus, provided that drought does not continue year after year, and trigger irreversible successional changes, the normal pattern may return. Although a few species, such as perlid Plecoptera, did not fare well during the drought, some tolerant and generalist species, such as lepidostomatid Trichoptera, increased both their abundance and range upstream. Obstructions to fish movements, such as weirs, prevented the full potential expansion of some fish species from being realised. It remains to be seen whether this will be reversed by more normal flow patterns. Long-term climate change in northern England is forecast to lead to wetter winters and drier summers (Environment Agency 1998), so new norms may be established, incorporating some of the features observed during the drought.

It was recognised at an early stage that it would be necessary to distinguish 'change' from 'damage', and therefore a project was initiated to formulate environmental evaluation criteria for water resource impact assessment. The initial stage of this was a literature review. Most of the papers found referring to fish were in relation to salmonids (Environment Agency 1998). The assignment of terms such as 'damage' and 'harm' depends on the perspective of the assessor, the timescale of the impact, and the sensitivity and 'value' of the impacted species, habitat or community. These difficulties in making assessments about environmental impact are, in part, due to the problem of classifying ecological communities (through to species and populations) on the basis of their 'value', sensitivity to drought impact and their resilience, i.e. ability to recover subsequently. Assigning judgmental terms to the observed changes in aquatic ecosystems is a problem in the absence of strong guidelines and detailed knowledge about interactions. Parameters such as rarity and conservation value are relevant and contribute to the debate, but are nevertheless influenced by natural as well as anthropogenic events. In addition to the problems of detecting ecological changes against high natural variability and making value judgments about their significance is the need to distinguish Drought Order impacts over and above those effects caused by natural drought and the normal abstraction and compensation flow regimes. These complexities mean that, in most cases, confident detection of harm by means of a monitoring programme initiated during the drought event is unlikely to be achievable.

Where serious water quality deteriorations occurred, these were due either to known pollution incidents or combinations of factors. It could be argued that low river flows limit the dilution available to absorb one-off pollution incidents, which therefore tend to be more marked in their effect. Very low dissolved oxygen levels recorded in the mid-reaches of the Calder during the drought were believed to be due to a combination of low river flows, a high proportion of treated sewage effluent, segregation of the river into a series of pounds separated by weirs, high temperatures and excessive weed growth. Separation of the effects of Drought Orders was not possible. Although a few water quality problems were encountered, water quality improvements were recorded in some instances and reflected YWS achieving improvements in effluent treatment at several sewage works.

Streams receiving compensation releases were of particular concern during the drought. Although only one fish rescue was necessary as a result of changed compensation releases under a Drought Order, the drought highlighted many anomalies in the reservoir flow compensation regimes operated by YWS. Some of the releases originated from bygone industrial demands, and are inappropriate for protecting the environment. In addition, some of the compensation arrangements result in higher levels of compensation water being released in the winter months, when it is least needed.

Fisheries survey techniques also have limitations, often in connection with the sampling methods, such that survey results may not reflect population changes adequately. Thus effects of the drought on fry may not be able to be examined until a later date, when the fish become fully vulnerable to capture by the survey methods commonly employed, but other influences in the intervening period may mask those of the drought. For example, good recruitment in drought conditions could be negated by a severe flood immediately afterwards, an occurrence that is thought to have affected some fish stocks following the 1975/1976 drought in Yorkshire.

One-off surveys in themselves cannot contribute to assessment of change but are useful in defining current status or providing a baseline for future work. To demonstrate change or impact, it is necessary to have a good basis with which to compare current data. This requires more than one previous sample, as impact needs to be shown to exist outside of the range of natural variability. Long-term data sets, especially for aquatic macroinvertebrates and fish, are lacking for many of the upper catchment and headwater streams influenced by compensation releases. A dilemma emerging from the drought studies and research elsewhere (Armitage *et al.* 1997) is that a high degree of focus on particular habitats is required, but baseline data may be required in relation to a number of possible impacts over a wide area. Thus both large-scale and detailed, and therefore expensive, monitoring programmes are required to cover most likely management questions about impacts. Appreciation of the scale, process rates and spatial integrity of the various aquatic systems in Yorkshire (and indeed elsewhere) is fundamental if effective baseline monitoring is to be implemented. Thus, although the long-term datasets collected during the drought may not have had immediate utility in defining impacts, they have been important in defining requirements for future monitoring.

15.9   Ecological lessons

(1) There was a lack of baseline data that would enable basic questions about effects of flow on ecology to be answered by monitoring instituted once the drought was in progress. Ecological responses are becoming better known, but need a comprehensive programme to account for the interactions. Natural variation is likely to be great, and other factors will complicate matters.

(2) Apart from the immediate surveillance needs during the drought, information from intensive, quantitative surveys was required. Focussed monitoring sites and methods needed to be identified in advance. Replicates and control sites should be found whenever possible, so that the effects of Drought Orders might be distinguished from those of natural drought and variations within existing regimes. Without such data, the utility of attempting long-term monitoring was questionable.

(3) Minimum flows specified in existing licences for controlling abstraction had not been derived in a consistent manner from considerations of relevant ecological effects.

(4) Community involvement could provide additional surveillance, but needed clear, rapid channels of communication of problems to facilitate timely investigation, e.g. a report of thin trout ascending streams for spawning was received too late for investigation.

(5) Effective strategies for maintenance of water quality, including trigger levels for actions based on observation of fish and indications from chemical determinands were successfully formulated and implemented.

(6) Losses and damage that are acceptable under normal flow regimes might become unacceptable with reduced flows and increased abstractions, e.g. entrainment losses at intakes. Consideration needed to be given to actions to ensure free movement of fish to suitable areas and to means of reducing vulnerability of fish to increased damage and mortality by predators and anglers.

(7) Realistic models of physical habitat and water quality were desirable to facilitate extrapolation to flows beyond previous experience and thus likely considerations for ecology.
(8) Following the recovery process may help to distinguish 'change' from 'damage'.

15.10   Management lessons

(1) A severe drought and associated need for Drought Orders was required before there was commitment to assessment of effects on aquatic ecology and suitable mitigation measures. Things happen quickly, so procedures need to be in place, including conditions to apply to Drought Orders.
(2) The opportunity of a large scale natural experiment should be grasped to learn as much as possible and get interested parties to pay for this whilst it could be clearly in their interests. Public pressure to protect the environment can be a powerful force, but needs to be harnessed.
(3) Lack of strategic planning for drought allowed the situation to deteriorate to a state where public health considerations had to override ecology, so persons interested in fisheries must question matters apparently outside their direct involvement.
(4) Consultations with interested parties need to occur before formal procedures begin.
(5) Agreements need to be watertight and to have formal commitment.
(6) Contingency plans need to be in place at an early stage to counteract increased risks of damage and to rescue fish stocks, if necessary. These should include lists of contacts among interested parties and other means of communication. Some requisite mitigation measures may not be immediately apparent and thus contingency measures and funds should be established to deal with adverse effects proved at a later stage.
(7) Without baseline data, a monitoring programme beyond surveillance during a drought may not be immediately worthwhile.
(8) Agreement about the definition of environmental damage will assist formulation of appropriate surveillance and monitoring programmes, responses to interested parties and mitigation measures.

References

Armitage P.D., Cannan C.A. & Symes K.L. (1997) Appraisal of the use of ecological information in the management of low flows in rivers, *R&D Technical Report W72*. Bristol: Environment Agency, 97 pp.
Environment Agency (1997) *Second Interim Report on the Environmental Impacts of the Drought on Yorkshire's Rivers 1995–1996*. York: Environment Agency, North East Region, 140 pp + tables and figures.

Environment Agency (1998) *Environmental Evaluation Criteria for Water Resource Impact Assessment*. Swindon: Sir William Halcrow and Partners Ltd, 78 pp + figures and appendices.

Everard M. (1996) The importance of periodic droughts for maintaining diversity in the freshwater environment. *Freshwater Forum* **7**, 33–50.

Poff N.L. & Ward J.V. (1989) Implications of stream flow variability and predictability for lotic community structure: a regional analysis of streamflow patterns. *Canadian Journal of Fisheries and Aquatic Science* **46**, 1805–1818.

Uff J. (1996). *Water Supply in Yorkshire – Report of the Independent Commission of Inquiry*. London: Uff, 159 pp.

Chapter 16
Impacts of the Ok Tedi copper mine on fish populations in the Fly River system, Papua New Guinea

S. SWALES, B.S. FIGA, K.A. BAKOWA and C.D. TENAKANAI

Environment Department, Ok Tedi Mining Limited, PO Box 1, Tabubil, Western Province, Papua New Guinea (e-mail Swales.Stephen.SS@bhp.com.au)

Abstract

The Ok Tedi copper mine in Papua New Guinea discharges waste rock and tailings into the Ok Tedi river in the headwaters of the Fly River system. Biological monitoring of the river system recorded approximately 100 fish species representing 32 families. Fish catches exhibited temporal and spatial variability, but since the commencement of mine operations monitoring has revealed significant reductions in fish catches at most riverine sites. However, no significant declines in fish catches have been recorded in the lower Fly or delta areas. Although catches in some floodplain off-river water bodies have also declined, these changes are thought to be associated with the effects of natural climatic phenomena, particularly El Niño droughts, introduced species and increased commercial and artisanal fishing. It is suggested that loss of fish habitat through increased river bed sedimentation due to the input of mine wastes is likely to be one of the major causes of the decline in fish catches in the Ok Tedi and middle Fly. However, other factors such as reduced water quality, introduced species and overfishing may also be involved.

Keywords: Mining, biological monitoring, fish populations.

16.1   Introduction

The Ok Tedi copper and gold mine is situated in the upper catchment of the Fly River in Western Province, Papua New Guinea (PNG), close to one of its largest tributaries, the Ok Tedi (Ok means river in the language of the local Yongom people). The mine produces approximately 80 000 t d^{-1} of waste tailings and 121 000 t d^{-1} of mined waste rock which are discharged directly into the Ok Tedi/Fly River system. (Because of the high rainfall and geological instability of the region, construction of a tailings dam was not feasible, and the mine has been operating without waste retention).

Ok Tedi Mining Ltd has a statutory requirement to protect the river system and to ensure that the mining operation does not cause unacceptable damage. The company commenced environmental monitoring of the Fly River system in 1981 and this now extends along the length of the river, from the headwaters down to the river delta and the Gulf of Papua. The extensive long-term monitoring programme means that the Fly River system is one of the most intensively studied tropical river systems. The main environmental impacts from mining operations arise from the introduction of large amounts of sediment into the river system, from waste rock and tailings, which results in elevated levels of total suspended solids (TSS) and increased river bed aggradation. The key water quality parameters affected by mine operations are TSS, particulate copper and dissolved copper.

In this chapter, the results of biological monitoring in the Ok Tedi/Fly River system and the possible effects of waste discharges from the Ok Tedi mine on the fish populations are discussed.

16.2 Study area

The Fly River system is the largest river, in terms of water flow, in Australasia. With a mean annual discharge of $\sim$6000 m^3 s^{-1} the Fly is similar in size to the Niger and Zambesi Rivers in Africa and the Danube in Europe (Welcomme 1985). However, with a catchment area of only 76 000 km^2 the Fly outranks all the world's major rivers in terms of the run-off per unit catchment area. This is due to the very high rainfall in the region, ranging from 10 000 mm yr^{-1} in the upper catchment near the mine to around 3000 mm yr^{-1} near the coast.

The Fly River, and its major tributary the Strickland River, flow for over 1200 km from their source in the Western Highlands of PNG down to the Gulf of Papua (Fig. 16.1). In its meandering course to the coast, the Fly falls only 20 m over the 800-km river length between the town of Kiunga and its estuary. Much of the Fly catchment, particularly in the middle and upper reaches, consists of dense, primary tropical rainforest. In the middle Fly area, tropical swamp forest thrives in the floodplain wetlands, while further downstream the drier climate gives rise to open savanna forest and grasslands.

The Fly River system supports the most diverse fish fauna in the Australasian region, with 128 recorded native freshwater species representing 33 families. Seventeen species are known only from the Fly basin, and thirty or more are known only from the Fly and one or more of the large rivers in central-southern New Guinea (Roberts 1978). The fish populations in the Fly are characterised by the large size of some species, the abundance of endemic species and the dominance by groups that are poorly represented in other parts of the world (including 17 species of catfish in the families Plotosidae and Ariidae). In most other ways the composition of the freshwater fish fauna is largely determined by its position in the Australasian zoogeographical zone (Roberts 1978; Coates 1993). The fish fauna of the Fly River conforms closely to the relationship between river basin area and fish species number recorded by Welcomme (1985) (Fig. 16.2). The river supports large populations of barramundi, *Lates calcarifer*, which provide a limited commercial fishery in the river and estuary. Artisanal subsistence fisheries also exist at most villages along the length of the Fly.

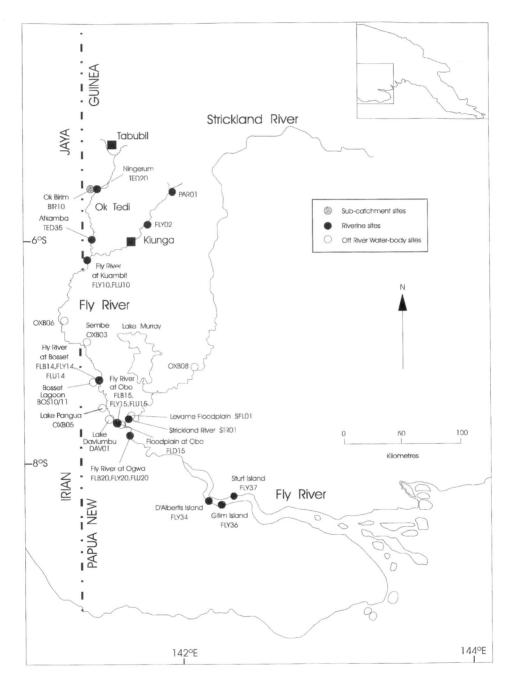

Figure 16.1 Locations of fish monitoring sites in riverine and floodplain areas of the Ok Tedi/Fly River system

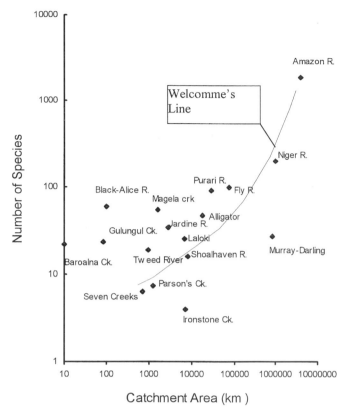

Figure 16.2 Relationship between river basin area and fish species richness, showing the position of the Fly River in relation to other Australasian and world rivers (modified after Welcomme (1985) and Bishop & Forbes, (1991))

The primary biological habitats of the Fly catchment are high-gradient streams in upland regions, lowland riverine habitats with and without associated wetlands, and the lakes and oxbow cut-offs associated with the wetlands of the middle Fly and Strickland River reaches. The river channels throughout the system are generally of low biological productivity, primarily due to high water turbidity and instability of the bed material. The off-river habitats are generally confined to the middle and lower Fly and the Lower Strickland areas. These habitats are highly productive areas and are the main feeding areas for fish in the associated river channels.

16.3 Methods

Fish populations were routinely monitored at over 30 sites throughout the Ok Tedi/ Fly River system (Fig. 16.1) using standardised methodologies, including gill-netting, seining, fish toxicants (rotenone) and electric fishing. However, a standardised gill-netting procedure forms the basis of most routine fish sampling at sites in the main river channel and floodplain off-river water bodies. A set of 13 gill-nets (stretched mesh size 2.5–17.5 cm) was set at each site for 24 h and checked at dawn, dusk and the end of the sampling period. Fish were identified to species, measured (fork or total length) to the nearest 1 mm and weighed (to the nearest g). Sampling at most sites was conducted on a monthly or quarterly basis.

16.4   Results

16.4.1   *Fish fauna – composition and species richness*

Approximately 100 fish species, representing 32 families, were recorded from the Ok Tedi/Fly River system in the biological monitoring programme since sampling commenced in the early 1980s. The freshwater fish fauna was dominated by catfish of the families Ariidae (16 species) and Plotosidae (ten species), groups which are rare elements in the freshwater fish faunas of other regions of the world (Roberts 1978).

The mean number of species recorded each year from sites in the main river (~20 species) has remained relatively constant (Fig. 16.3). However, in the Ok Tedi the number of recorded species has declined markedly in recent years falling from around 15 in 1983–1993 to just three species in 1996.

16.4.2   *Temporal and spatial trends in fish catches*

Fish catches in the Ok Tedi and Fly River, and the associated off-river water bodies, have shown considerable temporal and spatial variation. However, since sampling commenced in the early 1980s fish catches at many riverine sites, particularly in the Ok Tedi, have declined considerably. Catches at Ningerum (TED20) in the Ok Tedi and Kuambit (FLY10) in the middle Fly declined over the period of monitoring by

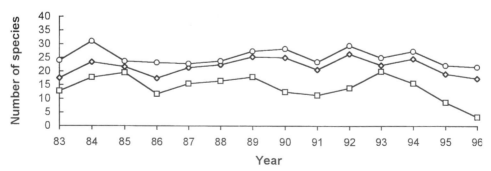

Figure 16.3 Mean number of fish species recorded at river channel sites since the commencement of sampling (◇ all riverine sites; □ Ok Tedi sites only; ○ all sites except Ok Tedi)

89% and 73%, respectively (Fig. 16.4). However, no significant decline in fish catches was recorded at riverine sites in the lower Fly (below Everill junction) or in the estuary.

Fish catches in many off-river water bodies on the floodplain of the middle Fly continue to be high, but also show considerable inter and intra-annual variability. Although catches at many sites declined in species diversity and biomass over time, such changes are thought to be due largely to climatic phenomena, particularly El

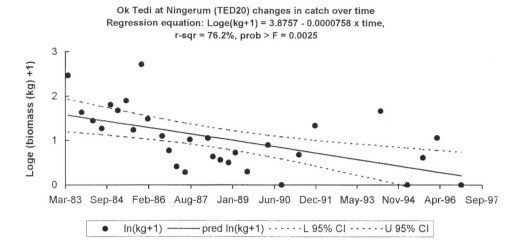

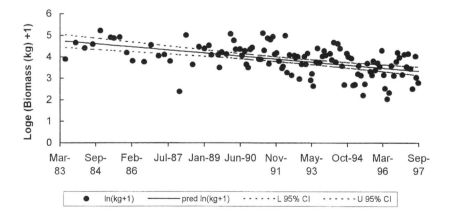

Figure 16.4 Temporal changes in gill-net catches and estimated linear regression (solid line) with 95% confidence limits (dashed line) at Ningerum (TED 20) in the Ok Tedi and Kuambit (FLY 10) in the middle Fly River

Niño droughts, which cause large areas of the floodplain to dry out periodically, with adverse consequences for fish stocks through loss of habitats and changes in water quality. There is, as yet, no evidence that declines in fish stocks in any off-river water bodies are due to the effects of the mine.

Mean total fish biomass and species number were much reduced at sites in the Ok Tedi and upper Fly compared to sites in the middle and lower Fly (Fig. 16.5). Reductions in fish catches at riverine sites downstream of the mine were partly due to decreases in catches of barramundi, which have declined considerably in catches at sites in the middle Fly over recent years (Fig. 16.6).

16.5   Discussion

The long-term monitoring of the biological impacts of the Ok Tedi mine on the Ok Tedi/Fly River system suggest that mining operations may be associated with significant changes to the aquatic ecosystem. Significant declines in fish catches at many sites over the period of sampling were observed. These coincide with the period of operation of the mine and the discharge of mine wastes into the Ok Tedi/Fly River system (Smith & Hortle 1991; Smith & Morris 1992).

The declines in fish catches downstream of the mine are suggestive of adverse environmental impacts arising from the discharge of mine-derived effluents into the headwaters of the river system. However, the evidence which is currently available is largely circumstantial and there is little information on cause and effect. Although a range of bioassay and toxicity studies has been carried out over recent years, no acute toxic effects of mine discharges on water quality and aquatic life have been detected (Smith *et al.* 1990). A recent study showed no evidence of toxicity of copper to algae in the Fly River system and the evidence which is available suggests that most of the copper in the system is bound to dissolved organic carbon and is not bioavailable (Stauber 1995).

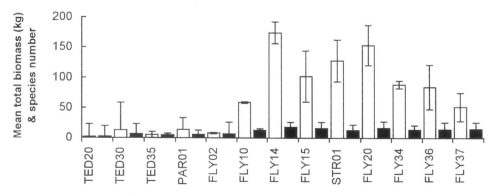

Figure 16.5   Mean estimates of total biomass (solid bar) and diversity (open bars) of fish catches at sites in the Ok Tedi, Fly and Strickland Rivers

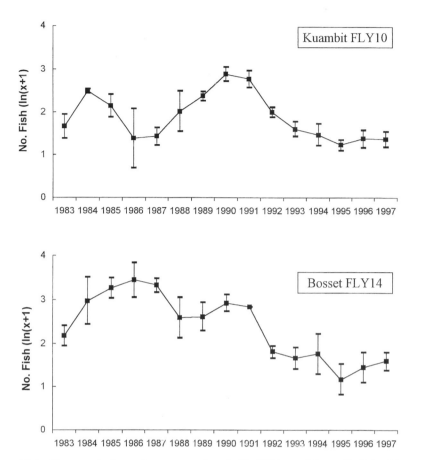

Figure 16.6 Temporal plots of mean catches (with 95% confidence limits) of barramundi *Lates calcarifer* at river channel sites in the middle Fly and Strickland River

A range of possible impact mechanisms may be associated with the declines in fish catches in the system. Although there is no evidence that changes in water quality through the input of mine wastes are directly toxic to aquatic life, monitoring of metal levels in fish tissues from samples taken from a range of sites in the Ok Tedi and Fly River demonstrated elevated levels of copper, lead, zinc and cadmium in fish flesh, liver and kidney (Swales *et al.*, in press). However, there is currently no evidence that the metal levels recorded in these samples are indicative of lethal or sub-lethal effects of mine wastes on fish survival. Studies have been initiated to investigate any possible effects of elevated metal levels due to mine waste input on fish histopathology.

Other possible causes for the declines in fish catches in the system include the increased suspended sediment concentration and increased rate of river bed aggradation. Although TSS concentrations have increased several-fold over pre-mine levels, there is no evidence that these levels are toxic to fish and other aquatic life. Most rivers in New Guinea are naturally turbid due to the high rainfall and land instability and it is thought that fish and other aquatic life are adapted to the high sediment loads. The dominance of Ariid and Plotosid catfish in the fish fauna of many rivers is thought to be indicative of the natural adaptation of the fish fauna to high sediment loads (Roberts 1978).

The extent of bed aggradation from pre-mine levels varies from over 6 m in parts of the Ok Tedi to around 2–4 m in the middle Fly. Although there is no direct evidence for adverse effects on fish and other aquatic fauna, it is likely that such high levels of sedimentation will result in considerable loss and damage to fish habitat and the elimination of other aquatic life. Consequently, it is likely that in severely impacted areas, such as the Ok Tedi, physical loss and damage to habitats is probably the main reason for the declines in fish catches observed. In other areas, such as the middle Fly, the role of river-bed aggradation in reduction of fish catches is less clear.

In addition, other anthropogenic and natural environmental factors may also be involved in the recorded declines in fish catches in the system. For example, the extent of commercial and artisanal fishing in the middle Fly River has increased over recent years and it is likely that these fisheries may have a substantial impact on fish stocks, particularly popular species such as barramundi, although there is little available information. Also, in recent years two species of introduced fish, the climbing perch, *Anabas testudineus,* and the walking catfish, *Clarias batrachus*, have become widespread and abundant in areas of the Fly River system, with unknown consequences for native fish stocks (I.D. Storey personal communication).

It is well established that riverine fish populations exhibit large natural variations in abundance associated with natural environmental factors, such as climatic changes associated with El Niño droughts, which have major effects on the extent of floodplain inundation in the Fly River (Winemiller 1996). Since the life cycle of many riverine fish is closely associated with the annual cycle of flooding, variations in river flow may have major consequences for fish population dynamics. Off-river water bodies on the Middle Fly floodplain provide important habitats for fish spawning and recruitment, and provide both adult and juvenile fish with valuable feeding and refuge areas. However, there is currently no evidence that mine operations are adversely affecting the floodplain ecosystem, and off-river water bodies support diverse and abundant fish populations.

Acknowledgements

We would like to thank the present and former members of the Environment Department, Ok Tedi Mining Ltd, for their assistance in the biological monitoring programme. The assistance provided by fisheries technicians in the Biology Section, particularly Kenneth Kambanei, Origen Mogelu, Mabi Dukawa, Cornelius Sare and Phillip Atio, is gratefully acknowledged. Permission from Ok Tedi Mining Ltd to publish this study is gratefully acknowledged.

References

Bishop K.A. & Forbes M.A. (1991) The freshwater fishes of northern Australia. In C.D. Haynes, M.G. Ridpath, M.A.J. Williams (eds) *Monsoonal Australia – Landscape, Ecology and Man in the Northern Lowlands*. Rotterdam: A.A. Balkema, pp. 79–107.

Coates D. (1993) Fish ecology and management of the Sepik-Ramu, New Guinea, a large contemporary tropical river basin. *Environmental Biology of Fishes* **38**, 345–368.

Roberts T.R. (1978) An ichthyological survey of the Fly River in Papua New Guinea with descriptions of new species. *Smithsonian Contributions to Zoology* **281**, 72 pp.

Smith R.E.W. & Hortle K.G. (1991) Assessment and prediction of the impacts of the Ok Tedi copper mine on fish catches in the Fly River system, Papua New Guinea. *Environmental Monitoring and Assessment* **18**, 41–68.

Smith R.E.W. & Morris T.F. (1992) The impacts of changing geochemistry on the fish assemblages of the Lower Ok Tedi and Middle Fly River, Papua New Guinea. *The Science of the Total Environment* **125**, 321–344.

Smith R.E.W., Ahsanullah M. & Batley G.E. (1990) Investigations of the impact of effluent from the Ok Tedi copper mine on the fisheries resource in the Fly River, Papua New Guinea. *Environmental Monitoring and Assessment* **14**, 315–331.

Stauber J.L. (1995) Toxicity testing using marine and freshwater unicellular algae. *Australian Journal of Ecotoxicology* **1**, 15–24.

Swales S., Storey A.W., Roderick I.D., Figa B.S., Bakowa K.A. & Tenakanai C.D. (in press) Biological monitoring of the impacts of the Ok Tedi copper mine on fish populations in the Fly River system, Papua New Guinea. *The Science of the Total Environment*.

Welcomme R.L. (1985) River Fisheries. *FAO Fisheries Technical Paper* **262**. Rome: FAO, 330 pp.

Winemiller K.O. (1996) Dynamic diversity in fish assemblages of tropical rivers. In M.L. Cody & J.A. Smallwood (eds) *Long-term Studies of Vertebrate Communities*. London: Academic Press, pp. 99–133.

Chapter 17
Riverine fish stock and regional agronomic responses to hydrological and climatic regimes in the upper Yazoo River basin

D.C. JACKSON and Q. YE

Mississippi State University, Department of Wildlife and Fisheries, Box 9690, Mississippi State, MS 39762, USA (e-mail: djackson@CFR.MsState.edu)

Abstract

The upper Yazoo River basin (UYRB) is a river–floodplain ecosystem in Mississippi (USA). Riverine hydrological and regional climatic regime relationships to agriculture (cotton, soybeans) and principal riverine fish stocks in the UYRB were determined. Climatic factors influenced fish stocks less than did hydrological factors. Lag periods (1–2 years) between flooding and stock responses by bigmouth buffalo *Ictiobus cyprinellus*, smallmouth buffalo *Ictiobus bubalus* and channel catfish *Ictalurus punctatus* reflected time requirements for recruitment. Channel catfish recruitment was also enhanced by warmer temperatures for a given year. Flathead catfish *Pylodictis olivaris* catches increased under stable stream flow conditions during seasonal low flow periods. No models could be developed for blue catfish *Ictalurus furcatus*. Climatic factors (e.g. date of last spring frost and degree days) were principal influences on cotton and soybean production. Floods resulted in reduced area planted (primarily cotton), but had little influence on crop yield and total annual production.

17.1   Introduction

The upper Yazoo River basin (UYRB) is a river–floodplain ecosystem located in the interior delta region (Mississippi River alluvial valley) of western Mississippi (USA). Historically, this region has been highly productive regarding wildlife and fisheries resources (Smith 1954; US Fish and Wildlife Service 1979). The floodplains of the UYRB are extensively developed for row-crop agriculture, primarily for cotton and soybeans. More than one-third of the cotton and soybean production in Mississippi is from the UYRB (Mississippi Agricultural Statistics Service 1995).

Extensive flood control programmes have developed throughout the entire Yazoo River watershed since 1928, primarily to protect and expand agricultural lands (Jackson *et al.* 1993). Specific projects have included river channelisation, construction of reservoirs and levees, river channel dredging and snagging (i.e. removal of instream large woody debris), and clearing of riparian vegetation.

Despite these efforts, crop lands in the UYRB can still be threatened by flooding, and especially from storm run-off during the growing season when the main tributary streams of the upper Yazoo River approach or exceed bankfull stage as a result of upstream reservoir releases (US Army Corps of Engineers 1991). To address these situations, maintenance and expansion of flood control activities continue in the UYRB. However, there is concern regarding potential negative impacts to riverine fisheries that could result from these activities (Jackson & Jackson 1989; Jackson 1991; Jackson *et al.* 1993).

With multiple use of floodplain river ecosystem resources as the guiding orientation, a study was conducted on the UYRB at a regional (basin-wide) level of resolution to address fisheries and agricultural interests. The objective of the study was to determine the inter-relationships of riverine hydrological and regional climatic regimes to relative abundances, assemblage composition and structure of principal catfish (*Ictaluridae*) and buffalofish (*Catostomidae*) stocks in UYRB rivers, and to agricultural area planted, per cent planted area harvested, yield and annual production of principal agricultural crops (cotton and soybeans) in the UYRB.

17.2   Materials and methods

This study utilised five years of riverine fish stock assessments (1990–1994) and 31 years (1964–1994) of agricultural, climatic and hydrological data from the UYRB. The fish stock data were collected in cooperation with the Mississippi Department of Wildlife, Fisheries and Parks (Jackson *et al.* 1995). Agronomic data were obtained from the Mississippi Agricultural Statistics Services. Climatic data were obtained from the National Oceanic and Atmospheric Administration (NOAA). Hydrological data were provided by the US Army Corps of Engineers, Vicksburg, Mississippi.

17.2.1   *Fish stock characteristics*

Fish stock data incorporated hoopnet catches obtained downstream from headwater flood control dams for five principal tributaries of the Yazoo River: the Coldwater, Little Tallahatchie, Tallahatchie, Yalobusha and Yocona rivers, all located in the UYRB (Fig. 17.1). Rivers were divided into major sections (e.g. upstream, midstream, downstream) for sampling purposes. Each major section was subsequently divided into 1-km sub-units (stream reaches) which served as experimental units. Sampling effort was evenly distributed among major sections of the respective rivers.

Hoop nets had front hoop diameters of 1.04 m and bar mesh size of 3.81 cm and were fished from January–August each year of the 5-year period. On each sample date, an attempt was made to set 10 nets (1990–1992; 1994) or five nets (1993), distributed at approximately 100-m (1990–1992; 1994) or 300-m (1993) intervals, in the randomly selected stream reach of the major section.

Nets were set on the bottom of the river with openings facing downstream and positioned in a manner considered to maximise catch efficiency (i.e. within an

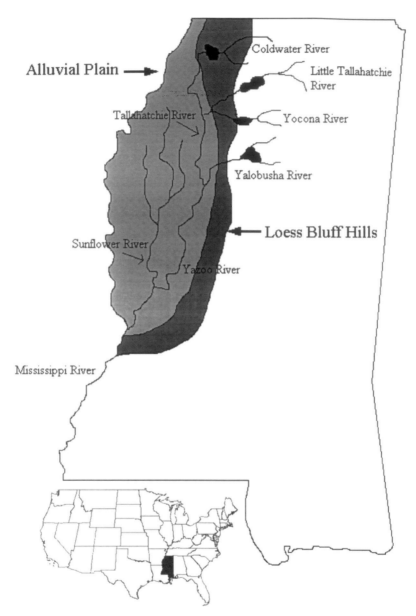

Figure 17.1   The upper Yazoo River basin (shaded) and principal tributary rivers of the Yazoo River, Mississippi, USA

interval, sites having steep banks, eddies and/or instream structure were selected, if available) (Jackson & Jackson 1989). During the 5-year study, 2878 overnight hoopnet sets were conducted in the UYRB.

All fish collected were identified to species (Robins *et al.* 1991), tagged and individually measured (total length, mm) and weighed (g) before being returned to their respective capture sites. Blue catfish (*Ictalurus furcatus*, N = 293), channel catfish (*Ictalurus punctatus*, N = 789), flathead catfish (*Pylodictis olivaris*, N = 619), bigmouth buffalo (*Ictiobus cyprinellus*, N = 288) and smallmouth buffalo (*Ictiobus bubalus*, N = 1739) were the principal exploitable species targeted in the study, and contributed (% by weight) the most to the total catch (Jackson *et al.* 1995). Catch per unit of effort (CPUE) was expressed as kg net-night^{-1}. Catch composition was assessed using species-specific per cent contribution, by weight (kg), of the total combined ictalurid and catostomid species catch. Proportional stock density (PSD) for ictalurid and catostomid species were determined, as proposed by Gabelhouse (1984).

17.2.2 *Agronomic characteristics*

Agronomic data were collected for 15 UYRB counties that contain farm lands considered most threatened by flooding of the river system (US Army Corps of Engineers 1991). Area planted (km^2), planted area harvested (%), yield (t km^{-2}) and annual production (*t*) of cotton and soybeans were the principal agricultural descriptors for the study. Data for area planted and for per cent planted area harvested for soybeans were only available for the period 1974–1994. All other agricultural data were available for the period 1964–1994.

17.2.3 *Hydrological characteristics*

Hydrological data for the five UYRB rivers were the mean of monthly maximum, minimum and average relative water stage (% of bankfull level) and their standard error during the high flow period (January–April) and low flow period (July–October) for each year, number of days above flood stage per year, and metre-days above flood stage per year. The metre-days value combines the effects of height and duration of flooding, and was determined by adding the height above flood stage over all the days of the flood, by year, for each river. Relative water stages, flood days and flood metre-days of the whole upper Yazoo River system were generated by averaging the flow/flood values among the five rivers. Maximum and yearly mean area flooded (km^2) data for each tributary river basin were only available for 1964–1989. Equations were therefore developed using these data to estimate area flooded for 1990–1994 by regressing absolute water stage (m) and flow discharge (m^3 s^{-1}) variables against maximum and yearly mean area flooded. An assumption was made that geotopographical characteristics in the UYRB did not change significantly during the period 1989–1994. Basin-wide maximum and mean area flooded were calculated by adding areas associated with each river basin to address relationships between fishes and the height and duration of flood events (*sensu* Welcomme 1976).

Climatic characteristics

Climatic data for the entire UYRB were obtained by combining data from weather stations located in the watershed of each river. The annual mean of daily maximum, minimum and average air temperature (°C), degree days above 18.3°C per year, total annual rainfall (cm), the number of frost-free days per year, last spring frost date, and first autumn frost date were considered as primary climatic factors. Both the last spring frost date and the first autumn frost date were adjusted to the number of days from 1 January. Only annual means of daily maximum, minimum and average air temperature, and degree-days per year were incorporated in model selection as potential descriptors for fish stock characteristics.

17.2.5   *Analyses*

The five rivers studied combine to form an integrated system, and have similar instream, riparian, and watershed characteristics subject to the same prevailing climatic, hydrological, and anthropogenic (primarily agricultural) influences (Jackson *et al.* 1993). Therefore a basin-wide (UYRB) level of resolution was utilised in multiple linear regression analyses (Myers 1990) to determine relationships of fish stock and agronomic characteristics with hydrological and climatic regimes. With regard to fish stock characteristics, total and principal species CPUE, catch composition and PSD were dependent variables; hydrological and climatic factors were independent variables. A lag time of 0–2 years was included in hydrological regimes to account for the time between fish spawning and first capture (Holčik & Bastl 1977; Risotto & Turner 1985).

For agronomic characteristics, area planted (km^2), planted area harvested (%), yield (t km^{-2}) and annual production (*t*) of principal agricultural crops (cotton and soybeans) were dependent variables in multiple linear regression models, and independent variables were selected climatic factors and flood-associated factors (number of days above flood stage, metre-days above flood stage, and maximum and mean area flooded per year), which were considered the most influential regarding agricultural enterprises. A lag of 0–2 years was also included for flood regimes as they related to agriculture. To demonstrate the dynamic function of climatic and hydrological regimes on crop production enterprises, models were developed based on 31 years of data (1964–1994) and 5 years of data (1990–1994). This last period corresponded with the duration of fish sampling.

Data were transformed, as required, according to partial regression plots and residual plots for non-linear relationships, as well as to address assumptions of the statistical analyses (Afifi & Clark 1984). Commonly performed transformations were square, square root, and logarithmic. Mallow's *Cp* and adjusted r^2 were used to select independent variables that best described the fisheries data. Mallow's *Cp* expresses variance and bias, and is useful as a criterion for discriminating between models (SAS Institute 1988; Myers 1990). If two or more independent variables were selected for a model, standardised regression coefficients were calculated to compare the relative importance of the regressors in describing dependent variables (Steel & Torrie 1980). The Statistical Analysis System (SAS Institute 1988) PC version was used for data analyses. Significance was set at $P < 0.05$ for all statistical tests.

17.3   Results

17.3.1   *Relationships between fish stock characteristics and hydrological and climatic characteristics*

Fish assemblages in UYRB rivers, when considered collectively as a single fish stock, apparently have functional connections to the hydrological dynamics of this integrated floodplain–river ecosystem. Evidence of these connections was revealed from the collective (all species combined) annual catch rates (CPUE: kg net-night^{-1}) tracking overall annual flow regimes of the system (Fig. 17.2).

At a species-specific level of resolution, however, it was difficult to discern relationships describing individual catostomid and ictalurid CPUE using hydrological and climatic data collected for this study. Indeed, the only significant CPUE model that could be generated in this regard was for smallmouth buffalo. In this regard, smallmouth buffalo CPUE was positively related to both the previous year's maximum relative water stage during the seasonal low flow period (SXL1), and the total flood metre-days occurring two years prior to the catch (FMD2) ($r^2 = 0.99$; Table 17.1). Standardised partial regression coefficients (b') suggest that SXL1 was more than twice as important as FMD2 in describing CPUE of smallmouth buffalo.

Channel catfish PSD was negatively related to mean daily minimum temperature ($r^2 = 0.99$, $P = 0.001$), suggesting that years with warmer average temperatures were conducive to recruitment of proportionally more smaller channel catfish into the stock compared to years with cooler average temperatures. No significant stock density (PSD) models could be generated for the other four principal target fish species.

Catch composition (% by weight of catch) of flathead catfish was negatively related to the standard error of mean relative water stage during the low flow period (SEL) ($r^2 = 0.84$, $P = 0.027$). Higher flood metre-days occurring 2 years before the catch (FMD2) were positively related to the proportion of the overall catch attributable to bigmouth buffalo ($r^2 = 0.85$, $P = 0.027$). The number of flood metre-days during the previous year was positively related to the proportion of the overall catch contributed by channel catfish ($r^2 = 0.78$, $P = 0.012$). No significant catch composition models could be generated for blue catfish and smallmouth buffalo.

17.3.2   *Relationships between agriculture characteristics and climatic and hydrological characteristics*

i. Long-term perspective (1964–1994)

Cotton production (yield expressed as t yr^{-1}) in the UYRB tended to be more stable across years than soybean production (Fig. 17.3). Models generated for agricultural enterprise indicated considerable unexplained variability for area planted in cotton (COTAP), and for soybean yield (SOYYD) during 1964–1994 in the UYRB; $r^2 = 0.42$ and 0.67, respectively (Table 17.2). However, both models, and all the

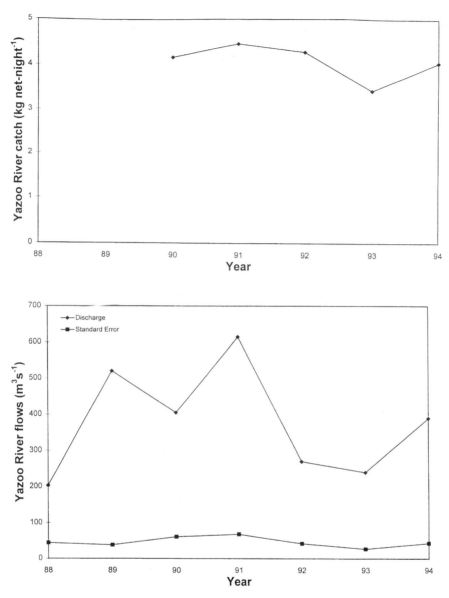

Figure 17.2   Collective fish catches (all fish species combined; 1990–1994) and flows of the Yazoo River system, Mississippi, USA (1988–1994)

variables within these two models, were highly significant ($P < 0.01$). Large values for last spring frost date (LSFD) tended to be the most significant factor related to reduced area planted in cotton. Other climatic variables, as well as variables describing hydrological regimes of the rivers, did not influence area planted for cotton, nor did they significantly influence cotton yield.

The area planted in soybeans (SOYAP) was apparently not influenced by the climatic and hydrological regime variables addressed by this study, because no

Table 17.1 Multiple regression models describing relative abundance (CPUE: catch per unit of effort, kg hoopnet-night^{-1}), proportional stock density (PSD: *sensu* Gabelhouse 1984), and catch composition (COMP: per cent by weight) of five principal fish species captured from the upper Yazoo River basin, Mississippi (USA) during 1990–1994

		Standardised partial regression coefficient			
Models[a]	N (yr)	b'_1	b'_2	r^2	P
CPUE-SM = −1.088 + 0.020 SXL1 + 0.0002 FMD2	5	1.23	0.51	0.99	0.015
PSD-CF = 449.631 − 37.020 TMIN	5	–	–	0.99	0.001
COMP-CF = 10.463 + 0.014 FMD1	5	–	–	0.78	0.048
COMP-FH = 57.181 − 11.859 SEL	5	–	–	0.84	0.027
COMP-BM = 6.250 + 0.014 FMD2	5	–	–	0.85	0.027

BF = blue catfish; CF = channel catfish; FH = flathead catfish; BM = bigmouth buffalo; SM = smallmouth buffalo. SXL = average monthly maximum water stage during low flow period (1 July–31 October); SXL1 = previous year's average monthly maximum water stage during low flow period; FMD1 = previous year's flood-metre days; FMD2 = flood-metre days 2 years prior to sample; TMIN = average minimum temperature (C); SEL = standard error of monthly mean water stage during low flow period.
[a]No models could be developed for blue catfish.

significant models could be generated using these data. However, soybean yield (SOYYD) was negatively related to degree-days (DD), and positively related to annual area flooded two years before harvest (AF2). Standardised partial regression coefficients indicated that AF2 was 0.68 times as important as DD in describing soybean yield.

From the long-term perspective, no significant models could be generated by the Stepwise Procedure for per cent planted area harvested for either crop. This result suggests that there is significant variability in factors influencing successful agricultural enterprise in the UYRB after crops are planted. Furthermore, the models generated suggested that climate and riverine hydrology have minimum influence on long-term agricultural enterprise descriptors in the UYRB.

ii. Short-term perspective (1990–1994)

The only significant multiple regression models developed, relating agricultural enterprise (1990–1994) in the UYRB with hydrological and climatic regimes, were for COTAP and SOYPH. During the 5-year period, COTAP was negatively related to annual mean precipitation ($r^2 = 0.86$, $P = 0.024$) (Table 17.2). Annual mean precipitation was positively, and highly, correlated with the maximum and mean area flooded ($r = 0.98$ and 0.97, respectively).

Flood-metre days during the previous year (FMD1), and the average maximum air temperature (TMAX) of the current year, accounted for 99.9% of the variance in

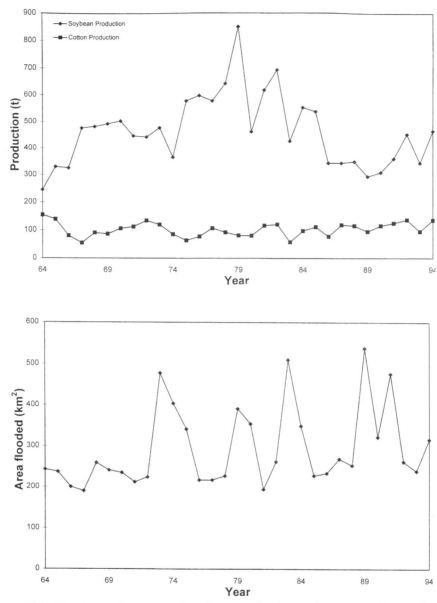

Figure 17.3 Total annual cotton and soybean production, and corresponding total annual area flooded in the upper Yazoo River basin, Mississippi, USA (1964–1994)

SOYPH. SOYPH was negatively related to FMD1, and positively related to TMAX ($P = 0.001$). Standardised partial regression coefficients suggested that FMD1 is approximately 3.6 times as important as TMAX in estimating SOYPH. However, during the 1990–1994 period, no significant ($P < 0.05$) model could be generated to relate any of the climatic and riverine hydrological regime variables to yield or total production of either crop.

Table 17.2 Multiple regression models describing agriculture enterprises for cotton and soybeans in the upper Yazoo River basin, Mississippi (USA) for the periods 1964–1994 and 1990–1994

Models[c]		Standardised partial regression coefficient			
	N (yr)	b'_1	b'_2	r^2	P
1964–1994					
COTAP = 3131.212 − 16.548 LSFD	31			0.42	0.009
SOYYD = 42.670 − 0.243 DD	31	0.974	0.662	0.67	0.001
+ 0.156 AF2					
1990–1994					
COTAP = 2057.527 − 2.408 PRE	5			0.86	0.024
SOYPH = 0.429	5	1.047	0.293	0.99	0.001
− 0.00017 FMD1 + 0.0256 TMAX					

COTAP = annual area planted to cotton (km^2); SOYYD = annual yield of soybeans (tons/km^2); SOYPH = per cent area planted in soybeans eventually harvested; LSFD[a] = last spring frost date; DD[b] = annual degree days; AF2 = area flooded 2 years prior to sample; PRE = annual precipitation (cm); FMD1 = previous year's flood-metre days; TMAX = average maximum air temperature (°C).
[a]Last spring frost date: number of days from the beginning of the calendar year until last spring minimum temperature (0°C) occurred.
[b]Degree days: one degree day is accumulated for each degree that the daily mean temperature is above 18.3°C.
[c]No significant models could be generated for cotton % planted area harvested, yield (t km^{-2}) and annual production (*t*); and for soybean area planted and annual production.

17.4 Discussion

The dynamics of fish populations in rivers and their associated fisheries tend to exhibit linkages to flow regimes (Jackson & Davies 1988a, 1988b; Jackson 1991) and climatic/weather characteristics (Cushing 1982; Jackson & Davies 1988b). These relationships can be particularly strong with respect to floodplain river ecosystems (Welcomme 1976, 1985, 1986; Goulding 1980; Junk *et al.* 1989; Jackson 1993). In addition to utilising main channel habitats, fishes exploit the spatially complex floodplain during seasonal inundation for spawning and nursery habitat, and for refuge and feeding (Risotto & Turner 1985; Bayley 1989; Ward & Stanford 1989). Human exploitation of fisheries resources in these systems is influenced by fish production processes occurring prior to exploitation and to river conditions extant during exploitation (Holčik & Bastl 1977).

Characteristics of riparian vegetation coupled with the dynamics of overbank flooding ultimately determine much of the productive potential for fisheries in floodplain river ecosystems through input of allochthonous organic materials, as well as nutrients of terrestrial origin (Vannote *et al.* 1980; Junk *et al.* 1989). Processing of allochthonous detritus as well as autochthonous production are stimulated by flooding and become the principal energetic foundations supporting the fish populations (Bayley 1989, 1995; Thorp & Delong 1994; Sparks 1995). Flooding can also introduce snags (large woody debris), which in turn provide important instream attachment substrates for invertebrates (Benke *et al.* 1985) as well as habitat for stream fishes (Gorman & Karr 1978; Benke *et al.* 1985; Insaurralde 1992).

Integrating fisheries conservation and management with other socio-economic uses of the respective river–floodplain ecosystem's resources (e.g. forestry, agriculture, urban/industrial development) requires not only an assessment of regional human culture and value systems (e.g. Jackson 1991; Brown *et al.* 1996), but also rigorous assessment of compatibility when potentially conflicting uses occur. In the UYRB, establishing priorities regarding natural resources management and agriculture is the foundation for tremendous efforts, as well as challenges, in the bio-political arena (Jackson & Jackson 1989; Jackson *et al.* 1993). Flood control to protect agricultural enterprise is a central issue. In this last regard, row crop production of cotton and soybeans is the critical concern because these two crops dominate agricultural operations throughout the UYRB (Jackson *et al.* 1993).

This study addressed both the dynamics of riverine fish stocks and those of row crop agriculture (cotton and soybeans) throughout the UYRB. Because flooding is a principal concern to both dimensions of the study, and because climatic regimes were hypothesised to influence agricultural enterprise as well as fisheries dynamics, climate and hydrology were established as principal categories for independent variables, while response (dependent) variables were established as characterisations of regional fisheries and agriculture.

17.4.1   *Fish stock dynamics*

The UYRB is an integrated, temperate, floodplain–river ecosystem. Stock assessment of riverine fisheries resources in the region have documented relatively abundant, well-structured fish populations conducive for exploitation by recreational, subsistence and artisanal (small-scale commercial) fisheries (Jackson *et al.* 1995; Ye 1996; Cloutman 1997; Cloutman *et al.* 1998). The US Army Corps of Engineers (1991) determined that the abundance of exploitable fishes in the principal rivers within the UYRB was similar to that of the lower White River, Arkansas (USA), a system considered to have an excellent fishery (Baker *et al.* 1989).

This study indicated that from a fishery resource perspective (i.e. more fish and larger fish are desirable), channel catfish, bigmouth buffalo and smallmouth buffalo exhibited positive relationships between flooding and fish stock characteristics. Temporal lag periods (years) between flooding and stock responses reflected time requirements for recruitment processes of these three species. Flathead catfish catches, however, tended to be associated more with variation in flow than to magnitude, with proportionally more flathead catfish captured under stable low flow conditions than under more variable, high flow regimes. Stable low flow conditions are likely beneficial to large predatory fishes (e.g. flathead catfish) seeking prey in these rivers. Blue catfish stock dynamics tended to be refractory to the influences of hydrological regime at an annual level of resolution.

In lightly exploited river fisheries, such as the UYRB (Brown *et al.* 1996; Cloutman & Jackson 1996; Cloutman 1997; Cloutman *et al.* 1998), abundance and structure of the adult (i.e. exploitable) component of fish stocks can be influenced more by characteristics of seasonal low flow regimes in the system than by annual flood regimes (Jackson *et al.* 1995). For example, in this study, relative abundance (CPUE) of smallmouth buffalo was directly and positively related to the previous year's water stage (% bankfull) during the summer low flow season. Although reduced stream flow during the seasonal low flow period may be beneficial to piscivores (as mentioned above regarding flathead catfish), higher stream flow during this season would probably increase survival of fishes vulnerable as prey. With regard to smallmouth buffalo CPUE, maximum water stage during the low flow period was a more important determinant than was flooding. Consequently, it is likely that future fisheries conservation concerns for rivers in the UYRB will focus on instream flow issues as much, if not more, than on flood issues.

With the exception of channel catfish, climatic variables had little influence on fish stock dynamics in the UYRB during this study. For channel catfish, however, temperature was the primary factor related to stock structure. A single variable, the average minimum temperature, explained 99% of the variation in PSD for this species. Rutherford *et al.* (1995) reported that growth increments of channel catfish in the lower Mississippi River were positively related only to length of the growing season (number of days > 15°C) and attributed this to favourable production of fish food items (primarily invertebrates) during extended warm environmental conditions.

17.4.2   *Row crop agriculture dynamics*

Mississippi is located within North America's humid subtropical climate region, with temperate winters and long hot summers. A high atmospheric pressure phenomenon known as the 'Bermuda High' typically occurs during summer and strongly influences precipitation in the region during this season. Atmospheric circulation associated with the Bermuda High produces subsiding air over the region, creating strong inversions, and stability, and consequently periods of drought (Pote & Wax 1986). Most flooding in Mississippi's UYRB occurs during winter and early spring, but flood events can occur during any month of the year.

During the period 1964–1994, most flooding in the UYRB occurred prior to the spring planting season. The current analyses indicated that, during this 31-year period, no factor associated with flood events adversely influenced production of cotton and soybeans. However, with regard to soybeans, the amount of area flooded two years prior to a crop was positively related to soybean yield. From a long-term perspective therefore, the data suggest that flooding may benefit agricultural enterprises associated with soybean production.

Shorter temporal resolution tended to reflect different patterns. During the 5-year period corresponding to the riverine fish stock assessments in the UYRB (1990–1994), high precipitation, coupled with flood events, was negatively related to area planted in cotton and the per cent of the area planted in soybeans that was eventually harvested. Subsequently, it is easy to understand public perception that flooding adversely impacts agricultural practices in the UYRB. However, low correlation coefficient values for the models generated indicated that flooding during this 5-year period did not significantly affect the overall yield of cotton and soybeans in the region. Conversely, cotton yield was positively correlated with maximum area flooded during the same year. This relationship could be attributed to elevated soil moisture (a by-product of flooding) during these years.

Drought is a common feature of Mississippi's climate; especially during the growing season (Pote & Wax 1986). Water deficits may reduce the growth of cotton, decrease the total number of potential fruiting points, exert negative effects on fruiting and flowering, and consequently result in less cotton production (Jordan 1986). Subsequently, enhanced soil moisture associated with flooding can increase yield (as revealed by the models). At a regional level of resolution, the UYRB had relatively stable total cotton crop production, even when the area planted to cotton was reduced due to flooding. From a longer temporal perspective (31 years of data) late spring frost (not flooding) was the only significant ($P = 0.009$) factor which caused reduction in area planted in cotton in the UYRB, while 58% of the variance remained unexplained by the model.

With regard to soybeans, the only long-term climatic relationship revealed was a negative correlation between soybean yield (t km^{-2}) and degree-days. However, high accumulative temperature, in conjunction with inadequate water, can be disastrous to soybeans during growth and reproductive stages (Abara 1986). Under conditions of high temperature and low rainfall, evapo-transpiration is accelerated. Excessive evapo-transpiration can deplete soil moisture, thereby inducing severe plant stress; ultimately resulting in yield reduction (Critchfield 1983). Flooding may not only contribute to alluvial deposits on crop lands, but can also increase soil moisture by recharging aquifers, which, from a long-term perspective, can ultimately lead to enhanced soybean yield. Soybean yield was positively correlated with area flooded two years preceding the year of harvest. However, a low correlation coefficient value for this model indicated considerable unexplained variability. Numerous factors other than climate and hydrological regimes (e.g. soil type and condition, cultivars, cultivation practices) affect agricultural enterprise interactively, and can sometimes result in plants responding in unexpected ways (Hodges *et al.* 1993).

17.4.3   *Integrated fisheries and agricultural perspective*

In the UYRB, abiotic factors associated with enhanced row crop agricultural production and yield were also associated with enhanced fish catches from the rivers. The collective relative abundance (CPUE) of the five fish species addressed tended to be positively correlated with yield of cotton and soybeans ($r = 0.90$; $r = 0.49$, respectively) during the five years of simultaneous collection of row crop and fish data. These relationships suggest that fisheries and row crop agronomic enterprises in the UYRB are not necessarily mutually exclusive.

Climatic influences, which are beyond human control, apparently have at least as much influence on agriculture as do hydrological influences from rivers in the UYRB. Since regional, and particularly riverine, hydrology provides opportunity for human control of fluvial processes, anthropogenic manipulation of the rivers in the UYRB has received considerable attention, often to the detriment of the fisheries (US Fish and Wildlife Service 1979; Jackson & Jackson 1989; Jackson 1993).

Flood regimes in the UYRB are fundamental determinants of structural and functional dynamics of the ecosystem from a regional perspective. As natural occurrences, flood pulses are often so predictable and long-lasting that flora, fauna and human societies have adapted to take advantage of them (Sparks 1995). Subsequently, if flooding does not significantly impact (as this study demonstrated), and in the long term might benefit (as this study suggests), crop production in the UYRB, then justification for large-scale, traditional flood control projects in the region should be subject to question, particularly when that justification focuses on protection of row crop agriculture for cotton and soybeans.

Acknowledgements

This chapter was approved for publication as article WF124 of the Mississippi Forest and Wildlife Research Centre, Mississippi State University.

References

Abara O.I.C. (1986) Stochastic analysis of irrigation strategies for soybean production in the Mississippi delta. PhD dissertation, Mississippi State University, 104 pp.

Afifi A.A. & Clark V. (1984) *Computer-aided Multivariate Analysis.* Belmont, CA: Wadsworth, 458 pp.

Baker J.A., Killgore K.J., Kasul R.L. & Sanders L.B. (1989) Fisheries investigations on the lower White River, Arkansas, *Technical Report EL-89-9.* Vicksburg, MI: US Army Corps of Engineers, Waterways Experiment Station, 54 pp.

Bayley P.B. (1989) Aquatic environments in the Amazon Basin, with an analysis of carbon sources, fish production and yield. In D.P. Dodge (ed.) Proceedings of the International Large River Symposium. *Canadian Special Publication of Fisheries and Aquatic Sciences* **106**, 399–408.

Bayley P.B. (1995) Understanding large river-floodplain ecosystems. *Bioscience* **45**, 153–158.

Benke A.C., Henry R.L. III, Gillespie D.M. & Hunter R.J. (1985) Importance of snag habitat for animal production in southeastern streams. *Fisheries* **10**(5), 8–13.

Brown R.B., Toth J.F. & Jackson D.C. (1996) Sociological aspects of river fisheries in the delta region of western Mississippi, *Completion Report, Federal Aid project F-108.* Jackson, MI: Mississippi Department of Wildlife, Fisheries and Parks, 33 pp.

Cloutman D.G. (1997) Biological and socio-economic assessment of stocking channel catfish in the Yalobusha River, Mississippi. PhD dissertation, Mississippi State University, 184 pp.

Cloutman D.G. & Jackson D.C. (1996) Biological and socio-economic assessment of supplemental stocking of catchable-size channel catfish in the Yalobusha River, *Annual Report, Federal aid Project F-111*. Jackson, MI: Mississippi Department of Wildlife, Fisheries and Parks, 48 pp.

Cloutman D.G., Chisam. C.A. & Jackson D.C. (1998) Biological and socio-economic assessment of supplemental stocking of catchable-size channel catfish in the Yalobusha River, *Annual Report, Federal Aid Project F-111*. Jackson, MI: Mississippi Department of Wildlife, Fisheries and Parks, 49 pp.

Critchfield H.J. (1983) *General Climatology*. Englewood Cliffs, NJ: Prentice-Hall, 453 pp.

Cushing D.H. (1982) *Climate and Fisheries*. New York: Academic Press, 373 pp.

Gabelhouse D.W. Jr (1984) A length-categorization system to assess fish stocks. *North American Journal of Fisheries Management* **4**, 273–285.

Gorman G.T. & Karr J.R. (1978) Habitat structure and stream fish communities. *Ecology* **59**, 507–515.

Goulding M. (1980) *The Fishes and the Forest*. Berkeley, CA: University of California Press, 280 pp.

Hodges H.F., Reddy K.R., McKinion J.M. & Reddy V.R. (1993) *Temperature Effects on Cotton*. Mississippi: Mississippi Agricultural and Forestry Experiment Station, 15 pp.

Holčik J. & Bastl I. (1977) Predicting fish yield in the Czechoslovakian section of the Danube River based on the hydrological regime. *Internationale Revue der Gesamten Hydrobiologii* **62**, 523–532.

Insaurralde M.S. (1992) Environmental characteristics associated with flathead catfish in four Mississippi streams. PhD dissertation, Mississippi State University, 152 pp.

Jackson D.C. (1993) Floodplain river fish stock responses to elevated hydrological regimes in unimpacted stream reaches and stream reaches impacted by clearing, dredging and snagging. *Polskie Archiwum Hydrobiologii* **40**, 77–85.

Jackson D.C. & Davies W.D. (1988a) The influence of differing flow regimes on the tailwater fishery below Jordan Dam, Alabama. *Proceedings of the Annual Conference of the Southeastern Association of Fish and Wildlife Agencies* **40** (1986), pp. 37–46.

Jackson D.C. & Davies W.D. (1988b) Environmental factors influencing summer angler effort on the Jordan Dam tailwater, Alabama. *North American Journal of Fisheries Management* **8**, 305–309.

Jackson D.C. & Jackson J.R. (1989) A glimmer of hope for stream fisheries in Mississippi. *Fisheries* **14(3)**, 4–9.

Jackson D.C., Brown A.V. & Davies W.D. (1991) Zooplankton transport and diel drift in the Jordan Dam tailwater during a minimal flow regime. *Rivers* **2**, 190–197.

Jackson D.C., Brown-Peterson N.J. & Rhine T.D. (1993) Perspectives for rivers and their fishery resources in the Upper Yazoo River Basin, Mississippi. In L.W. Hesse, C.B. Stalnaker, N.G. Benson & J.R. Zuboy (eds) *Restoration Planning for the Rivers of the Mississippi River Ecosystem, US Fish and Wildlife Service Biological Report 19*. Washington, DC: US Department of the Interior National Biological Survey, pp. 255–265

Jackson D.C., Ye Q., Stopha M.E., Rhine T.D. & Brown-Peterson N.J. (1995) Riverine fisheries resources assessments in the Upper Yazoo River Basin, *Completion Report, Federal Aid Project F-94*. Jackson, MI: Mississippi Department of Wildlife, Fisheries and Parks, 107 pp.

Jordan W.R. (1986) Water deficits and reproduction. In J.R. Mauney & J.McD. Stewart (eds) *Cotton Physiology*. Memphis, TN: The Cotton Foundation, pp. 63–72.

Junk W.J., Bayley P.B. & Sparks R.E. (1989) The flood pulse concept in river-floodplain systems. In D.P. Dodge (ed.) Proceedings of the International Large River Symposium. *Canadian Special Publication of Fisheries and Aquatic Sciences* **106**, 110–127.

Mississippi Agricultural Statistics Service (1995) Mississippi Agricultural Statistics 1986–1995, Supplement 29. Jackson, MI: Mississippi Department of Agriculture and Commerce, 107 pp.

Myers R.H. (1990) *Classical and Modern Regression with Applications*. Boston, MA: Wadsworth, 488 pp.

Pote J.W. & Wax C.L. (1986) Climatological aspects of irrigation design criteria in Mississippi, *Technical Bulletin 138*. Mississippi: Mississippi Agricultural and Forestry Experiment Station, 16 pp.

Risotto S.P. & Turner R.E. (1985) Annual fluctuation in abundance of the commercial fisheries of the Mississippi River and tributaries. *North American Journal of Fisheries Management* **5**, 557–574.

Robins C.R., Bailey R.M., Bond C.E., Brooker J.R., Lochner E.A., Lea R.N. & Scott W.B. (1991) *A List of Common and Scientific Names of Fishes from the United States and Canada*, fifth edition. Bethesda, MA: American Fisheries Society, Special Publication 20, 183 pp.

Rutherford D.A., Kelso W.E., Bryan C.F. & Constant G.C. (1995) Influence of physicochemical characteristics on annual growth increments of four fishes from the lower Mississippi River. *Transactions of the American Fisheries Society* **124**, 687–697.

SAS Institute, Inc. (1988) *SAS/STAT User's Guide*. Cary, NC: SAS Institute, 1686 pp.

Smith F.E. (1954) *The Yazoo River*. New York: Rinehart, 361 pp.

Sparks R.E. (1995) Need for ecosystem management of large rivers and their floodplains. *Bioscience* **45**, 168–182.

Steel R.G.D. & Torrie J.H. (1980) *Principles and Procedures of Statistics*, second edition. New York: McGraw-Hill, 633 pp.

Thorp J.H. & Delong M.D. (1994) The riverine productivity model: an heuristsic view of carbon sources and organic processing in large river ecosystems. *Oikos* **70**, 305–308.

US Army Corps of Engineers (1991) *Supplement 1 to the Final Environmental Impact Statement on the Operation and Maintenance (of) Arkabutla Lake, Enid Lake, Grenada Lake, and Sardis Lake, Mississippi*. Vicksburg, MI: US Army Corps of Engineers, Vicksburg District, 123 pp.

US Fish & Wildlife Service (1979) *The Yazoo Basin: An Environmental Overview*. Jackson, MI: Jackson Area Office, Fish and Wildlife Service Planning Aid Report to the US Army Corps of Engineers, 39 pp.

Vannote R.L., Minshall G.M., Cummins K.W., Sedell J.R. & Cushing C.E. (1980) The river continuum concept. *Canadian Journal of Fisheries and Aquatic Sciences* **37**, 130–137.

Ward J.V. & Stanford J.A. (1989) Riverine ecosystems: the influence of man on catchment dynamics and fish ecology. In D.P. Dodge (ed.) Proceedings of the International Large River Symposium. *Canadian Special Publication of Fisheries and Aquatic Sciences* **106**, 56–64.

Welcomme R.L. (1976) Some general and theoretical considerations on the fish yield of African rivers. *Journal of Fish Biology* **8**, 351–364.

Welcomme R.L. (1985) River fisheries. *FAO Fisheries Technical Paper* **262**. Rome: FAO, 330 pp.

Welcomme R.L. (1986) The effects of the Sahelian drought on the fishery of the central delta of the Niger River. *Aquaculture and Fisheries Management* **17**, 147–154.

Ye Q. (1996) Riverine fish stock and regional agronomic responses to hydrologic and climatic regimes in the upper Yazoo River basin. PhD dissertation, Mississippi State University, 99 pp.

Section V
Rehabilitation of river fisheries

Chapter 18
An assessment of anthropogenic activities on and rehabilitation of river fisheries: current state and future direction

M.C. LUCAS

Department of Biological Sciences, University of Durham, South Road, Durham DH1 3LE, UK
(e-mail: m.c.lucas@durham.ac.uk)

G. MARMULLA

FAO, Inland Water Resources and Aquaculture Service, Viale delle Terme di Caracalla, 00100 Rome, Italy

Abstract

River fish communities provide an important resource, as well as being elements of lotic ecosystems. Damage to river fisheries has been caused by a wide range of problems including overexploitation, pollution, hydraulic engineering, changes in land use, introduced species, and increased predation. However, the importance of these factors varies greatly on a scale between catchments and continents. The time-scale of damage by these factors is also variable between catchments. Damage to river fisheries is often the result of several factors and tends to necessitate an integrated approach for successful rehabilitation measures. Identification of principle factors causing damage, and appropriate solutions, is therefore catchment specific. Generally, with appropriate resources, fisheries scientists and managers are now in a position to identify the principle causes of damage to fisheries in any river system.

In the last decades, a great variety of technologies have been developed to enable rehabilitation of river fisheries, although there is a need to maintain and develop innovation in the variety of approaches used, and ensure better cost-benefit evaluation of rehabilitation methods through case studies. The greatest challenge for fishery experts is to expand their understanding of socio-economic applications to fisheries, and to use this as a basis for arguing the case for prevention/minimisation of damage in new schemes on rivers, and for appropriate rehabilitation works and monitoring. To do this, training of fisheries specialists in this field must be expanded, and the dialogue with politicians, planners and the public must be improved.

Keywords: Pollution, overfishing, river regulation, impoundment, rehabilitation, connectivity, species introduction, predation, migration, sustainable fisheries, cost-benefit.

18.1   Introduction

Globally, fish in river catchments provide a major resource for humans, especially where dense human populations are living close to rivers, both in developed and developing countries. According to the World Conservation Monitoring Centre

(Groombridge and Jenkins 1998), the most stressed catchments are to be found in South Asia, the Middle East as well as in western and north-central Europe, whereas the least stressed are in the north-western part of North America. Worldwide, among 151 river basins, 30 have been identified as supporting high aquatic biodiversity on the basis of fish family richness and vulnerability to future pressures (Table 18.1). Within floodplain and inland regions of developing countries, subsistence fisheries often provide a large component of dietary protein (Ruddle 1982). For example, Mali only produces freshwater fish, almost exclusively through capture fisheries. In Mali, the mean annual freshwater capture fish production was around 63 000 t for the period 1984–1994 and reached 133 000 t, 111 910 t and 99 550 in 1995, 1996 (FAO 1998) and 1997 (FAO 1999a), respectively. Table 18.2 provides insight into fish consumption data (FAO 1998) of some selected developing countries in which almost the whole fish production, or at least a substantial proportion, comes or is thought to come mainly from inland capture fisheries in rivers or floodplains. Commercial river fisheries are common in developed areas, especially within the

Table 18.1   Thirty high priority river basins (modified from Groombridge and Jenkins 1998). These are the 30 river basins that support high biodiversity and are most vulnerable to future pressures

Asia	Africa	South America
Ca	Gambia	Magdalena
Cauvery	Niger	Paranà
Chao Phraya	Nile	Parnaiba
Ganges-Brahmaputra	Senegal	Sao Francisco
Godavari	Volta	Uruguay
Indus		
Irrawaddy		
Krishna		
Ma		
Mahanadi		
Mekong		
Narmada		
Pahang		
Penner		
Perak		
Salween		
Sittang		
Song Hong (Red)		
Tapti		
Tembesi-Hari		

Table 18.2 Food balance of fish and fishery products in live weight and fish contribution to protein supply of some selected developing countries (partly extracted and modified from FAO 1998). Almost the whole fish production of these countries, or at least a substantial proportion, comes or is thought to come mainly from inland capture fisheries in rivers or floodplains

Country	Year	Inland capture fish production (t)	Total fish production (t)	Inland capture fisheries/ total fish production (%)	Non-food uses (t)	Imports (t)	Exports (t)	Total food-fish supply (t)	Population (thousands)	Per caput supply (total food-fish supply) (kg)	Per caput supply (country's inland capture fisheries only)[a] (kg)	Total food-fish[b]/ animal proteins (%)
AFRICA												
Burkina Faso	1995	8000	8000	100.0		6192	3	14189	10479	1.4	0.8	6.2
Central African Republic	1995	12900	13300	97.0		500		13800	3273	4.2	3.9	9.5
Chad	1995	90000	90000	100.0	49600	669		41069	6335	6.5	6.4	20.6
Mali	1995	132900	133000	99.9		1934	1000	133934	10795	12.4	12.3	21.9
Sudan	1995	40000	45000	88.9		1706	100	46606	26707	1.8	1.5	2.2
ASIA												
Cambodia	1995	72499	112510	64.4			22816	89694	10024	9.0	7.2	35.3
Laos	1995	25850	40250	64.2		4545		44795	4882	9.2	5.3	37.7

[a]Calculated for the purpose of this chapter.
[b]Calculated from total food-fish supply.

northern hemisphere, but also in other regions such as eastern South America (Novoa & Ramos 1990). Increasingly, recreational fisheries are becoming important both within developed and developing countries (FAO 1999b), especially as the amount of leisure time grows (Cornelly & Brown 1991). Fish communities are, of course, also key components in almost all freshwater ecosystems, representing the majority of biomass at upper trophic levels. Moreover, fish play an important role in nutrient and energy transfer to upland systems (Krine *et al.* 1997), and perhaps also for nutrient flux in lowland systems (Lucas *et al.* 1998). River fish also provide a key food source for many terrestrial or semi-aquatic predators such as otter *Lutra lutra* (L.) and osprey *Pandion halieatus* (L.).

Productive, historically sustained fisheries have existed on many rivers. In a wide variety of cases, decline or complete loss of these fisheries has been symptomatic of damage to the fish stocks. In some cases such declines can directly be attributed to overfishing, for example sturgeon (Acipenseridae) in the Amur River (Krykhtin & Svirskii 1997), but often declines in historical fisheries have been indicators of more general damage to the river ecosystem. In many circumstances, failure of commercial or recreational fisheries has been the first warning signal of progressive ecosystem damage, usually caused by humans; for example, the decline and failure of salmon fisheries on the Rhine (de Groot 1991). Many rivers in the developed world reached their worst state of damage from the early to mid-twentieth century, while significant anthropogenic effects on previously pristine rivers in the developing world are only now being detected. In the meantime there has been substantial progress made in the science of rehabilitation of river catchment environments, and in several developed countries the last few decades have seen the political and social changes necessary to bring about rehabilitation of previously damaged rivers, albeit with varying degrees of success.

The FAO Code of Conduct for responsible Fisheries (FAO 1995), and the related technical guidelines in support of the implementation of the Code, e.g. *FAO Technical Guidelines for Responsible Fisheries No. 6: Inland Fisheries* (FAO 1997), provide the agreed basis for the promotion of long-term conservation and sustainable use of fisheries resources. The provisions of the Code comprise, *inter alia*, the introduction of measures for depleted resources and those resources threatened with depletion, and set out that every effort be made to ensure that resources and habitats critical to the well-being of such resources which have been adversely affected by fishing or other human activities are restored. Conservation and management measures, amongst which also fall rehabilitation measures for inland fisheries, are explicitly mentioned and encouraged.

Although the term 'restoration' is often applied to such practices (e.g. 'River restoration: the physical dimension', *Aquatic Conservation: Marine Freshwater Ecosystems* special issue, vol. 8, pp. 1–264), the term 'rehabilitation' is preferred, since most projects aid the recovery of river systems, rather than restore them to their pristine state (Boon 1998). Bradshaw (1996) gave an account of definitions of various of the relevant terms. In this paper the main causes of damage to river fisheries, and the rehabilitation measures available to be employed are summarised (reviewed in greater detail by Cowx 1994; Cowx & Welcomme 1998). The success of these approaches is evaluated, and the areas and issues where progress is still to be made are examined.

18.2   Damage to river fish stocks

The causes of damage to river fisheries are many and varied. The most important sources of damage vary in time and in space; for example, deoxygenation typically occurs in the lower reaches of the river and during the warmest season. In many cases, factors causing damage are inter-related and have feedback effects. Figure 18.1 illustrates the range of factors having impacts on river fisheries, and some of their interactions. Several of the major factors are discussed below.

18.2.2   *Overfishing*

The effects of overfishing are most commonly observed for species which are of high economic value or vulnerable, especially – but not exclusively – long-lived species which are slow to mature and recruit. This is particularly the case for fish such as sturgeon, many of the 26 species of which are threatened or endangered. In Central and Eastern Europe and Asia, the high value of caviar, has provided a source of good income, set against the recent economic upheavals in many of these countries, and has, in addition to the permitted exploitation, resulted in excessive illegal fishing (e.g. Krykhtin & Svirskii 1997). Migratory fish using the main river channel are highly susceptible to fishing gears, and can be intercepted at known times of the year during migration. Overfishing has been a significant factor in the historical demise of many migratory salmonid stocks, especially those with constrained run timings. Brett (1986) elegantly expressed the susceptibility of diadromous stocks to overfishing:

'As a fishery, salmon are ideal. They comb the ocean for its abundant food, convert this to delectable flesh, and return regularly in hordes to funnel through a limited number of river mouths exposing themselves to the simplest method of capture – a gill net or seine net.'

However, even the causes of collapse of famous fisheries such as the Fraser River sockeye salmon *Oncorhynchus nerka* (Walbaum) remain unclear, despite extensive research and debate (Smith 1994). It was initially felt that reduced catches in the early 1900s could be attributed to the increased catches, combined with the final insult of the 1917 obstruction of the now notorious Hell's Gate by the actions of the Canadian Northern Railway construction efforts, severely impeding upstream migration. In contrast, Thompson later laid the blame on the declines resulting from piecemeal damage to the Fraser system, from the damming of Lake Quesnel in 1899, to a series of landslips resulting from construction activities by the Canadian Northern Railway, of which the Hell's Gate landslips were the most famous. Ricker (1954) felt that while obstruction at Hell's Gate might be a factor, the effects of

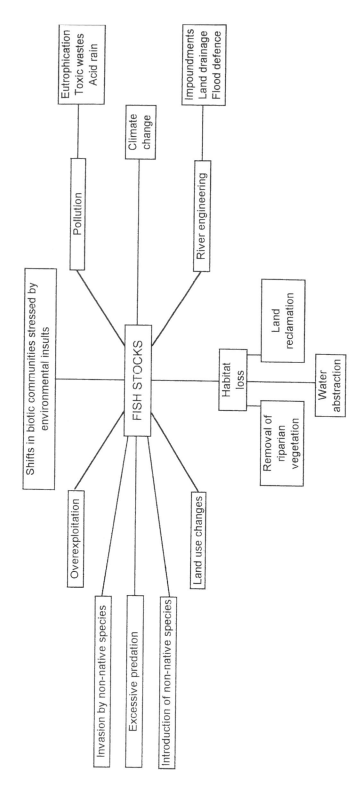

Figure 18.1 Factors causing damage to riverine fish stocks (modified from Cowx 1994)

overfishing were complicated and underestimated, and went on to demonstrate the importance of spawner–recruit relationships in determining future stocks. Recent studies, while acknowledging the intense fishery, have examined the physiological stress of migration past Hell's Gate and other obstructions (Hinch *et al.* 1996), and changes in the Pacific climate on productivity of sockeye salmon (Beamish *et al.* 1997) as factors influencing the performance of the stocks. Attempts to restore Fraser River salmon stocks to levels approaching those that occurred in the early twentieth century have incorporated a range of commercial and recreational fishery limits, spawning and nursery habitat improvement, and attempts to remove obstructions to migration, especially at Hell's Gate (Smith 1994; Beamish *et al.* 1997). These have met with reasonable success, although recent suggestions of sustainably doubling the Fraser River sockeye salmon spawning stock have been regarded as unlikely to succeed and of dubious cost–benefit value (Henderson & Healey 1993).

18.2.2 *Pollution*

Many rivers in developed countries have suffered the effects of a variety of kinds of chronic and acute pollution. Many rivers in industrialised areas suffered from the effects of oxygen-demanding industrial wastes and sewage and conservative pollutants such as heavy metals (e.g. Philippart *et al.* 1988). These rivers and their fisheries became degraded, associated particularly with the loss of diadromous species requiring rich oxygen, such as Atlantic salmon *Salmo salar* L. and shads, e.g. *Alosa alosa* (L.). Catchments such as the Rhine, Meuse and Thames almost completely lost their runs of these species. Political initiatives and technological advances since the mid-twentieth century have enabled great reductions in the pollutant loadings in such rivers and have improved water quality to such an extent that salmon rehabilitation programmes are underway in these and similar rivers (e.g. Hughes & Willis, Chapter 5).

Chronic, low-level input of pollutants, often from a wide area probably represents an at least equally challenging problem for the long-term sustainability of fisheries as acute, gross pollution, in that detecting and managing the sources of such pollution is a much greater task. Recently, there has been great concern over the long-term health of lowland river fish populations due to the presence of oestrogen mimicking chemicals (Purdom *et al.* 1994) and Jobling *et al.* (1998) recently showed that more than 60% of male roach *Rutilus rutilus* (L.) from English lowland rivers were hermaphrodites.

However, a much greater problem for lowland river fish stocks is that of eutrophication. The accelerated enrichment of rivers due to excessive inputs of nutrients, principally nitrogen and phosphorus due to man's actions, has damaged the ecology of many European and American lowland rivers. Major point source input of nutrients from sewage, as well as increased atmospheric deposition and intensive agriculture, have stimulated major changes in lowland river ecology: increased algal productivity and incidence of algal blooms, especially of Cyanobacteria, high turbidity, die-off of submerged macrophytes, reduced habitat availability and complexity, high

BOD and oxygen depletion during decomposition of algae, resulting in substantial fish kills (Harper 1992). The fish stocks in these waters have become limited to a few species characteristic of slow-flowing, eutrophic water, cyprinids, e.g. bream *Abramis brama* (L.) and roach. High nutrient concentrations in, and related low oxygen contents of, the interstitial water can constitute bottlenecks in restoration of salmonids after successful structural river rehabilitation. Recently, there has been a great deal of research in the use of biomanipulation in lakes, principally through removal of planktivorous fish to reduce predation pressure on algae-eating zooplankton, and benthivorous fish to limit release of phosphorus from sediments, in order to aid the reversal of gross eutrophication in fresh waters such as the Norfolk Broads and Dutch lakes, e.g. Lake Wolderwijd (Grimm & Backx 1994). However, since rivers have a better flushing ability in comparison to lakes, fewer active attempts have been made to abate river eutrophication, although legislative measures have been taken to reduce nutrient inputs.

Not all rehabilitation work involving attempts at shifting trophic status have sought to reduce nutrient levels – Stockner and Macisaac (1996) have deliberately added nutrients to nursery streams and lakes in British Columbia to increase productivity of juvenile sockeye salmon by 60%! In this case it was found that reduction of salmon runs resulted in nutrient limitation from a lack of the decaying carcasses of post-spawners, resulting in a lower carrying capacity for juveniles.

While eutrophication is the main chronic pollution problem for fish in lowland areas, upland stocks have suffered from surface water acidification. Acid precipitation in Europe and North America is a phenomenon which was greatly exacerbated by the Industrial Revolution, but which has mostly taken decades to result in substantial surface water acidification, particularly in catchments with base-poor geological characteristics (Howells 1990). Such effects have often been exacerbated by changes in land use such as large-scale coniferous afforestation, enhancing capture of acid particles and condensate. The changes in plant and animal communities, frequently including loss of salmonid stocks, are difficult to reverse, and are problems which are not solved quickly. The main rehabilitation technique of liming has been of variable success between studies, and is expensive. Conifer-free buffer strips along tributary streams can limit acid flush and aluminium leaching and enhance invertebrate and fish production. To this end, major changes in the forest formation (i.e. back to a higher proportion of broad-leaved trees, as was common for many parts of Europe before afforestation with conifers) of the catchment basin of the River Sieg (North Rhine-Westphalia, Germany) were claimed to be required to take place in addition to the rehabilitation of the aquatic environment in the context of the restoration of the Sieg's salmon stock (Marmulla 1992). In the long term, reduction of atmospheric emissions of sulphur- and nitrogen-containing gases, remains the most effective technique for rehabilitation of fish stocks affected by surface water acidification, in terms of cost–benefit value, but requires extensive international cooperation, long lead times, and substantial reductions (Howells 1990). Therefore, it is of concern, for freshwater fisheries in eastern Asia, that the acid deposition as a consequence of the industrial development in the region has increased dramatically, threatening salmonid stocks in, for example, Japan, but probably also much more widely.

18.2.3 *River engineering and hydrology*

The form of many river habitats, especially in large catchments, has altered greatly in the last few hundred years, largely because of engineering works designed to store water, produce electricity, prevent flooding, allow irrigation, enable navigation and a host of other uses (Welcomme 1994). This has dramatically altered the hydrology and patterns of sediment transport and deposition in such rivers. Impoundment and channelisation, frequently in the context of changes in agricultural practices, i.e. increases of intensive agriculture in the lower catchment, and deforestation in the upper catchment, are perhaps the main factors affecting both the upper and lower reaches of rivers. For lowland areas, in most cases in the developed world, the braiding of interconnecting channels over the floodplain has largely been lost. Furthermore, due to flood prevention measures the link between the river channels and the floodplain has mostly been broken, and the floodplain lakes are rarely inundated. Concerns have also been expressed about the possible effects of water transfer schemes between catchments, in affecting the ecology of receiving rivers (Davies *et al.* 1992).

A further example for the high degree of modification of river hydrology is the global total of reservoirs which is 60 000 in number, 400 000 km^2 in water surface and some 6500 km^2 in volume (Avakyan & Lakovleva 1998). It is remarkable that reservoir storage has attained about seven times the standing stock of water in rivers (Vorosmarty *et al.* 1997).

Although the effects of obstructions on the passage of migratory salmonids has received by far the greatest attention, with the development of a wide range of fish passage facilities to mitigate, and in some cases aid rehabilitation, of salmonid migrations (e.g. Beach 1984), more emphasis has recently been placed on ensuring the free movement for other fish species, including lowland river species (DVWK 1996; Cowx & Welcomme 1998; Jungwith *et al.* 1998). A positive example in this respect is the restoration of the allis shad, *Alosa alosa* (L.), population of the River Garonne, France (Larinier *et al.* 1994). The significance of engineered obstructions in limiting natural migratory behaviour of many lowland river fish species has, until recently, probably been greatly underestimated (Lucas & Batley 1996; Lucas *et al.* 1998). Undoubtedly the collection and analysis of long-term catch per unit effort data sets for fisheries on rivers such as the Nidd (Axford 1991) have been, and will continue to be, of great value in correlating probable changes in local fish stocks with events affecting the river's ecology. For example, such techniques will be important in evaluating the effects of removal of a weir, known to inhibit upstream migration of cyprinid fish on this river (Lucas & Frear 1997), in the near future. In the United States, the removal of Edwards dam in Maine is the first example for a federal government 'dismissing' an operational dam with a licence pending renewal (Anonymous 1997).

River engineering, in greatly reducing the natural seasonal inundation of floodplains and restricting natural movements of fishes, appears to be largely responsible for the declines in productivity and diversity of fish communities, and declines of fisheries, in temperate and tropical river systems (Philippart *et al.* 1988; Welcomme 1995; Cowx & Welcomme 1998). Paradoxically, attempts to control flooding in Bangladesh, in order to reduce damage to human populations, have resulted in reduced fish catches for some of the poorest people who rely on fish as a staple source of protein (Mirza & Ericksen 1996; Halls *et al.*, Chapter 14).

18.2.4   *Species interactions*

The physical and chemical changes to the river environment described above have major impacts on the fish communities present, but the fish communities, and the fisheries they support, have also been damaged by species interactions, in some cases directly by introduced or escaped exotic species, and in other cases due to changes in competitor or predator–prey relationships. Fish species such as rainbow trout *Oncorhynchus mykiss* (Walbaum) and brown trout, *Salmo trutta* L., introduced to catchments in Australia and New Zealand have probably been responsible for the decline in native species of stream-dwelling galaxiid (Tilzey 1976). In Australia, carp *Cyprinus carpio* (L.), introduced to the Murray-Darling River in the 1800s have become dominant in some areas, possibly at the expense of native species, but do not appear to have been responsible for high levels of turbidity (Brumley 1991).

Fishermen have, on occasion, been responsible for the persecution of fish-eating predators, often with poor evidence of causality between low fish abundance and predator numbers, or worse, between low catches and predator numbers. Nevertheless, recent spreads in the distribution and abundance of sawbill ducks *Mergus* spp. and cormorant *Phalacrocorax carbo* (L.) have generated genuine concern with regard to their possible effects on fish catches. Overall, breeding populations of cormorant in Britain have increased by 3% per year over the last three decades, and continue to do so, but inland-nesting cormorants have increased by 47% per year (Kirby *et al.* 1996). Cormorants especially, with their greater daily food requirement, and their ability to eat fish up to 0.5 kg, have been suggested as being responsible for substantial fish mortality and negative effects on fisheries at a number of sites throughout Europe. However, while it is clear that cormorants have the capacity to cause substantial impacts on fisheries, evidence from natural river fisheries (Suter 1995) and put-and-take stillwater fisheries (Callaghan *et al.* 1998) suggest that fish predation by cormorants is mostly a compensatory source of mortality and is therefore not likely to affect angler catches. There remains some debate about this, but it is indisputable that commercial fisheries bear a significant cost in stocking fish that are removed by cormorants and hence are unavailable to the angler.

18.3   Rehabilitation of rivers

Rehabilitation methods are almost as many and varied as the forms of damage to fisheries. Just as it is rare for a river fishery to suffer only one form of damage, it is rare for one rehabilitation method to be the best approach for achieving the greatest degree of restoration. However, prioritisation of the key cause(s) of damage and appropriate remedial action can have a dramatic effect, as demonstrated by the rapid increase in species diversity in rivers such as the lower Thames, following dramatic improvements in water quality. Generally, a planned, integrated approach of rehabilitation, tailored to the specific river's ecology, hydrology and chemistry is the most appropriate way of achieving successful rehabilitation. The range of possible rehabilitation methods is outlined in Figure 18.2. Cowx (1994) recommended the use of impact assessment methodology before and during implementation of a rehabilitation scheme, in order to predict and assess the feedback effects of the scheme on the system. To date there is limited evidence of such procedures being followed, but there is a growing number of publications recording the success or failure of river rehabilitation projects with regard to fisheries, although dominated by consideration of salmonids (e.g. Berg & Jørgensen 1994; Kelly & Bracken 1998; Scruton *et al.* 1998). However, the records, especially those demonstrating success, have to be scrutinised with care and analysed critically; there is a risk of bias as many projects need positive results to survive. Therefore, standard indicators should be agreed upon and used.

The various methods of rehabilitation of fisheries for use in rivers are well documented (e.g. Cowx & Welcomme 1998). It is worth considering some general points in relation to technical aspects of rehabilitation of rivers for fish. Until recently, water quality was regarded as perhaps the primary factor requiring attention for rehabilitation, and this is still the case in some polluted rivers. However, the consensus now in whole catchment rehabilitation is that the foremost priority is to maintain longitudinal connectivity within the system, this being essential not only for long-distance migrants such as salmon, but also for the free movement and migration of all species within the range of suitable habitats (Cowx & Welcomme 1998). Beyond this, maintenance of high water quality in the upper reaches, usually dominated by pollution-sensitive species such as salmonids, is of as high importance as maintenance of lateral connectivity in the same river stretch. Now that water quality is generally improving in many rivers, greater emphasis should also be paid to restore access to side channel and floodplain habitats in the middle and lower reaches. It is proposed that additional requirements such as maintaining sediment transport and habitat reinstatement are variable in character according to the specific areas of river being targeted.

18.4   Conclusions

In examining causes of damage to river fisheries and their rehabilitation several broad conclusions may be made:

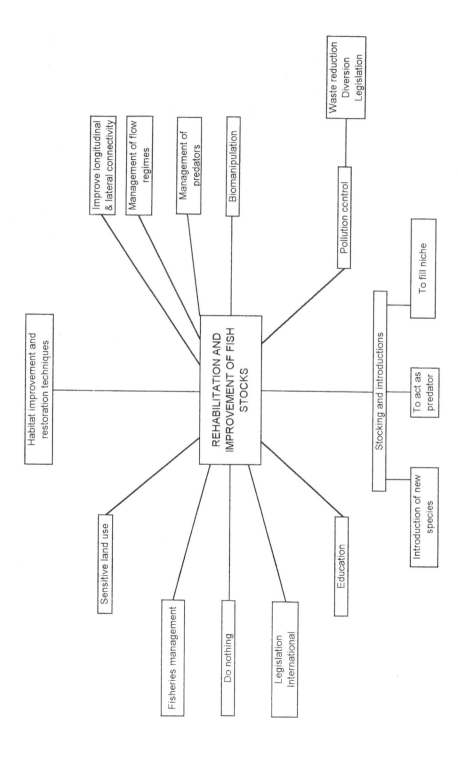

Figure 18.2 Methods appropriate for the rehabilitation of river fish stocks. Normally an appropriate rehabilitation programme will require the integration of several techniques. (Modified from Cowx 1994)

(1) It is generally possible to identify and prioritise problems impacting upon fisheries and aquatic communities, and to carry out appropriate technical solutions to these problems.
(2) Socio-economic aspects of mitigation and rehabilitation (including finance) are currently a more problematic feature than technical aspects, and present the greatest challenges to development and maintenance of aquatic ecosystems and fisheries.

It is clear that a great deal of progress has been made in understanding the causes of damaged river fisheries. Not least amongst these have been the major advances made in the ability to quantitatively or semi-quantitatively sample the fish populations of river environments. This, coupled with detailed environmental measurements and powerful computational and statistical techniques, makes it possible to correlate changes in fish abundance and diversity with environmental factors. Together with laboratory and field-based experiments demonstrating effects of specific factors on fish biology and fisheries, factors most likely to be causing damage can be identified. But before fully promoting such optimism, it is worth revisiting the Fraser River sockeye salmon fishery and noting that after nearly 100 years of study by many scientists, the primary cause of the crash in salmon stocks is still not established, and it still has not been possible to rehabilitate the fishery even close to its former status.

The limitations that fisheries experts currently suffer, in terms of their capabilities and opportunities for dealing with the socio-economic problems of maintaining and improving the state of river fisheries, remain a major problem. A low economic value is placed on fish and aquatic communities in comparison to factors such as use of water for domestic and industrial purposes, as a route for waste disposal, and for flood defence. It is, perhaps, unreasonable to expect the integrity of aquatic communities and the fisheries they may provide to be valued more highly than factors which have even greater direct effects on human life, such as flooding or drought, but improvement in the valuation of aquatic communities and fisheries in qualitative and quantitative terms is needed for the benefit for both developed and developing countries. Improvement in qualitative valuation can be achieved by dialogue with the public and politicians; for example, by demonstrating the importance of healthy fish populations for supporting populations of charismatic megafauna such as otters. In addition, positive effects could be achieved through better environmental education, which should include the allaying of public ignorance and prejudices with respect to fish (e.g. 'fish are slippery, ugly and stink') and to arouse peoples' interests in a better understanding and knowledge of 'ordinary' river fish (e.g. visitor centres and public observation chambers at fish passes; thematic exhibitions etc.).

Quantitative valuation requires the use of analytical techniques such as cost–benefit analysis, with either financial or relative units, and should be employed much more frequently as a tool by fisheries managers in seeking to highlight the potential damage which might be caused by a proposed project on a river, in suggesting less damaging alternatives, and in arguing the case for adequate funds to provide mitigation or rehabilitation of rivers for fish wherever impairment is unavoidable.

However, it has to be kept in mind that provision of even the best mitigation can never fully compensate for losses of functions or structures. This is especially true for fish passes. Therefore, the know-how of the construction of fish passes to allow migration through an obstacle must not lead to the (wrong) conclusion that multiplication of weirs and dams can be done without causing any harm (Larinier *et al.* 1994). Analytical techniques are already being used in a pro-active fashion in a few cases. For example, economic analysis of the value of coastal wetland areas in Florida for recreational and commercial fisheries has lead to increased pressure for state purchase of more coastal land to preserve it from development, to safeguard the habitat, fish stocks and fisheries (Bell 1997).

Furthermore, it is increasingly felt that boundaries for management units, including those responsible for the implementation of structural and physico-chemical improvements of riverine habitats (i.e. often the Water Authorities or equivalent, but not fisheries!) within agencies, or at higher hierarchical level, e.g. ministries (including fisheries, wildlife, forest management, and parks and recreation) should be redrawn along watershed and eco-region lines. Contrary to previous independent division management plans, the design and implementation of the new integrated management plans will require input and contributions from managers in all relevant units. These new plans will thereby become more comprehensive and ecosystem-specific, as they will require a great deal of communication, debate, cooperation and collaboration among the specialised agency personnel who were previously acting more independently (FAO 1999b).

The socio-economic objectives relating to minimising damage to river and other freshwater fisheries, and improving the opportunities for rehabilitation schemes can be summarised as follows.

(1) Modern integrated management should aim at decisions which do not put fisheries resources at unnecessary risk, as prevention of loss of potential is much more preferable to rehabilitation, for obvious reasons, including costs.

(2) There should be evaluation and differentiation between those problems which are 'perceived' by members of the community and those which are genuine, and their relative magnitude and priority should be defined.

(3) There should be more and improved cost–benefit evaluation in arguing the case, in the first instance, for more and better conservation as well as less damaging schemes, and for mitigation and rehabilitation work. There is a need to improve and standardise the evaluation methodologies appropriate to the scale of the project.

(4) Training of fisheries specialists in socio-economic assessment of aquatic resources, including fisheries, should be improved by encouraging multi-disciplinary knowledge and 'new blood'.

(5) There should be dialogue with politicians and the public in order to promote the value of fish and aquatic communities.

(6) Objectives should be set in conjunction with allied groups, taking account of political and economic realities.

(7) According to the scale of the problem, how and by what means of resource funding, the objectives can be achieved should be identified.

(8) A phased, multiple-objective, flexible approach for rehabilitation studies should be adopted. Progress and the best approach at each stage should be continually assessed. Effective communication with allies and with the public should be maintained.

(9) Post-rehabilitation assessment should be improved, to prove that the work has been effective; and if not, determine why. This should include appropriate budgeting when seeking funds, and incorporation of objective assessment methods, such as control areas and appropriate indicator groups. Technical development in this area is a priority. With respect to post-rehabilitation improvements, legal aspects also have to be taken into consideration, as does much of the know-how to deal with the results of a reassessment technically, i.e. the implementation of improvements does not necessarily mean that the required measures can be carried out.

(10) There should be effective dissemination of results, successes and failures, to colleagues, politicians and the public by use of appropriate media and measures.

Fisheries managers need to be more pro-active in their interaction with other water resources personnel (Sambrook & Cowx, Chapter 13). Planners and engineers still do not actively solicit the input of fisheries specialists at an early stage when potentially damaging schemes are being considered, or when alterations to existing schemes might allow rehabilitation work. Fisheries specialists therefore need to interact with other disciplines, and need to be enabled to do so, to be made aware of, and participate, at the carliest stage of possible projects (Sambrook & Cowx, Chapter 13). Since most of the factors causing problems for fish communities are beyond our immediate control, fisheries staff need to broaden and strengthen their cause by interacting and making alliances with other interested parties, in seeking to advance the importance of properly functioning aquatic ecosystems, to limit damage to aquatic habitats, and to promote rehabilitation.

Acknowledgements

We are grateful to all those at the HIFI *Ecology and Management of River Fisheries* meeting and elsewhere for discussion and comments which have contributed to the production of this chapter. We acknowledge with thanks the valuable and constructive comments by Dr J. Kapetsky, Senior Fishery Resources Officer, FAO.

References

Anonymous (1997) Environmentalists claim victory as FERC orders removal of US dam. *International Water Power and Dam Construction*, September 1997.

Avakyan A.B. & Lakovleva V.B. (1998) Status of global reservoirs: the position in the late twentieth century. *Lakes and Reservoirs: Research and Management* **3**, 454–52.

Axford S. (1991) Some factors affecting catches in Yorkshire rivers. In I.G. Cowx (ed.) *Catch Effort Sampling Strategies*. Oxford: Blackwell Science, pp. 143–153.

Baras E., Lambert H. & Philippart J.C. (1994) A comprehensive assessment of the failure of *Barbus barbus* spawning migration through a fish pass in the canalized River Meuse (Belgium). *Aquatic Living Resources* **7**, 181–189.

Beach M.H. (1984) Fish pass design criteria for the design and approval of fish passes and other structures to facilitate the passage of migratory fishes in rivers. *MAFF Fisheries Technical Report* **78**, 46 pp.

Beamish R.J., Neville C.E.M. & Cass A.J. (1997) Production of Fraser River sockeye salmon (*Oncorhynchus nerka*) in relation to decadal-scale changes in the climate and ocean. *Canadian Journal of Fisheries and Aquatic Sciences* **54**, 543 554.

Bell F.W. (1997) The economic valuation of saltwater marsh supporting marine recreational fisheries in the south east United States. *Ecological Economics* **21**, 243–254.

Berg S. & Jørgenson J. (1994) Stocking experiments with 0+ eels (*Anguilla anguilla* L.) in Danish streams: post-stocking movements, densities and mortality. In I.G. Cowx (ed.) *Rehabilitation of Freshwater Fisheries*. Oxford: Fishing News Books, Blackwell Science, pp. 314–325.

Boon P.J. (1998) River restoration in five dimensions. *Aquatic Conservation: Marine and Freshwater Ecosystems* **8**, 257–264.

Bradshaw A.D. (1996) Underlying principles of restoration. *Canadian Journal of Fisheries and Aquatic Sciences* **53** (Suppl. 1), 3–9.

Brett J.R. (1986) Production energetics of a population of sockeye salmon, *Oncorhynchus nerka*. *Canadian Journal of Zoology* **64**, 555–564.

Brumley A.R. (1991) Cyprinids of Australasia. In I.J. Winfield & J.S. Nelson (eds) *Cyprinid Fishes: Systematics, Biology and Exploitation*. London: Chapman & Hall, pp. 264–383.

Callaghan D.A., Kirby J.S., Bell M.C. & Spray J. (1998) Cormorant *Phalacrocorax carbo* occupancy and impact at stillwater game fisheries in England and Wales. *Bird Study* **45**, 1–17.

Cornelly N.A. & Brown T.L. (1991) Net economic value of the fresh-water recreational fisheries of New-York. *Transactions of the American Fisheries Society* **120**, 770–775.

Cowx I.G. (1994) Strategic approach to fishery rehabilitation. In I.G. Cowx (ed.) *Rehabilitation of Freshwater Fisheries*. Oxford: Fishing News Books, Blackwell Science, pp. 3–10.

Cowx, I.G. & Welcomme, R.L. (eds) (1998) *Rehabilitation of Rivers for Fish*. Oxford: Fishing News Books, Blackwell Science, 260 pp.

Davies B.R., Thoms M. & Meador M. (1992) An assessment of the ecological impacts of inter-basin transfers, and their threats to river basin integrity and conservation. *Aquatic Conservation: Marine and Freshwater Ecosystems* **2**, 325–350.

DVMK (1996) *Fischaufstieg sanlangen: Bemessung, Gestaltung, Funktionskontrolle. Deutscher Verband für Wassenwirtschaft und Kulturban e.V. (DVMK)*. DVMK – Merbletter, Heft 232/ 1996, 110 pp.

Food and Agriculture Organization of the United Nations (FAO) (1995) *Code of Conduct for Responsible Fisheries*. Rome: FAO, 41 pp.

FAO (1997) *FAO Technical Guidelines for Responsible Fisheries No. 6: Inland Fisheries*. Rome: FAO, 36 pp.

FAO (1998) Fish and fishery products: world apparent consumption statistics based on food balance sheets (1961–1995), *FAO Fisheries Circular No.* **821**, Revision 4. Rome: FAO, 253 pp.

FAO (1999a) Capture production. *Yearbook of Fishery Statistics, 1997, Vol. 84*. Rome: FAO, 715 pp (in press).

FAO (1999b) Review of the state of world fishery resources: inland capture fisheries, *FAO Fisheries Circular No.* **942**. Rome: FAO, 54 pp.

Grimm M.P. & Backx J.J.G.M. (1994) Mass removal of fish from Lake Wolderwijd, The Netherlands. Part I: Planning and strategy of a large-scale biomanipulation project. In I.G. Cowx (ed.) *Rehabilitation of Freshwater Fisheries.* Oxford: Fishing News Books, Blackwell Science, pp. 390–400.

Groombridge B. & Jenkins M. (1998) *Freshwater Biodiversity: A Preliminary Global Assessment.* Oxford: World Conservation Press, 104 pp & 14 maps.

Groot S.J. de (1991) Decline and the fall of the Rhine salmon observed in the light of a possible rehabilitation. In D. Mills (ed.) *Strategies for the Rehabilitation of Salmon Rivers.* London: Atlantic Salmon Trust, Institute of Fisheries Management, Linnean Society of London, pp. 147–153.

Harper D. (1992) *Eutrophication of Freshwaters.* London: Chapman & Hall, 327 pp.

Henderson M.A. & Healey M.C. (1993) Doubling sockeye salmon production in the Fraser River – is this sustainable development? *Environmental Management* 17, 719–728.

Hinch S.G., Diewert R.E., Lissimore T.J., Prince A.M.J., Healey M.C. & Henderson M.A. (1996) Use of electromyogram telemetry to assess difficult passage areas for river-migrating adult sockeye salmon. *Transactions of the American Fisheries Society* 125, 253–260.

Howells G. (1990) *Acid Rain and Acid Waters.* New York: Ellis Horwood, 212 pp.

Jobling S., Nolan M., Tyler C.R., Brighty G. & Sumpter J.P. (1988) Widespread sexual disruption in wild fish. *Environmental Science and Technology* 32: 2498–2506.

Jungwith M., Schmutz S. & Weiss S. (eds) *Fish Migration and Fish Bypasses.* Oxford: Fishing News Books, Blackwell Science, 438 pp.

Kelly F.L. & Bracken J.J. (1998) Fisheries enhancement of the Rye Water, a lowland river in Ireland. *Aquatic Conservation: Marine and Freshwater Ecosystems* 8, 131–144.

Kirby J.S., Holmes J.S. & Sellers R.M. (1996) Cormorants *Phalacrocorax carbo* as fish predators – an appraisal of their conservation and management in Great Britain. *Biological Conservation* 75, 191–199.

Krine T.C., Goering,J.J. & Piokowski R.J. (1997) The effect of salmon carcasses on Alaskan freshwaters. In A.N. Milner & M.W. Oswood (eds) *Freshwaters of Alaska: Ecological Synthesis.* New York: Springer, pp. 179–204.

Krykhtin M.L. & Svirskii V.G. (1997) Endemic sturgeons of the Amur River: Kaluga, *Huso dauricus,* and Amur sturgeon, *Acipenser schrenkii. Environmental Biology of Fishes* 48, 231–240.

Larinier M., Porcher J.P., Travade F. & Gosset C. (1994) Passes à poissons: expertise, conception des ouvrages de franchissement, *Collection Mise au Point, Conseil Supérieur de la Pêche,* Paris, France, 336 pp.

Lucas M.C. & Batley E. (1996) Seasonal movements and behaviour of adult barbel *Barbus barbus,* a riverine cyprinid fish: implications for river management. *Journal of Applied Ecology* 33, 1345–1358.

Lucas M.C. & Frear P. (1997) Effects of a flow-gauging weir on the migratory behaviour of adult barbel, *Barbus barbus,* a riverine cyprinid. *Journal of Fish Biology* 50, 382–396.

Lucas M.C., Thom T.J., Duncan A.D. & Slavík O. (1998) Coarse fish migration: occurrence, causes and implications, *R&D Technical Report W152.* Bristol: Environment Agency, 147 pp.

Lucas M.C., Mercer T., Batley E., Frear P.A., Peirson G., Duncan A. & Kubecka J. (1998) Spatio-temporal variations in the distribution and abundance of fishes in the Yorkshire Ouse system. *Science of the Total Environment* **210/211**, 437–455.

Marmulla G. (1992) Überprüfung der Sieg als Lachsgewässer – Abschlußbericht Phase I. – Landesanstalt für Fischerei Nordrhein-Westfalen, Kirchhundem-Albaum, 121 pp.

Mirza M.Q. & Ericksen N.J. (1996) Impact of water control projects on fisheries resources in Bangladesh. *Environmental Management* 20, 523–539.

Novoa D. & Ramos F. (1990) Commercial fisheries in the Orinoco river – its current regulations. *Intersciencia* 15, 486–490.

Philippart J.C., Gillet A. & Micha J.C. (1988) Fish and their environment in large European river ecosystems. The River Meuse. *Sciences de l'Eau* **7**, 115–154.

Purdom C.E., Hardiman P.A., Bye V.J., Eno N.C., Tyler C.R. & Sumpter J.P. (1994) Estrogenic effects of effluents from sewage treatment works. *Chemistry and Ecology* **8**, 275–285.

Ricker W.E. (1954) Stock recruitment. *Journal of the Fisheries Research Board of Canada* **11**, 559–623.

Ruddle K. (1982) Traditional integrated farming systems and rural development- the example of ricefield fisheries in Southeast-Asia. *Agricultural Administration* **10**, 1–11.

Scruton D.A., Anderson T.C. & King L.W. (1998) Pamehac Brook: a case study of the restoration of a Newfoundland, Canada, river impacted by flow diversion for pulpwood transportation. *Aquatic Conservation: Marine and Freshwater Ecosystems* **8**, 145–158.

Smith T.D. (1994) *Scaling Fisheries: the Science of Measuring the Effects of Fishing, 1855–1955.* Cambridge: Cambridge University Press, 392 pp.

Stockner J.G. & Macisaac E.A. (1996) British Columbia lake enrichment programme – 2 decades of habitat enhancement for sockeye salmon. *Regulated Rivers: Research and Management* **12**, 547–561.

Suter W. (1995) The effect of predation by wintering cormorants *Phalacrocorax carbo* on grayling *Thymallus thymallus* and trout (Salmonidae) populations – 2 case studies from Swiss rivers. *Journal of Applied Ecology* **32**, 29–46.

Tilzey R.D.J. (1976) Observations on interactions between indigenous Galaxiidae and introduced Salmonidae in the Lake Eucumbene catchment, New South Wales. *Australian Journal of Marine and Freshwater Research* **27**, 551–563.

Vorosmarty G.J., Sharma K.P., Fekete B.M., Copeland A.H., Holden J., Marble J. & Lough J.A. (1997) The storage and ageing of continental runnoff in large reservoir systems of the world. *Ambio* **26**, 210–219.

Welcomme R.L. (1994) The status of large river habitats. In I.G. Cowx (ed.) *Rehabilitation of Freshwater Fisheries.* Oxford: Fishing News Books, Blackwell Science, pp. 11–20.

Welcomme R.L. (1995) Relationships between fisheries and the integrity of river systems. *Regulated River: Research and Management* **11**, 121–136.

Chapter 19
Defining and achieving fish habitat rehabilitation in large, low-gradient rivers

P.B. BAYLEY
104 Nash Hall, Oregon State University, Corvallis OR 97331, USA

K. O'HARA and R. STEEL
Environmental Advice Centre Ltd, The Bothy, Waen Farm, Village Road, Nercwys, Mold CH7 4EW, UK

Abstract

Rehabilitation of a system towards a restored state with more natural functions requires knowledge of what natural processes existed initially. Large, low-gradient, temperate rivers, typified by Huet's bream zone in Europe, have mostly lost their natural habitats and the hydrological regimes that maintained them. If restoration of some of the natural support systems for fish communities for reasons of conservation and/or low-maintenance fisheries is required, existing role models of optimal habitats for all life stages are probably lacking, because of restriction of biota to the main channel, excessive water regulation and riparian vegetation alteration. Consequently, it is believed that attempts to find optimal habitats in existing systems will more probably find marginal habitats. Equally, attempts to replicate instream physical habitat structures used in smaller rivers are usually inappropriate from a maintenance viewpoint (due to hydraulic stress and fine substrate effects) and were naturally absent or infrequent in the main channel.

Evidence from tropical and a few temperate systems under restoration suggests that the larger the river and the lower the gradient, the more critical habitats and sources of production lie in a variety of off-channel habitats in the floodplain. Therefore, we advocate that restoration of larger rivers should follow long-term adaptive management experiments using a diversity of off-channel habitat types under different hydrological regimes that are connected to the main channel for part or all of the year. Evaluation should initially focus on local habitat preferences and productivity in these restored or simulated off-channel areas, to demonstrate desirable features to be replicated on a larger scale sufficient to benefit river fish populations.

Keywords: Fisheries, floodplain, hydrology, off-channel, restoration.

19.1   Introduction

Restoration, as a desirable goal, can be simply defined as facilitating the return of a system to one governed by natural functions (Gore 1985) and should logically tend towards zero maintenance costs. Even partial restoration may result in systems with lower maintenance costs and an environment that increases some natural processes, and encourages locally adapted, indigenous species. The term rehabilitation (Gore &

Shields Jr 1995) is applied to such a process, which would, for example, exclude the recovery of a species by enhancing processes that were not considered part of a pristine system. If full restoration is not realistic, which for cultural and demographic reasons is the case for most, if not all, large river systems, in order to rehabilitate part of a system, a concept of the structure of the original system, and how it functioned and reacted to natural disturbance, is still needed (Sparks *et al.* 1990).

Although concepts of temperate river–floodplain systems are limited by largely descriptive information (Forbes 1895; Antipa 1928) or by inferential challenges based on the few intact tropical systems (Welcomme 1979, 1985), there are several common functions and features that low-gradient, river–floodplain systems appear to possess (Junk *et al.* 1989; Bayley 1995) which can be used as guidelines for partial or full restoration (Bayley 1991; Trexler 1995). However, these systems are so structurally and dynamically complex, that prediction of what will occur where, and at what frequency, will always be challenging and unpredictable at small scales. Therefore, any restoration process should be regarded as a long-term experiment subject to evaluation and redirection (Walters 1986), even though spatial replication of independent floodplains will normally not be feasible. This process is most germane when physical alterations are designed to speed up the response.

This chapter focuses on physical habitat restoration, with emphasis on fish and other aquatic fauna, and discusses the desirability of concentrating restoration activities on the main river channel or in the floodplain. Only low-gradient systems characterised in their natural state by meanders or anastomosed reaches (Amoros *et al.* 1987) and associated with mature floodplains dominated by fine sediments are considered. This corresponds to types in the 'low-energy cohesive floodplain' Class C of Nanson and Croke (1992), and may, to a more limited extent, correspond to their Class B3. Although much altered during recent centuries, the 'bream' zone of European rivers (Huet 1959) is considered to correspond, but may not be limited, to Class C floodplain types. Corresponding systems in North America are characterised by Centrarchidae species that use floodplain lentic systems, when available, for all or part of their life cycles (Groen & Schmulbach 1978; Trexler 1995). Most of the tropical floodplains in savanna or humid environments considered by Welcomme (1979, 1985) also apply.

19.2   Traditional, instream habitat restoration

The generally successful habitat development work in American trout streams (Hunt 1988) influenced much of the thinking and subsequent activity in temperate streams, and parallel work increased trout (Mills (1980) in Mann 1988) and coarse fish (Swales & O'Hara 1983) biomass through instream structures in the UK. However, the success rate was often overstated because insufficient attention was paid to distinguishing between the attraction effect of a structure and demographic change in the fish population (Fausch *et al.* 1995). The generally positive effects of large woody debris (LWD) on measures of several fish properties in North American streams is less frequently documented (Fausch & Northcote 1992; Flebbe & Dolloff 1995) than the effects of LWD on increasing pool sizes and frequency and other habitat diversity features (Andrus *et al.* 1988; McIntosh *et al.* 1994; Ralph *et al.* 1994; Richmond & Fausch 1995). However, large-scale effects of removal of habitat and physical diversity in streams can be marked and indisputable (Cowx *et al.* 1986).

19.3 Main channels

Large, low-gradient river channels associated with fine alluvium tend to be uniformly deep, even though pools with relatively static water are not common in unimpounded reaches. Rivers in naturally forested basins normally carry considerable quantities of woody debris, as they still do in the Amazon basin and used to do in temperate regions (Maser & Sedell 1994). Such instream structure is transient, becoming buried, deposited in off-channel areas, or carried downstream, and therefore requires constant recruitment from the watershed upstream (Andrus *et al.* 1988; Maser & Sedell 1994).

LWD is, or was, undoubtedly important for fish in large river channels, although depth itself provides protection from some aerial predators. Log piles provide protection from aquatic predators and resting places for migratory characins in the Amazon River (Alcides Gomes da Silva, personal communication). Some invertebrate production is dependent on sunken logs (Nilsen & Larimore 1973), but an estimate based on historical log accumulations in the Mississippi indicated a low biomass of 0.007 g m^{-2} of main channel water surface area (Junk *et al.* 1989). In sandy river channels lacking floodplain habitats, LWD can provide a relatively large proportion of secondary production (Benke *et al.* 1985).

Triska's reconstruction (1984) of the Red River in the Lower Mississippi system revealed a notable accumulation of LWD in the nineteenth century. However, quantities in the main channel were minor compared to those in the large, off-channel lakes that were created and maintained by LWD that was swept on to the floodplain.

Littoral zones of large river channels are important, and may be vital in the absence of floodplain habitats. However, compared to the volume or area in the main channel they are limited, similar to a lake with low shoreline development except that erosion boundaries will normally be less biologically productive than most littoral zones in lakes or the moving littoral of floodplains (Junk *et al.* 1989). Shoreline length can and should be increased by restoring side channels, but the benefits are marginal on a per unit area basis compared with the variety and productivity of floodplain habitats accessible to aquatic fauna (Junk *et al.* 1989; Bayley 1995; Sparks 1995).

A common response to calls for habitat rehabilitation, often in the name of restoration, has been to attempt to replicate procedures in smaller rivers or find some substitute that is more feasible from an engineering point of view. There are strong engineering, legal and financial disincentives to simulate the recruitment and flow of

LWD in rivers of deforested basins, and several decades are required to restore forested riparian zones to provide natural recruitment of LWD from upstream (Andrus *et al.* 1988). Even if benefits of LWD as fish microhabitat may be independent of whether the structure is transient or not, maintaining fixed, natural structures is increasingly problematic as channels become larger, because practical considerations rule out structures that are quickly buried or swept downstream (Lyons & Courtney 1990).

An alternative has evolved from the use of wing dykes or dams to direct and maintain direction of flow and minimum navigation depths (Schnick *et al.* 1981). Early versions of such structures provided sediment traps that were instrumental in converting inshore and off-channel habitats to terrestrial habitats, an effect that was very marked in the Lower Missouri River (Morris *et al.* 1968). Wing dykes have since been modified with notches to reduce sediment trapping, and their rationale has conveniently evolved into satisfying a need for providing habitat for particular fish and invertebrates (Hurley *et al.* 1987; Newcomb 1989). Whether such developments increase biomass levels and production rather than merely attract fish may not be important if the objective and product is more satisfied fishermen. However, it is inappropriate to finance such structures under the name and spirit of restoration or rehabilitation, and they should instead be regarded as examples of enhancement (NRC 1992). Such practices, in combination with other structures that maintain a constant, confined channel (Brookes 1988; Petts 1989) will often increase downgrading of the river bed and thereby make rehabilitation projects that provide access to floodplains require more expensive flow modifications.

Although there are localised situations, often in shallower water, where hard structures can be maintained, it is predicted that rehabilitation or enhancement efforts in main channels will, at best, marginally increase fish production compared with similar investments that provide off-channel aquatic habitats. Moreover, extensive, permanent maintenance of a main channel in a single position may render genuine rehabilitation projects infeasible.

19.4   Aquatic habitats in the floodplain

In regulated systems with established terrestrial conversions of floodplains, there are significant impediments to floodplain or off-channel restoration to a scale that would measurably improve fish populations and communities systemwide. This is especially true in the most regulated rivers in which hydrological modifications are necessary to maintain seasonal access to off-channel lentic habitats with minimum maintenance (Gore & Shields Jr 1995). Access to former floodable areas and isolated water bodies is further impeded by downgrading as mentioned previously (Brookes 1988).

Considering these fiscal and cultural impediments, it is remarkable that attempts at off-channel rehabilitation with a long-term ecological perspective have proceeded at all. However, projects involving significant floodplain restoration have been initiated on the River Danube (Heiler *et al.* 1995; Tockner & Schiemer 1997), Kissimmee River in Florida (Toth *et al.* 1993; Trexler 1995), Upper Mississippi (Lubinski & Gutreuter 1993; Sparks 1995), and Rhine river–floodplains (van Dijk *et al.* 1995), as well as many smaller projects connecting specific water bodies to main rivers.

19.4.1 *Fish responses*

The foregoing floodplain rehabilitation projects provide opportunities for understanding previously unrecognised elements of fish ecology that are important but not sufficient for optimal plans of new floodplain projects. Most studies of fish ecology have concentrated on environments that are hydrologically stable, at least where perennial surface water connections exist. Conversely, the degree of adaptation and opportunism that fish show in a variable hydrological environment is only slowly being understood in temperate regions, because most systems there are strongly regulated and/or access to off-channel lacustrine habitats is denied.

It is useful to be aware of hints of how fish behaved before system modification. Minnows, *Phoxinus phoxinus* (L.), have recently been observed to prefer shallow inshore habitats of rivers during winter (P. Garner unpublished data). If this preference, observed in an altered environment with water regulation, were to be interpreted as a guide for approaching optimal conditions in a rehabilitated environment, it would be misleading. Flooding naturally occurs during the winter in the UK, and Izaac Walton (Walton & Cotton 1994) reported that during the seventeenth century, ' . . . the Minnow or Penk: he is not easily found and caught till March or April, for then he appears first in the river; nature having taught him to shelter and hide himself, in the winter, in ditches that be near to the river . . . '. Hints such as this one from less-modified environments may provide more appropriate guides for rehabilitation than detailed studies in heavily modified environments.

When options of lotic and lentic environments are seasonally available, traditional ecological classifications of fish are seen in a different light. In addition to limnophilic fish, rheophilic fish species have successfully located and utilised gravel pits connected to the Lower Rhine river and young stages are more numerous than in the main channel (Staas & Neumann 1993). The ability of native fish to locate specific floodplain habitat connections has been demonstrated in the Colorado River (Valdez & Wick 1983), and young coho salmon, *Oncorhynchus kisutch* (Walbaum), can locate minor inflows from tributaries during main channel flood stages, which lead them to off-channel pools (Peterson 1982).

The construction of off-channel pools connected to the Great Ouse River in England is intended to benefit the recruitment of juvenile roach, *Rutilus rutilus* (L.), and bream, *Abramis brama* (L.), (Mann 1988). Preferences of these and other cyprinids for lacustrine microhabitats in the regulated main channel (Garner 1996) hint at the productive potential of restored off-channel pools in a system currently constrained by limited habitat diversity (Copp 1991). Considering that the recruitment of coarse fish species in the UK is temperature limited (Mills & Mann 1985), an additional advantage of warmer spring and summer temperatures would be expected in many off-channel pools where cool groundwater flows do not counteract direct warming.

Although one expects such rehabilitation attempts to be most beneficial to low-gradient reaches corresponding to the bream zone in Europe, there are hints that localised effects may occur in higher gradient rivers. In the River Wey in Surrey, UK, a small lake of 1 m maximum depth was connected by a narrow ditch to a section of the river that corresponds to the barbel, *Barbus barbus* (L.), zone. Intensive electric fishing produced samples of young chub, *Leuciscus cephalus* (L.), that were consistent with them entering the lake (which was not stocked with that species), growing and exiting (R. Steel, personal observation). The young chub, ranging from 110–250 mm long, were mostly congregated close to the reeds fringing this shallow lake.

The facultative behaviour of riverine fish is demonstrated by their ability to inhabit flooded areas that have not been accessible for decades (P. Bayley, unpublished data, Willamette River). Although they were not former floodplain lakes, gravel-mined pits provide an excellent opportunity for seasonal or perennial use by fish (Staas & Neumann 1993), provided that some shallow littoral zones are created and that connections with the main river are maintained by seasonal main channel flow regimes and/or tributaries. It can be argued that these would constitute enhancement projects. However, in the short term, demonstrable benefits from the voluntary use by riverine fish are facilitated, and in the long term, natural vegetation succession and reworking of the pools and surrounding floodplain during major floods may occur, or be encouraged.

19.4.2 *Hydrological constraints*

A major restoration challenge is the facilitation of sufficient deregulation, or simulated flow variation, of the river. Connecting off-channel pools to a highly regulated river may provide benefits to fish fauna, but will not benefit from the recycling of organic matter from a moving littoral (Junk *et al.* 1989; Bayley 1995). The action may need more investment in maintenance, such as keeping the connecting channel open and forestalling a tendency to return to a terrestrial habitat (Tockner & Schiemer 1997). It is reasonable to argue that low-maintenance restoration or rehabilitation of floodplain habitats is not possible without some degree of seasonal discharge variation that corresponds to regimes to which the fauna have adapted.

Connecting existing or created floodplain pools raises the issue of geographic permanence of restored off-shore habitats. Given a restored or simulated hydro-logical regime, no feature in a floodplain typically remains in one location on a scale of decades, even if the project received excellent direction from geomorphologists and plant ecologists. The movement or disappearance of habitats may be an embarrassment in a project demonstrating certain ecological benefits. It is known that significant fishery and biodiversity potential exists at large scales in intact floodplain systems (Welcomme 1985; Bayley 1995), but there is no guarantee of proportional benefits in the limited areas for which the resources to restore are available (Bayley 1991). Therefore, more investment in maintenance needs to be budgeted to maintain specific ecological goals in limited areas. Normally this will be required anyway for the maintenance of seasonal flow variation, and to achieve this some of the existing flow regulation structures upstream may need to be retained.

Although most rehabilitation projects need to increase flood stages seasonally (usually in spring and early summer), changes imposed by navigation dams on the Upper Mississippi and Illinois Rivers permit seasonal flooding to the extent that artificial levees and dam operation allow. However, too much water is retained in the lower reaches of main channel pools during the summer, preventing the stabilisation of soft substrates and the regeneration of moist soil vegetation (Sparks 1995). Therefore a major goal of hydrological restoration in those systems is to expose more floodplain by lowering water levels during July and August.

19.4.3   *Evaluation*

Because restoration or rehabilitation projects in large river–floodplains are not routine or predictable, each project should be regarded as an experiment and evaluated accordingly. It is very expensive, if not impossible, to quantify the benefit of restoration attempts at current scales on fish populations in the river system as a whole. However, it is less expensive to measure the changes in biodiversity, population densities, structure and production in an off-channel lentic habitat prior to and following its connection to the river, and to monitor seasonal movements into and out of the restored habitat. It is essential to continue this monitoring for sufficient years to encompass a broad range of annual hydrographs, and to continue until vegetation succession, sediment transport and behavioural patterns of fish populations are understood sufficiently to make predictions with acceptable risk. Such information can then serve as a guide for future site restoration within the range of conditions encountered.

Unfortunately, common practice fails to regard the project as an experiment. Subsequently, administration ignores the need to evaluate and/or attempts to make accurate predictions. In the latter case, this results in significant resources being spent on detailed modelling and predictions based on the original, highly modified system prior to starting the restoration process. Such resources are better spent evaluating conditions before and during the process based on a simple, flexible model that can be subsequently parameterised to the degree that the accuracy of the evaluation process requires.

19.5   Conclusion

It is understandable that reported successes, as measured by fish or invertebrate responses, in the rehabilitation of stream reaches through the addition of hard structures or large woody debris have prompted similar attempts in large river channels. Moreover, the task of the institutional, political and professional organisation to pass the threshold of a significant floodplain restoration is daunting, but it is being overcome. It is suggested that the following questions should be asked and dealt with before major habitat construction projects in main channels, that include fish or fisheries as benefits, are attempted.

(1) What is the long-term cost of maintenance or replacement of structures, after accounting for the likelihood of damage, downstream transport, or burial following a major flood?
(2) How will the structure change the channel and hydrology and will those changes impede future, more comprehensive restoration programmes?
(3) What is the cost of an effective evaluation of the impact on fish populations and other biota of interest? Can it even be achieved?
(4) Are there alternative off-channel restoration opportunities, not necessarily in the same system, that may provide superior benefits and lower long-term costs?

Floodplain habitat restoration is recommended to the extent that appropriate hydrological features can be restored or simulated (Bayley 1991), usually in that order of preference. At the same time, it should be noted that development of projects that are tailored to particular fish species, or even fish in general, usually carry more risk unless greater start-up and maintenance costs are budgeted. Projects that cater for a variety of conservation-related interests and ecological services spread the risk, reduce maintenance in the long term, and place the project closer to the restoration ideal on the restoration–rehabilitation–enhancement spectrum.

To make good choices there is a need to consider which systems have appropriate or manageable hydrological variations that justify the effort of land acquisition or easement, as well as the considerable investment in multidisciplinary and multi-institutional, professional activity and involvement with stakeholders. These latter costs tend to be underrated, as is their value when planning other enterprises. Much needs to be learnt from the few existing floodplain restoration projects, from the institutional as well as from the ecological and evaluation viewpoints (Walters 1997).

References

Amoros C., Roux A.L., Reygrobellet J.L., Bravard J.P. & Pautou G. (1987) A method for applied ecological studies of fluvial hydrosystems. *Regulated Rivers: Research and Management* **1**, 17–36.

Andrus C.W., Long B.A. & Froehlich H.A. (1988) Woody debris and its contribution to pool formation in a coastal stream 50 years after logging. *Canadian Journal of Fisheries and Aquatic Sciences* **45**, 2080–2086.

Antipa G.P. (1928) Die biologischen Grundlagen und der Mechanismus der Fischproduktion in den Gewässern der unteren Donau. *Academie Roumaine, Bulletin de la Section Scientifique* **11**, 1–20.

Bayley P.B. (1991) The flood pulse advantage and the restoration of river-floodplain systems. *Regulated Rivers: Research and Management* **6**, 75–86.

Bayley P.B. (1995) Understanding large river-floodplain ecosystems. *Bioscience* **45**, 153–158.

Benke A.C., Henry II R.L., Gillespie D.M. & Hunter R.J. (1985) Importance of snag habitat for animal production in southeastern streams. *Fisheries* **10**, 8–13.

Brookes A. (1988) *Channelized Rivers: Perspectives for Environmental Management.* Chichester: Wiley, 326 pp.

Copp G.H. (1991) Ecology of aquatic habitats in the Great Ouse, a small regulated lowland river. *Regulated Rivers: Research and Management* **6**, 125–134.

Cowx I.G., Wheatley G.A. & Mosley A. (1986) Long-term effects of land drainage works on fish stocks in the upper reaches of a lowland river. *Journal of Environmental Management* **22**, 147–156.

Fausch K.D. & Northcote T.G. (1992) Large woody debris and salmonid habitat in a small coastal British Columbia stream. *Canadian Journal of Fisheries and Aquatic Sciences* **49**, 682–693.

Fausch K.D., Gowan G., Richmond A.D. & Riley S.C. (1995) The role of dispersal in trout population response to habitat formed by large woody debris in Colorado mountain streams. *Bulletin Français de la Pêche et de la Pisciculture* **337–339**, 179–190.

Flebbe P.A. & Dolloff C.A. (1995) Trout use of woody debris and habitat in Appalachian wilderness streams of North Carolina. *North American Journal of Fisheries Management* **15**, 579–590.

Forbes S.A. (1895) *Illinois State Laboratory of Natural History, Champaign, Illinois. Biennial Report of the Director, 1893–1894.* Illinois Fish Commissioner's Report for 1892–1894, pp. 39–52.

Garner P. (1996) Microhabitat use and diet of 0+ cyprinid fishes in a lentic, regulated reach of the River Great Ouse, England. *Journal of Fish Biology* **48**, 367–382.

Gore J.A. (1985). Introduction. In J.A. Gore (ed.) *The Restoration of Rivers and Streams.* Boston, MA: Butterworth, pp. vii–xii.

Gore J.A. & Shields F.D. Jr (1995) Can large rivers be restored? *Bioscience* **45**, 142–152.

Groen C.L. & Schmulbach, J.C. (1978) The sport fishery of the unchannelized and channelized middle Missouri River. *Transactions of the American Fisheries Society* **107**, 412–418.

Heiler G., Hein T., Schiemer F. & Bornette G. (1995) Hydrological connectivity and flood pulses as the central aspects for the integrity of a river-floodplain system. *Regulated Rivers: Research and Management* **11**, 351–361.

Huet M. (1959) Profiles and biology of Western European streams as related to fish management. *Transactions of the American Fisheries Society* **88**, 155–163.

Hunt R.L. (1988) A compendium of 45 trout stream habitat development evaluations in Wisconsin during 1953–1985, *Technical Bulletin 162.* Madison, WI: Department of Natural Resources, 80 pp.

Hurley S.T., Hubert W.A. & Nickum J.G. (1987) Habitats and movements of shovelnose sturgeons in the upper Mississippi River. *Transactions of the American Fisheries Society* **116**, 655–662.

Junk W.J., Bayley P.B. & Sparks R.E. (1989) The flood pulse concept in river-floodplain systems. *Special Publication of the Canadian Journal of Fisheries and Aquatic Sciences* **106**, 110–127.

Lubinski K.S. & Gutreuter S. (1993) Ecological information and habitat rehabilitation on the Upper Mississippi River. In L.W. Hesse, C.B. Stalnaker, N.G. Benson & J.R. Zuboy (eds) Restoration planning for the rivers of the Mississippi River ecosystem, *Biological Report 19.* Washington, DC: National Biological Survey, US Department of the Interior, pp. 87–100.

Lyons J. & Courtney C.C. (1990). A review of fisheries habitat improvement projects in warmwater streams, with recommendations for Wisconsin, *Technical Bulletin 169.* Madison, WI: Department of Natural Resources, 34 pp.

Mann R.H.K. (1988) Fish and fisheries of regulated rivers in the UK. *Regulated Rivers: Research and Management* **2**, 411–424.

Maser C. & Sedell J.R. (1994) *From the Forest to the Sea: The Ecology of Wood in Streams, Rivers, Estuaries, and Oceans.* Delray Beach, FL: St Lucie Press, 200 pp.

McIntosh B.A., Sedell J.R., Smith J.E., Wissmar R.C., Clarke S.E., Reeves G.H. & Brown L.A. (1994) Historical changes in fish habitat for select river basins of Eastern Oregon and Washington. *Northwest Science* **68**, 36–53.

Mills C.A. & Mann R.H.K. (1985) Environmentally induced fluctuations in year class strength and their implications for management. *Journal of Fish Biology* **27** (Suppl. A), 209–226.

Morris L.A., Langemeier R.N., Russell T.R. & Witt A. Jr (1968) Effects of main stem impoundments and channelization upon the limnology of the Missouri River, Nebraska. *Transactions of the American Fisheries Society* **97**, 380–388.

Nanson G.C. & Croke J.C. (1992) A genetic classification of floodplains. *Geomorphology* **4**, 459–486.

Newcomb B.A. (1989) Winter abundance of channel catfish in the channelized Missouri River, Nebraska. *North American Journal of Fisheries Management* **9**, 195–202.

Nilsen H.C. & Larimore R.W. (1973) Establishment of invertebrate communities on log substrates in the Kaskaskia River, Illinois. *Ecology* **54**, 366–374.

National Research Council (NRC) (1992) *Restoration of Aquatic Ecosystems – Science, Technology, and Public Policy.* Washington, DC: National Research Council, National Academy Press.

Peterson N.P. (1982) Immigration of juvenile coho salmon (*Oncorhynchus kisutch*) into riverine ponds. *Canadian Journal of Fisheries and Aquatic Sciences* **39**, 1308–1310.

Petts G.E. (1989) Historical analysis of fluvial hydrosystems. In G.E. Petts (ed.) *Historical Change of Large Alluvial Rivers: Western Europe.* Chichester: Wiley, pp. 1–18.

Ralph S.C., Poole G.C., Conquest L.L. & Naiman R.J. (1994) Stream channel morphology and woody debris in logged and unlogged basins of western Washington. *Canadian Journal of Fisheries and Aquatic Sciences* **51**, 37–51.

Richmond A.D. & Fausch K.D. (1995) Characteristics and function of large woody debris in subalpine Rocky Mountain streams in northern Colorado. *Canadian Journal of Fisheries and Aquatic Sciences* **52**, 1789–1802.

Schnick R.A., Morton J.M, Mochalski J.C. & Beall, J.T. (eds) (1981) Mitigation/enhancement handbook for the Upper Mississippi River System (UMRS) and other large river systems, *Technical Report E. Comprehensive Master Plan for the Management of the Upper Mississippi River System.* Minneapolis, MN: Upper Mississippi River Basin Commission, 702 pp.

Sparks R.E. (1995) Need for ecosystem management of large rivers and their floodplains. *Bioscience* **45**, 168–182.

Sparks R.E., Bayley P.B., Kohler S.L. & Osborne L.L. (1990) Disturbance and recovery of large floodplain rivers. *Environmental Management* **14**, 699–709.

Staas S. & Neumann D. (1993) Reproduction of fish in the lower River Rhine and connected gravel pit lakes. In J.A. Van-de-Kraats (ed.) *Rehabilitation of the River Rhine.* Tarrytown, NY: Elsevier, pp. 311–313.

Swales S. & O'Hara K. (1983) A short-term study of a habitat improvement programme on the distribution and abundance of fish stocks in a small lowland stream in Shropshire. *Aquaculture and Fisheries Management* **14**, 135–144.

Tockner K. & Schiemer F. (1997) Ecological aspects of the restoration strategy for a river-floodplain system on the Danube River in Austria. *Global Ecology and Biogeography Letters* **6**, 321–329.

Toth L.A., Obeysekera J.T.B., Perkins W.A. & Loftin M.K. (1993) Flow regulation and restoration of Florida's Kissimmee River. *Regulated Rivers: Research and Management* **8**, 155–166.

Trexler J.C. (1995) Restoration of the Kissimmee River: A conceptual model of past and present fish communities and its consequences for evaluating restoration success. *Restoration Ecology* **3**, 195–210.

Triska F.J. (1984) Role of wood debris in modifying channel geomorphology and riparian areas of a large lowland river under pristine conditions: a historical case study. *Verhandlungen Internationale Vereinigung fur theoretische und angewandte Limnologie* **22**, 1876–1892.

Valdez R.A. & Wick E.J. (1983) Natural versus man-made backwaters as native fish habitat. In V.D. Adams & V.A. Lamarra (eds) *Aquatic Resources Management of the Colorado River Ecosystem*. Ann Arbor, MI: Ann Arbor Science, pp. 519–536.

van Dijk G.M., Marteijn E.C.L. & Schulte-Wülwer-Leidig A. (1995) Ecological rehabilitation of the River Rhine: plans, progress and perspectives. *Regulated Rivers: Research and Management* **11**, 377–388.

Walters C.J. (1986) *Adaptive Management of Renewable Resources*. New York: Macmillan.

Walters C.J. (1997) Challenges in adaptive management of riparian and coastal ecosystems. *Conservation Ecology* [www.consecol.org/vol1/iss2/art1] 1: 1.

Walton I. & Cotton C. (1994) The complete angler (reprint of fifth edition first published in 1676). London: Senate, Studio Editions, 427 pp.

Welcomme R.L. (1979) *The Fisheries Ecology of Floodplain Rivers*. London: Longman, 317 pp.

Welcomme R.L. (1985) River fisheries. *FAO Fisheries Technical Paper* 262. Rome: FAO, 330 pp.

Chapter 20
Watershed analysis and restoration in the Siuslaw River, Oregon, USA

N.B. ARMANTROUT

Eugene District, Bureau of Land Management, PO Box 10226, Eugene, OR 97440, USA

Abstract

The Siuslaw River is an eighth order river flowing into the Pacific Ocean on the central Oregon coast. Historically it produced runs of coho salmon, *Oncorhynchus kisutch,* approaching 500 000 fish per year, as well as chinook, *O. tshawytscha,* and chum, *O. keta,* salmon and cutthroat, *O. clarki,* and steelhead, *O. mykiss,* trout. In the mid-1800s, the river and tributaries were dominated by large woody material and beaver dams that created a complex habitat. As a result of man's activities, including the removal of the large woody material, the available habitat declined. Loss of habitat and intense fishing pressure contributed to a decline in salmonid runs to less than 5% of historical numbers. Watershed analysis was used, using historical information, field sampling and remote sensing, to describe the basin hydrology, climate, geomorphology and vegetation.

The Siuslaw River lacks a typical headwater, cutting through the Coast Range Mountains that are composed of uplifted marine silt and sand deposits. Gradients in the Siuslaw mainstem are low, with the steepest portion of the basin found in the adjoining slopes in the middle of the basin. Precipitation is seasonal and primarily rainfall during intense cyclonic winter storms. Stream flows are closely tied to precipitation events. Large, older conifers that could serve to provide stream channel structure have been largely removed. Watershed analysis was used in the Siuslaw to develop proposals for stream restoration that included in-channel structures, conversion of riparian vegetative communities to a higher percentage of conifer, improving access for fish, and restoring hydrological processes. Evaluation to date has documented an increase in habitat complexity and increased use of project areas by juvenile salmonids.

20.1   Introduction

The Siuslaw River is an eighth order river (Strahler 1957) originating south of Eugene, Oregon, USA, and flowing 177 river km through the Coast Range Mountains to drain into the Pacific Ocean on the central Oregon coast. The river drains a basin of 1753 km^2. The Siuslaw has been connected to both the Willamette River to the north and east, and to the Umpqua River to the south. The current elevation difference at the Willamette and Siuslaw Divide is only about 3 m.

The Bureau of Land Management (BLM) and US Forest Service (USFS) together manage about half of the basin. Private industrial forest companies, private citizens and the State of Oregon own the rest. The predominant commercial activity in the basin is the management and harvest of forest products.

The basin supports runs of anadromous and resident fish. The most important economically are the salmonids, including coho, *Oncorhynchus kisutch*, chinook, *Oncorhynchus tshawytscha,* and chum, *Oncorhynchus keta,* salmon and cutthroat, *Oncorhynchus clarki,* and steelhead trout, *Oncorhynchus mykiss.* Other anadromous fish include sturgeon, *Acipenser transmontanus,* and Pacific lamprey, *Lampetra tridentatus.* Native resident fish are predominantly cyprinids, sculpins (*Cottus* spp.), and brook lamprey, *Lampetra richardsoni.* Several species of centrarchids and catfishes have been introduced into the basin but have remained in Triangle Lake and Hult Pond and 'have not become established in the mainstem or tributaries.

The river is a major producer of coho salmon, with harvesting reaching a peak in 1899 of 462 000 fish (Oregon Department of Fish and Wildlife 1997). Recent runs have declined to less than 1% of this total (Oregon Department of Fish and Wildlife, personnel communication). The other anadromous salmonids have also declined in numbers, but only the chum salmon shows the same degree of decline as the coho salmon. Coho is now listed as a threatened species, and chum, steelhead and cutthroat have all been recently considered for listing as threatened or endangered species.

Declines in runs are attributed to a variety of factors including habitat alteration, fishing pressure, planting of hatchery fish originating from other basins, predation, and ocean conditions. Of these factors, BLM has influence only on freshwater habitat. Aquatic habitat deterioration on lands administered by BLM is similar to that which has occurred in the rest of the Siuslaw basin and in adjoining river basins. The changes have altered the hydrological processes and greatly simplified channel complexity and stability. The loss of aquatic habitat and decline in salmonid runs has been recognised for nearly a century, but it is only within the past 30 years that management of aquatic habitat has been changed in an effort to rehabilitate salmonid runs.

Restoration of aquatic habitat on public lands in the Eugene District was first initiated in 1969, and has continued sporadically since. Restoration efforts have accelerated since the release of the Northwester Forest Plan (FEMAT) in 1993 (Forest Ecosystem Management Team 1993) and the Eugene Resource Management Plan (RMP) in 1995 (USDI 1995a). FEMAT and the Eugene RMP include aquatic habitat restoration as a part of an Aquatic Conservation Strategy. It also includes retention of riparian reserves along all streams, designation of key watersheds, a series of standards and guides for management of activities influencing the aquatic system, and preparation of a watershed analysis prior to initiation of activities that would markedly alter the aquatic system in a basin.

Watershed analysis is a systematic procedure for characterising watershed and ecological processes to meet specific management and social objectives, and is used to guide management prescriptions, including setting refining boundaries of riparian and other reserves, developing restoration strategies and priorities, and revealing the most useful indicators for monitoring environmental changes. Watershed analysis was to be completed on watersheds of 45–450 km^2 (FEMAT 1993). The process included upslope areas as well as the aquatic system.

Since adoption of the Aquatic Conservation Strategy by BLM and the USFS, approximately 90% of the basin is covered by completed watershed analysis, with the remaining 10% expected to be completed by 1999. In addition to the Federal watershed analysis process, private industrial forest companies have also prepared watershed analyses for part of their lands. The existing watershed analyses have not been compiled into a single document, although efforts in that direction are underway. However, all use similar information on aspects such as geology, hydrology and vegetation history.

20.2   Siuslaw watershed analysis

The following discussion summarises the basic features of the Siuslaw basin as found in the completed watershed analyses, and the use of this information in providing context for the planning, implementation and monitoring of aquatic habitat restoration by the Eugene District, BLM. Original sources of information in the watershed analyses are cited in USDA (1996), USDI (1995b, 1995c), USDI (1996) and Weyerhaeuser Corp. (1996).

Factors influencing aquatic habitats may be summarised in four categories: geomorphology, climate, vegetation and human influences (see Armantrout 1999). Geomorphology includes the soils, underlying geological type, erosion and deposition processes, and the basic landforms. The climate includes precipitation, hydrology and temperature profiles. Vegetation includes riparian and upslope vegetation. Human influences include land management, direct influences on the aquatic system and influences on aquatic biota.

20.2.1   *Geomorphology*

The Siuslaw River has its origins east of the Coast Range Mountains in the Lorane Valley south of Eugene, Oregon, which is bordered by low hills of basaltic origin. It cuts through the Coast Range, an actively uplifting area of marine silt and sand sedimentary layers running parallel to the coast. The sedimentary layers are clearly visible in the uplifted materials. The Siuslaw has been able to erode the material at a rate sufficient to compensate for the uplift, retaining a nearly constant elevation. Basaltic dikes are present in the Coast Range, formed by volcanic intrusions that approached the surface.

The Siuslaw River lacks typical headwaters. The uppermost stream channels flow off low hills bordering the Lorane Valley. The Siuslaw River itself forms from the union of the North Fork and the South Fork near the town of Lorane at an elevation of less than 200 m. As a result, the Siuslaw, with an elevation drop of less than 200 m along the 135 km to the estuary, has a low gradient along most of its length.

Tributaries flowing off the actively uplifting slopes adjoining the river are actively forming as the Coast Range uplifts and are steeper in gradient, especially in their upper reaches. The steepest areas are in the middle of the basin, the area of the most active uplift, rather than at the upper end, with slopes exceeding 70%.

Soils are shallow, perched on the adjoining slopes or accumulated in river valleys. Stability varies, but soils are most stable in the upper reaches of the basin and least stable on steep slopes in the middle portion of the basin. The most common natural erosion is from slope failures or from channel failures (Dietrich *et al.* 1993). Erosion occurs most frequently during intense winter storms. Erosion rates have increased as a result of timber harvesting activities and road building (Benda & Dunne 1987; Benda 1990; Bilby *et al.* 1989).

Delivery of sediments to the stream by erosive processes is greatest in the middle portion of the basin where elevations are steepest, and lowest in the upper reaches of the basin. The material delivered to the stream channel is predominantly silt and sand. Rocks derived from silt stone and sand stone typically disintegrate into the parent materials within a relatively short period of time after reaching stream channels. The basaltic dikes are the only source of competent rock in the basin; as a result cobble, rubble and boulder particles are most concentrated in the limited stream areas associated with those dikes.

Silt and sandstone materials dominate stream channels. Where gradients and structural materials are present, the sedimentary materials accumulate; where structure is absent, the materials flush from the channel, resulting in channel incision in the valley floor until it reaches the parent bedrock. Only two natural lakes are found in the basin. Both were formed by lateral movement of bedrock and associated materials that blocked the channel. Typically, lakes that form are quickly filled in by accumulated sediments or break apart (Worona 1993). Incision into valley sediments is thought to contribute to decreased water storage capacity in the basin (Masters *et al.* 1991; Schumm 1993).

20.2.2   *Climate*

The basin has a maritime climate moderated by its proximity to the Pacific Ocean. Over 80% of the precipitation falls as rain in the period November–February. Snow does fall in the winter months, but does not persist. Of particular importance are the winter cyclonic storms that produce more than 2.5 cm of rainfall in less than 24 h. These high energy storms produce peak stream flows and erosional events.

Patterns of precipitation in the Siuslaw basin are quite variable. The highest precipitation occurs in the middle portions of the basin in association with the highest elevations. Localised precipitation may exceed 250 cm yr^{-1}. The lowest precipitation, <100 cm yr^{-1} is in the upper reaches of the basin.

The underlying bedrock layers have low permeability. Groundwater storage is limited, being greatest in the slope soils and valley deposits. Because of the limited groundwater storage, stream flows are closely related to precipitation events, with highest flows in the winter and lowest in late summer. The onset of higher winter flows following the initiation of winter rainfall is delayed for days or weeks by the recharging of groundwater. Once the soils are saturated streams rise quickly during storm events and flooding may occur. In summer, stream levels decrease, with lowest flows in late August and September. In the driest years many of the tributaries lose all surface flow. Temperatures are also highest during the low flow periods. Because of the limited headwaters and low gradient, reduced stream cover, and extensive bedrock substrate temperatures in the larger tributaries and the mainstem may exceed 27°C.

Precipitation and temperatures vary considerably from year to year depending on the direction and intensity of the storm tracts. In recent years, the El Niño phenomenon caused decreases in winter storm intensity, with below average precipitation and above average temperatures, both of which are reflected in stream flows and water temperatures.

20.2.3   *Vegetation*

Douglas fir, *Pseudotsuga mensiezii*, western hemlock, *Tsuga heterophylla*, and western red cedar, *Thuja plicata*, are the dominant native tree species, replacing an earlier more arid community that flourished following the last ice age (Worona 1993). The conifer community has a history of major fires every 80–120 years (Teensma *et al.* 1991; Long 1996). Fires were caused by natural events, particularly lightning strikes and human activities. The fires would remove 30–100% of the forest stands, with the result that the forest was typically a mosaic of young, mature and old forests.

One exception is the upper valleys of the Siuslaw River and Lake Creek, a major tributary. In both these flat valleys, wetter areas predominated. The valleys were burned at regular intervals by the native Americans to preserve the natural grass, oak and pine communities that were present prior to the development of the Douglas fir conifer forest.

Most of the Siuslaw basin is managed forest. Only limited areas of disjoint older forests remain; the majority of the forests have been harvested two or three times over the past century and are in young age classes of Douglas fir, hemlock and cedars. Riparian areas were usually treated as part of the upslope forest and were only recently provided with different management guidelines.

Riparian areas along the more unconfined and lower gradient portions of the Siuslaw and its tributaries were frequently flooded as a result of abundant large woody material and beaver activity and consisted of a mosaic of older and younger conifers, open wet and marshy areas, and hardwoods. Large woody material entered the channel through wind throw and the migration of the stream channel. Major events, such as floods, fires and landslides contributed to the presence of large woody material and hardwoods; steeper, narrow valleys were dominated by conifer with only limited hardwoods.

Most of the conifer in the valley bottoms were harvested, replanted and are in young age classes, or were converted to other land uses. Beaver ponds and wet areas were drained either by ditches or by cleaning woody material from the stream channel, allowing water levels to lower. Harvesting of trees, construction of roads and railroads, and the development of agriculture fragmented the riparian communities. In larger valleys, grasses and brush commonly replaced trees. Conifers are present, but are a much smaller part of the riparian community. The dominant tree species are the hardwood species red alder, *Alnus rubra,* and big leaf maple, *Acer macrophyllum.*

Only limited numbers of conifer in riparian areas are of suitable size to be potential sources of woody material in the channels. Recent changes in management regulations are designed to provide future supplies of large woody material from the riparian area for channel structure, but it will require at least 100–150 years before most of these trees are of sufficient size to remain in place once they reach the stream channel.

20.2.4 *Anthropogenic changes*

Primary human activities in the Siuslaw basin are timber management and agriculture. The only sizeable community is at the mouth of the river; otherwise permanent human settlements are scattered in the basin, concentrating in the upper reaches and along flatter, broader valleys. Changes in the basin are most pronounced in vegetation patterns and transportation routes. Hult Pond, a 16 ha mill pond on upper Lake Creek, is the only reservoir in the basin. At one time there were a series of small millponds to store logs before milling, but all other dams have been removed. Forests remain the dominant vegetative community but are now in a mosaic of mostly younger age classes. Flatter valley vegetation in settled areas is predominantly grasslands and brush; where human settlements are absent, it is hardwood and conifer forest, brush and grass. Roads parallel all major streams and many smaller tributaries, cross hillslopes and run along ridge tops. Very few areas without roads exceeding 500 ha can be found in the basin. The roads are a major factor in fragmenting forests and riparian areas, and in blocking access for fish and other aquatic species.

Changes in the aquatic system from human activities are seen mostly in hydrological and sediment patterns and in channel characteristics. Road networks have the greatest impact on hydrology and sediment. Roads intercept groundwater and divert it into channels associated with roads and road crossings. Roads are also the major source of erosion, including channel and slope failure events. Roads and associated culverts also create barriers to the movement of fish and other aquatic organisms.

Large woody material and beaver dams were the dominant stream structure in basin stream channels. Populations of beaver decreased as a result of intensive trapping initiated nearly 200 years ago, but have partially recovered in recent years. Large, woody material was removed from the channel with few trees now available as replacement. Without structure to retain depositional materials, widespread downcutting of stream channels occurred. The lowered stream channels reduced channel complexity, reduced groundwater storage capacity of the system, and reduced the connectivity between the channel and riparian areas. Lowered groundwater, reduced stream shading and exposure of bedrock are also thought to contribute to elevation of water temperatures but too little data are available to document such a relationship.

Information on fish communities is available only for anadromous salmonids. The information shows a marked decline in coho and chum salmon runs with a lesser decline for other species. While loss of freshwater habitat contributed to these declines, other factors also played a major role. Insufficient information is available to determine changes in populations of other aquatic species. Self-sustaining populations of non-native fish species are found only in Triangle Lake and Hult Pond.

20.3   Siuslaw watershed restoration

Restoration activities by BLM to date have been implemented on lands managed by the Bureau of Land Management located in the middle and upper half of the basin. BLM has worked with private landowners and the Oregon Department of Fish and Wildlife (ODFW) to address opportunities on adjoining private lands. Cooperative projects are expected to increase as a result of the formation of a Siuslaw Watershed Council and passage of legislation granting BLM the authority to expend funds on private lands. The US Forest Service, who manages land in the lower part of the basin, has also implemented habitat restoration efforts. This discussion will cover the projects implemented by BLM.

BLM initiated habitat restoration projects in the basin in 1969. The number and size of the projects have increased since 1993, with projects improving 35 km of habitat and opening passage for fish to over 160 km of additional habitat. Early projects include blasting pools in bedrock and placement of boulders in stream channels. Subsequent projects also used gabions and small trees (<15 cm in diameter). An evaluation of these projects is summarised in Armantrout (1991). Evaluation of existing projects in the Siuslaw basin and information on projects in other stream basins are used in developing current project proposals (Anderson & Miyajima 1975; Anderson & Cameron 1980; Anderson 1981; Hall & Baker 1982; House *et al.* 1989; House *et al.* 1990; Frissell & Nawa 1992; Crispin *et al.* 1993; House 1996).

20.3.1   *Selection of restoration sites*

No formal criteria for project location or design have been developed, but there are some informal concepts used in developing and designing projects. Only limited reaches of habitat in good condition remain, all in medium to small streams, but these were used as a template for restoration projects. Historical conditions, summarised in the watershed analyses, are indicators of the types of habitat that were present before extensive modification began. This information is most complete for the largest streams in the basin. Even in reaches that are generally degraded, there may be sections that are providing good habitat (see Everest *et al.* (1985) for discussion of habitat). The factors creating and maintaining these sections of better habitat were also used in developing restoration plans. From the conditions and processes in the basin a number of informal criteria were developed for selecting sites.

(1) Until recently, projects were only on lands managed by BLM, although conditions on adjoining non-Federal lands were included in project planning.
(2) The emphasis is on sites where projects would restore habitat and more natural conditions to benefit not only the economically important salmonids but also other aquatic organisms and resources.
(3) Sites are selected using existing patterns of flow and deposition in selecting sites, working with the stream. Good habitats are maintained, while improving degraded areas.
(4) Projects are generally located in larger streams to provide more functional habitats and to attract more species. Projects in larger streams are often designed to influence tributary stream habitats and processes.
(5) Projects are designed for both short- and long-term benefits. Channel projects involve placing structure with varying anticipated longevity in the channel. For the longer term, restoration of natural conditions, particularly riparian vegetation of predominantly large conifer, is expected to provide greater habitat benefits.
(6) Cluster projects increase interaction among projects, improve hydrological and sediment processes, and improve habitat.
(7) Providing access for fish to a blocked stream will generally produce more benefits than rehabilitating the same stream.

20.3.2 *Restoration issues*

Restoration options are based on the identified problems in the basin and the opportunities for improvements. The watershed analysis process has as its primary purpose the identification of these issues. Based on the information generated by the watershed analyses in the Siuslaw River basin, a number of issues to be considered when designing aquatic habitat restoration projects were identified.

(1) Low gradients in the larger and intermediate size streams.
(2) Only limited amounts of competent rock and gravel for structure and spawning are delivered to the stream channel. The dominant sediment material delivered to the channel, up to 95%, is sand and silt.
(3) Stream channels were dominated by large woody material and beaver dams. Riparian areas, the primary source of the conifers that provided the large woody

material, are now nearly devoid of source materials. Beavers are present but in greatly reduced numbers and subject to trapping.

(4) With little or no structural material, stream channels have become simplified, with limited habitat for fish or other aquatic species.

(5) Groundwater storage is limited by underlying geology, and is greatest in the depositional valley materials. As a result of channel incision, the storage capacity has been reduced. This also contributes to higher peak flows during flood events and lower flows during periods of little or no precipitation.

(6) Water quality is generally good except for elevated temperature and associated low oxygen. Temperatures have increased as a result of removal of streamside shading and channel incision.

(7) Roads have altered sediment delivery patterns, although the rate varies by location in the basin. The roads also intercept groundwater and divert it into surface flow, increasing peak stream flows and reducing groundwater storage.

(8) No large reservoirs have been built in the basin. Natural barriers are present in the basin, but most barriers to fish migration in medium to large streams are due to roads and road crossings.

Based on the issues developed from watershed analysis, it was decided that the emphasis should be to restore hydrological and sediment processes. Prior to anthropogenic alteration, stream structure from large woody material and beaver activity controlled these processes. In the short term, there are no natural sources of large woody material. Riparian restoration projects were developed to create future sources of large woody material. In the meantime, stream structures of boulders and logs were placed in the channel to create structure. In the largest channel, the emphasis was on hydrological and sediment movement, with sites selected that could influence tributaries. In the smaller streams, where the response to structure is more immediate, additional emphasis was placed on providing channel complexity to create a diversity of habitats.

20.3.3 *Project design*

Project designs developed as part of the Siuslaw habitat restoration project can be summarised in four general categories. One or more categories may be carried out at a particular site. Other than road rehabilitation, none of the actions in the aquatic habitat restoration programme involve upslope management activities. Management of upslope federal lands is guided by the Northwest Forest Plan, which includes a series of forest reserves and management standards and guides. These guidelines also provide for riparian reserves, which require that activities in the riparian area maintain or improve the function of the riparian area and aquatic system. Management of private lands is controlled by local and state laws, which differ from federal management guidelines.

(1) *Passage*. The most common passage problem for anadromous fish and other aquatic species is culverts that are too steep or inaccessible. Replacement of problem culverts with ones that are passable or modification of the channel above and below the culvert to improve access into and through the culvert for fish and other aquatic species opens additional habitat. Blasting of pools has been used to open access past bedrock falls and chutes. Two natural falls have been by-passed using fish ladders, one built by the state and one by BLM, with a third ladder built by BLM to permit passage over Hult Pond Dam.

(2) *Roads and culverts*. Use of most roads, including forest roads, is shared with other agencies or with private landowners. This limits the opportunity to close and rehabilitate roads. Where opportunities exist, roads are closed, with some also being subsoiled and planted with trees. Water drainage associated with roads is controlled using water barring, ditching, and stabilisation to reduce sediment production and changes in hydrology. Culverts are replaced, re-aligned or remove to improve movement of water, sediment and debris.

(3) *Riparian restoration*. Based on unpublished inventories of riparian areas in mature and old growth forests, the District found that the percentage of conifer in unharvested riparian areas ranged from 70% to over 95%. The Eugene District fisheries programme has established 70% conifer in riparian areas as a management target. Riparian vegetation restoration involves removal of patches or strips of existing riparian vegetation, nearly all brush or red alder, after which conifer seedlings are planted in the prepared sites. Most of the openings are developed in conjunction with concurrent stream channel project work, using the equipment access routes as riparian restoration sites. Additional opening may be added. Planted sites are maintained for several years by controlling competing vegetation.

(4) *In-channel stream projects*. In-channel projects involve placement of logs and boulders in a series of structures that include single boulder and log placement, clusters, jetties, ramps and cascades (Figs 20.1–20.4). Existing log and boulder structures found in other locations in the basin are used as templates. The cascade design is based on naturally occurring cascades in the stream that create channel changes similar to those associated with log jams. Materials are placed with heavy equipment and key elements are anchored. The District has, in the past, used gabions, pool blasting and construction of small structures using small boulders and logs. Work was done primarily in smaller streams by hand, using volunteers. While these projects produce benefits they had only limited impacts on the aquatic system as a whole and, other than blasted pools, seldom survived longer than a decade (Armantrout 1991).

20.3.4 *Monitoring*

Prior to stream or riparian restoration, an inventory is completed on existing habitat. In selected habitats, sampling is done for existing fish communities using seine

Figure 20.1 Wood structure placed in smaller stream (note small particle substrate)

Figure 20.2 Log and boulder structure in larger river

netting, snorkelling and electric fishing. Spawning ground counts have been carried out for all anadromous species since 1984 at selected index areas in BLM streams. Many of the projects are implemented in index areas, providing before and after counts. Counts by BLM in non-project areas, on BLM land, and counts on non-BLM land by the Oregon Department of Fish and Wildlife permit comparisons of spawning population changes in project areas to overall spawning in the Siuslaw basin and coastwide (Armantrout 1990). During the period when most projects were implemented anadromous salmonid populations coast-wide decreased to historical lows. Spawning counts in individual project areas were higher than the counts for the Siuslaw basin as a whole, suggesting a favourable response from fish. Overall, the projects cannot be shown to significantly increase the runs of anadromous salmonids in the Siuslaw basin.

Figure 20.3 Single large log structure during high flows (note accumulated woody material)

Figure 20.4 Artificial cascade created with boulders and large logs

Sampling of juvenile salmonids demonstrated an increase in numbers associated with habitat projects (Armantrout 1988, 1991). These increases were seen for both summer and winter populations. Non-salmonids also responded favourably to habitat structures, particularly the dace (*Rhinichthys* sp.) and shiners (*Richardsonius balteatus*). Since 1994, BLM has operated a smolt trap on Wolf Creek, a major tributary of the Siuslaw, to assess changes in fish production as a result of projects. Numbers of smolts of all species of anadromous salmonid showed a generally upward trend.

All sites are photographed prior to and after installation of projects. Since 1996, digital photography has been used and photos entered directly into a geographically linked database. Comparison of photographs before and after are used to document general changes in stream channel characteristics and to monitor changes in the habitat projects. Smaller structures have a lower life span (Armantrout 1991), with boulders having a longer retention time than logs. Large boulder and log structures built since 1995 are still being evaluated, but to date approximately 95% of the structures are still in place and functioning after three flood events. Stream channel changes in response to habitat structures are primarily increased depth of channel, increased deposition of sediment, increased cover and complexity, and more frequently, over-flooding onto adjoining floodplains.

Riparian vegetation restoration is still in its early stages. Survival and growth of seedlings varied, but was greatest when canopy opening was most complete and competing vegetation vigorously controlled. Full establishment of conifers is expected in 10–20 years, with trees reaching a size sufficient to create in-channel structure requiring an additional 100–150 years.

20.4   Discussion

Watershed analysis provides a process for summarising basic properties and processes in the Siuslaw River basin and identifying potential project areas. The watershed analyses written since 1994 for the Siuslaw basin incorporate watershed plans on which earlier projects were undertaken. The watershed analyses prepared under the FEMAT guidelines did provide the broader landscape perspective for these projects. They also provided a more comprehensive view of the expected results from restoration activities. Of particular value was information on historical conditions in the basin. The watershed analysis documentation also proved helpful in working with other landowners in developing cooperative projects.

Project work carried out between 1969 and 1994 in smaller streams showed promise. Improvements in habitat conditions and fish numbers were documented at project sites. The most notable changes in physical features were an increase in pool percentage and depth, increased cover, and a shift in substrate from bedrock to smaller particle sizes. In streams where restoration projects were developed at several extended sites, the overall quality of habitat was improved significantly for that stream. For the Siuslaw basin as a whole, habitat improvement projects have occurred in less than 5% of stream kilometres, most of it concentrated in small to medium-size streams. Removal of barriers, particularly the construction of a fish ladder over Lake Creek Falls, which increased available habitat in the Siuslaw basin by more than 13%, has had the greatest impact on available habitat for anadromous salmonids.

Habitat projects have increased the connectivity between the stream channel and riparian areas. As anticipated, deposition of sediments, primarily of finer particle sizes, has increased. Data to date are inadequate to document any changes in hydrological processes or temperatures. Wetlands have increased in streamside areas, thought to be due to raised groundwater water levels, but no measurements have been made. Temperatures in Wolf Creek, where many projects have been clustered, did show declines of 2–5°C, but this may be due at least in part to higher flows and lower air temperatures in 1997 when the lower temperatures were recorded.

Evaluation of the impacts of project work on salmonids, particularly anadromous salmonids, is hampered by the multiple influences on the fish runs. In addition to habitat changes, the fish are influenced by harvest regulations, hatchery practices, freshwater and marine predators, and ocean condition changes, due to the influence of global climate change and the increasing frequency and intensity of El Niño events.

Spawning by anadromous salmonids in several of the streams with projects shifted from non-project reaches onto project sites. Juvenile salmonid numbers also increased in project sites. Increased use of project sites, increased juveniles in project sites, and a generally upward trend in smolts at the Wolf Creek fish trap are all indicative of increased salmonid production from projects. However, with the exception of passage projects, it was not possible to document significant changes in fish numbers for the overall runs in the Siuslaw basin. Declines in runs in most coastal Oregon streams in recent years, a result of ocean conditions, harvesting, and non-freshwater habitat influences, have had a more significant impact on the numbers of spawning adult salmonids than have the fish projects.

Current project work by the Eugene District BLM is extending to more substantial structures in the mainstem and larger tributaries where the influence of the structures on hydrological and sediment processes is expected to be greater than for work in smaller streams. Project work will continue in smaller streams, on roads and with culverts. The improvements are expected to be cumulative, resulting in an overall upward trend for habitat in the basin. Riparian restoration projects, involving a smaller area and a timeframe measured in decades, is not expected to contribute to the improvement of in-channel aquatic habitat until well into the twenty-first century, although it is expected to improve shading of the stream.

Overall habitat changes resulting from restoration efforts have been close to those predicted from watershed analysis. Evaluation of projects provides an opportunity to anticipate the results of future restoration projects. Information from watershed analysis provides a framework for the evaluation, and for planning for continued restoration.

Acknowledgements

The field work associated with the project development, construction and evaluation depended on the assistance of Leo Poole, Russel Hammer, Larry Folenius, of BLM, Student Conservation Association volunteers, and the Oregon Department of Fish and Wildlife.

References

Anderson J.W. (1981) Anadromous fish projects 1981. USDI Bureau of Land Management Coos Bay District. In T. J. Hassler (ed.) *Proceedings of the Propagation, Enhancement, and Rehabilitation of Anadromous Salmonid Populations and Habitat in the Pacific Northwest*

Symposium. Arcata, CA: California Cooperative Fisheries Research Unit, Humboldt State University, pp. 109–114.

Anderson J.W. & Cameron J.J. (1980) The use of gabions to improve aquatic habitat, *Technical Note 342*. Denver, CO: USDI, Bureau of Land Management, Denver Service Centre, 22 pp.

Anderson J.W. & Miyajima L. (1975) Vincent Creek fish rearing pool project, *Technical Note 274*. Denver, CO: USDI Bureau of Land Management, Denver Service Centre, 25 pp.

Armantrout N.B. (1988) Use of various types of habitat improvements by salmonids in winter. In B.G. Shepherd (ed.) Proceedings of the 1988 Northeast Pacific chinook and coho salmon workshop, 2–4 October 1988, Bellingham, WA: North Pacific Chapter of the American Fisheries Society, Penticton, BC, pp. 142–147.

Armantrout N.B. (1990) Index areas as population indicators. In T.J. Hassler (ed.) *Proceedings of the 1990 Northeast Pacific Chinook and Coho Salmon Workshop*. Arcata, CA: California Cooperative Fisheries Research Unit, Humboldt State University, pp. 71–99.

Armantrout N.B. (1991) Restructuring streams for anadromous salmonids. *American Fisheries Society Symposium* **10**, 136–149.

Armantrout N.B. (1999). Aquatic and riparian inventories using remote sensing. *UNESCO MAB Symposium: Fish and land/inland water ecotones, 22–24 May 1995, Zakopane, Poland*. Paris: UNESCO (in press).

Benda L. (1990) The influence of debris flows on channels and valley floors in the Oregon Coast Range, USA. *Earth Surface Processes and Landforms* **15**, 457–466.

Benda L & Dunne T. (1987) Sediment routing by debris flow. In R.L. Beschta, T. Blinn, G.E. Grant, G.G. Ice & F.J. Swanson (eds) *Erosion and Sediment in the Pacific Rim*. Wallingford, Oxon: IAHS Publication 20165, pp. 213–223.

Bilby R.E., Sullivan K. & Duncan S.H. (1989) The generation and fate of road-surface sediment in forested watersheds in southwestern Washington. *Forest Science* **35**, 453–468.

Crispin V., House R.A. & Roberts D. (1993) Changes in instream habitat, large woody debris, and salmon habitat after the restructuring of a coastal Oregon stream. *North American Journal of Fisheries Management* **13**, 96–102.

Dietrich W.E., Wilson C.J., Montgomery D.R. & McKean J. (1993) Analysis of erosion threshholds, channel networks, and landscape morphology using a digital terrain model. *Journal of Geology* **101**, 259–278.

Everest F.H., Armantrout N.B, Keller S.M., Parante W.D., Sedell J.R., Nickolson T.E., Johnston J.M. & Haugen G.M. (1985) Salmonids. In E.R. Brown (ed.) *Management of Wildlife and Fish Habitats in Forests of Western Oregon and Washington*, 20Rb-F & WL-192–1985. Portland, OR: USDA, Forest Service, Pacific Northwest Region, pp. 199–230.

Forest Ecosystem Management Team (FEMAT) (1993*) Forest Ecosystem Management: An Ecological, Economic, and Social Assessment*. Portland, OR: USDA Forest Service, USDC National Marine Fisheries Service, USDI Bureau of Land Management, USDI Fish and Wildlife Service, USDI Park Service and Environmental Protection Agency, IX sections, 990 pp.

Frissell C.A. & Nawa R.K. (1992) Incidence and causes of physical failure of artificial habitat structures in streams of western Oregon and Washington. *North American Journal of Fisheries Management* **12**, 182–197.

Hall J.D. & Baker C.O. (1982) Rehabilitating and enhancing stream habitat. 1. Review and evaluation, *General Technical Report PNW-138*. Portland, OR: USDA Forest Service.

House R.A. (1996) An evaluation of stream restoration structures in a Coastal Oregon stream, 19811993. *North American Journal of Fisheries Management* **16**, 272–281.

House R.A. & Boehne P.L. (1985) Evaluation of instream enhancement structures for salmonid spawning and rearing in a coastal Oregon stream. *North American Journal of Fisheries Management* **5**, 283–295.

House R., Crispin V. & Monthey R. (1989) Evaluation of stream rehabilitation projects in Salem District (1981–1988), *Technical Note T/N OR-6*. Portland, OR: USDI Bureau of Land Management, 344 pp.

House R.A., Anderson J., Boehne P. & Suther J. (eds) (1990) *Stream Rehabilitation Manual.* Corvallis, OR: Oregon Chapter, American Fisheries Society.

Long C.J. (1996) Fire history of the central Coast Range, Oregon: A *c.* 9000 year record from Little Lake. MA thesis, University of Oregon, Eugene, 147 pp.

Masters L.S., Burkhardt J.W. & Tausch R. (1991) The geomorphic process: Effects of base level lowering on riparian management. *Rangelands* **13**, 280–284.

Oregon Department of Fish and Wildlife (1997) *Siuslaw River Basin Fish Management Plan. Draft version, 13 June 1997.* Corvallis, OR: Oregon Department of Fish and Wildlife, 161 pp.

Schumm S.A. (1993) River response to baselevel change: Implications for sequence stratigraphy. *Journal of Geology* **101**, 279–294.

Strahler A.N. (1957) Quantitative analysis of watershed geomorphology. *Transactions of the American Geophysics Union* **38**, 913–920.

Teensma R.D., Rienstra J.T. & Yeiter M.A (1991) Preliminary reconstruction and analysis of changes in forest stand age classes of the Oregon Coast Range from 1850–1940, *Technical Note T/N OR-9*. Portland, OR: USDI Bureau of Land Management.

USDA Forest Service (1996) *Indian/Deadwood Watershed Analysis.* Corvallis, OR: Siuslaw National Forest, 111 pp, 14 append.

USDI (1995a) *Record of Decision and Resource Management Plan.* Eugene, OR: USDI, Bureau of Land Management, Eugene District, 263 pp.

USDI (1995b) *Wolf Creek Watershed Analysis.* Eugene, OR: USDI, Bureau of Land Management, Eugene District, 203 pp., 9 app.

USDI (1995c) *Lake Creek Watershed Analysis.* Eugene, OR: USDI, Bureau of Land Management, Eugene District, 298 pp.

USDI (1996) *Siuslaw Watershed Analysis.* Eugene, OR: USDI, Bureau of Land Management, Eugene District.

Weyerhacuser Corporation (1996) *Upper Siuslaw Watershed Analysis.* Springfield, OR: Weyerhaeuser Corporation, 350 pp.

Worona M.A. (1993) Late Quaternary vegetation and climate history of the central Coast Range, Oregon: A record of the last *ca.* 42,000 years from Little Lake. MS thesis, University of Oregon, Eugene, Oregon.

Chapter 21
Planning implications of a habitat improvement project conducted on a Newfoundland stream

M.C. VAN ZYLL DE JONG
Inland Fish and Wildlife Division, Department of Forest Resources and Agrifoods, Box 8700, Building 810 Pleasantville, St John's, Newfoundland, Canada, A1B 4J6 (e-mail: mdejong@wild.dnr.gov.nf.ca)

I.G. COWX
University of Hull International Fisheries Institute, Hull HU6 7RX, UK

D.A. SCRUTON
Marine Habitat Research Section, Science Branch, Department of Fisheries and Oceans, PO Box 5667, St John's, Newfoundland, Canada, A1C 5X1

Abstract

The success of habitat restoration or improvement projects is dependent on planning and design using the proper information. This information consists of an understanding of how aquatic ecosystems behave over a wide range of temporal and spatial scales. A case study of the effect of habitat restoration techniques on brook trout, *Salvelinus fontinalis* (Mitchill), and Atlantic salmon, *Salmo salar* L., in a northern Newfoundland stream illustrates the need for improved planning. The case study reveals deficiencies in current planning and design and suggests improvements to the process.

Keywords: Salmonids, planning, habitat improvement.

21.1 Introduction

The loss of habitat through environmental perturbations has led to the use of numerous habitat management methods to restore or increase carrying capacity for fish (Cowx 1994; Cowx & Welcomme 1998). Improvement of fish habitat is critical to remove bottlenecks to fish production in degraded streams. At present, planning and design of habitat improvement projects are usually based on qualitative observations and a conceptualised vision of pre-perturbation habitat. This has arisen because few studies have attempted to examine the pre- and post-rehabilitation status of the fish communities to quantify improvements and feed back the information to expand our knowledge and improve methodology. When scientific evaluations of habitat improvement projects have been conducted, they suggest that project planning and design are unable to provide the necessary inputs to fully explain the outcome of improvement measures (e.g. van Zyll de Jong *et al.* 1997).

To select the most suitable sites and the most appropriate technique for rehabilitation, one must be able to predict the outcome of activities with reasonable accuracy. It is critical that the correct limiting factor(s) are identified and the most suitable technique(s) employed over the appropriate temporal and spatial scales. In terms of management, these objectives require some kind of quantitative prediction of expected output. An appropriate strategic framework to planning and design of habitat improvement and restoration projects, techniques and objectives must be clearly delineated before projects are initiated. This type of planning process is fundamental for assessing the merits of different approaches, and their ability to produce desired habitat features and response of target species. This chapter illustrates habitat improvement project planning activities through a case study on Joe Farrell's Brook in Newfoundland, Canada. Inadequacies in planning processes and recommendations to improve planning practices are discussed.

21.2   Case study: Joe Farrell's Brook

The case study was based on a long-term evaluation of a major habitat improvement project on Joe Farrell's Brook, a second order tributary of the Salmon River (Main Brook Newfoundland; Fig. 21.1) (van Zyll de Jong *et al.* 1997). In this river, clear cut harvest and pulp wood transportation have resulted in the channelisation and alteration of stream hydrology in the area, and reduced the amount and diversity of fluvial habitat available for salmonid populations. The main objectives of the project were: (1) to evaluate the long-term effectiveness and stability of restoration procedures on both physical habitat and juvenile fish populations, especially brook trout, *Salvelinus fontinalis* (Mitchill); and (2) to provide a model for effective approaches to stream habitat improvement.

A pre-improvement study provided baseline information for an analysis of the factors most likely to be limiting the development of the salmonid fish communities. On the basis of this analysis, sites were identified for remedial procedures. Treatments included boulder clusters ($n = 4$), V-dams ($n = 4$) and halflog covers ($n = 2$) (Fig. 21.1). Boulder clusters were intended to increase habitat diversity, thereby increasing protective and rearing habitat for all age classes of juvenile salmonids. V-dams were designed to create a deep pool environment with roughened surface water to provide high quality cover area for larger brook trout, particularly during low flow periods. Halflog covers were intended to provide hiding–resting–security cover for yearling and older brook trout. A sub-basin control site was chosen to compare results obtained from the restored sites with an untreated, historically logged reach.

Biological and physical habitat variables were sampled at 10 sampling sites on Joe Farrell's Brook and one sub-basin control site on Eastern Pond Brook (Fig. 21.1) from 1 July to 14 August 1993, 27 June to 15 August 1994 and 30 June to 18 August 1995. All sites were 40 m long, measured as a meander length.

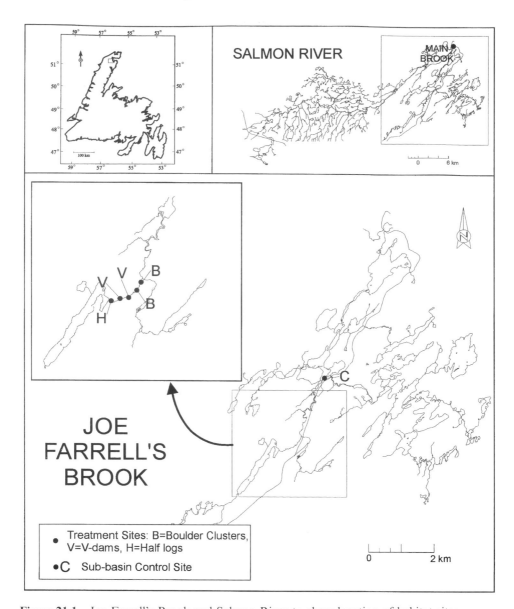

Figure 21.1　Joe Farrell's Brook and Salmon River to show location of habitat sites

Stream habitats measurements were collected at 10 cross-sectional transects, spaced every 4 m. Physical habitat parameters measured included stream gradient (%), depth (cm), width (m), substrate composition (modified Wentworth; after Gibson 1993), water velocity (cm s^{-1}), and cover types (%).

Fish were sampled from each 40-m section of stream enclosed with a 6-mm mesh size barrier net, with a Smith-Root VIII-A DC backpack electro-fisher. Population estimates were made using the depletion method (Zippin 1958). The maximum number of passes was based on the rate of decline in successive fishings with a minimum of four passes. Each salmonid was identified to species, fork length measured to the nearest mm, and weighed to the nearest 0.1 g. All population estimates and fish age and growth analyses were performed using MICROFISH 3.0 (van Deventer & Platts 1985).

Raw density data for each age class of each species and physical habitat variables were investigated for overall differences between treatments and sampling times using one-way analysis of variance (Duncan Multiple Range Test). Residuals were examined for normality, homogeneity and independence. Randomisation methods were used to compute type 1 error. Where there were significant differences (at the 5% level), product–moment correlation coefficients were used to elucidate relationships between changes in age class density with changes in physical habitat. All tests were judged significant at $P < 0.05$. All statistical analysis was performed using SYSTAT procedures (Anon 1996).

In boulder treatments one way analysis of variance showed that $0+$ and $1+$ salmon density increased significantly ($P < 0.05$) in both post-treatment years when compared to pre-treatment densities, while older age classes showed no significant ($P > 0.05$) change in density (Fig. 21.2). Brook trout, aged $0+$ increased significantly in the first post-treatment year, while older classes exhibited no density change ($P > 0.05$) (Fig. 21.2). V-dam treatments significantly increased the density ($P < 0.05$) of $0+$ salmon in the second post-treatment year compared to the first post-treatment and pre-treatment years (Fig. 21.2). There was no change ($P > 0.05$) in density between the first post-treatment and pre-treatment year. Aged $1+$ salmon densities for both post-treatment years were significantly higher ($P < 0.05$) than the pre-treatment period, while no change was evident for older age classes. Trout densities did not change significantly for any age class (Fig. 21.2). Half-log cover increased ($P < 0.05$) $0+$ salmon density for both post-treatment years when compared to the pre-treatment density, while all other age classes of salmon showed no change ($P > 0.05$) in densities (Fig. 21.2). Trout densities did not change significantly ($P > 0.05$) for any age class (Fig. 21.2). The combined effect of all rehabilitation efforts on Joe Farrell's Brook was examined in relation to control sites. Salmon aged $0+$, $1+$ and $3+$ demonstrated increased densities ($P < 0.05$) in the post-treatment years in response to habitat restoration, while all trout age classes and age $2+$ salmon did not change significantly ($P > 0.05$) at treatment sites when compared to the sub-basin control (Fig. 21.2).

One-way analysis of variance of the measured and calculated habitat variables showed significant differences ($P < 0.05$) between the 11 sites for velocity, stream volume, surface area, stream width, maximum depth, mean depth, coarseness rating, instream cover, overhanging cover, canopy cover, and pool, riffle and flat percentage in total, and significant differences ($P < 0.05$) between sampling periods for velocity, stream volume, maximum depth and mean depth.

Densities of young of the year salmon ($0+$) were negatively correlated with stream volume, and mean depth, and positively correlated with percentage pool (Table 21.1). One-year-old salmon ($1+$) densities were negatively correlated with stream volume and mean depth and positively correlated with velocity and overhanging

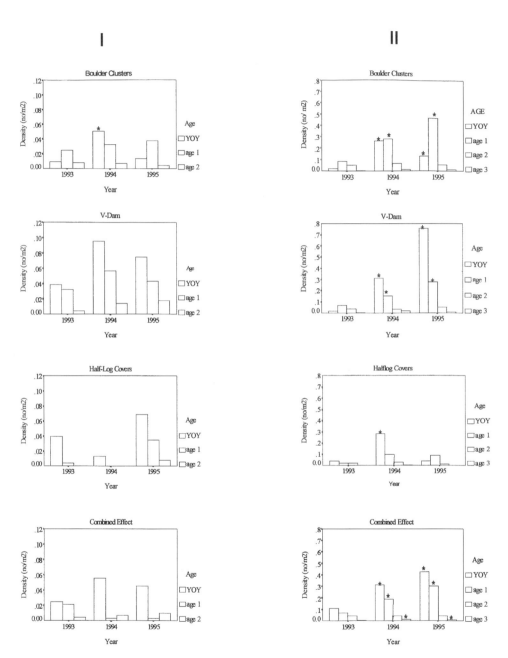

Figure 21.2 Mean density at age distribution of (I) brook trout and (II) Atlantic salmon in the period 1993–1995 for treatments and combined effects. Significant changes are indicated by an asterisk (*)

Table 21.1 Product–moment correlation coefficients for brook trout and Atlantic salmon age classes and habitat variables

	Surface area	Velocity	Volume	Width	Mean depth	Coarseness	Pool (%)	Flat (%)	Instream	Overhang	Canopy
Salmon 0+	-0.102	-0.268	-0.320	-0.107	-0.463	0.258	0.445	-0.071	0.207	0.201	0.256
Salmon 1+	-0.153	0.388	-0.478	-0.158	-0.652	0.229	0.171	-0.243	-0.227	0.385	-0.282
Salmon 2+	-0.091	0.494	-0.009	-0.102	-0.113	0.386	0.202	-0.219	-0.275	0.391	-0.422
Salmon 3+	-0.151	0.025	-0.167	-0.153	-0.212	0.236	0.218	-0.322	-0.157	0.448	-0.235
Trout 0+	-0.361	-0.126	-0.294	-0.360	-0.170	0.096	0.024	-0.250	0.005	-0.047	0.117
Trout 1+	-0.258	0.020	-0.041	-0.248	-0.054	0.363	0.556	-0.470	0.017	0.112	-0.011
Trout 2+	-0.223	-0.021	-0.124	-0.226	-0.075	0.144	0.283	-0.330	0.062	0.041	-0.065

Correlations ($P < 0.05$) are underlined.

cover (Table 21.1). Two-year-old salmon (2+) densities were negatively correlated with canopy cover and positively correlated with velocity, substrate coarseness and overhanging cover (Table 21.1). Three-year-old salmon (3+) densities showed no significant negative correlations but were positively correlated with overhanging cover (Table 21.1). Young of the year trout (0+) densities were negatively correlated with surface area and stream width, but no positive associations were found (Table 21.1). Densities of 1+ trout were positively correlated with coarseness rating and pool percentage and negatively correlated with flat percentage (Table 21.1). Densities of trout aged 2+ were not correlated with any habitat attribute (Table 21.1).

21.3   Discussion

Despite considerable attention to biological and physical factors during the project planning and design phase of the Joe Farrell's Brook rehabilitation scheme, unexpected results were observed. Instream structures, specifically V-dams and half-logs, were designed to improve habitat for juvenile brook trout. Although the structures created the desired habitat conditions, they were unsuccessful in increasing brook trout densities. Juvenile salmon density did, however, increase dramatically. The increase in Atlantic salmon density of all age classes was primarily as a result of the creation of greater habitat diversity. It is difficult to account for the failure to increase the density of brook trout. One hypothesis is the absence of overhanging and canopy cover or an inadequate combination of cover types. In addition, an accurate picture of the extent to which physical habitat alteration was responsible for changes in fish density cannot be given. Intra- and inter-specific competition plays a role in determining salmonid density, therefore factors other than measurable physical habitat may play a role in determining fish community structure (Gibson 1973; Heggenes & Saltveit 1990; Gibson *et al.* 1993).

Habitat manipulations need to be based on ecologically sound principles, where relations among fish populations and habitat factors have been demonstrated and knowledge of the potential capacity of a system to produce fish exists. Unfortunately, most degradation is ecologically undocumented, and in addition, pre-perturbation habitat is rarely documented. This lack of vital information presents a planning problem: how to determine the potential of the afflicted river after the event and the importance of factors limiting such potential. These considerations must be accounted for in the planning phase.

A strategic approach used to plan and develop the project is illustrated in Fig. 21.3. The first step in a logical approach to planning a project is to identify what the objectives of the enhancement scheme are and to relate these to the status of the fishery. Consequently, there is first a need for an inventory of the status of the fish stocks and habitat prior to any development scenario being formulated (Everest *et al.* 1991; Cowx 1994). In this case the inventory techniques of Buchanan *et al.* (1989) and Hamilton and Bergersen (1990) and were used to qualify and quantify habitat. A

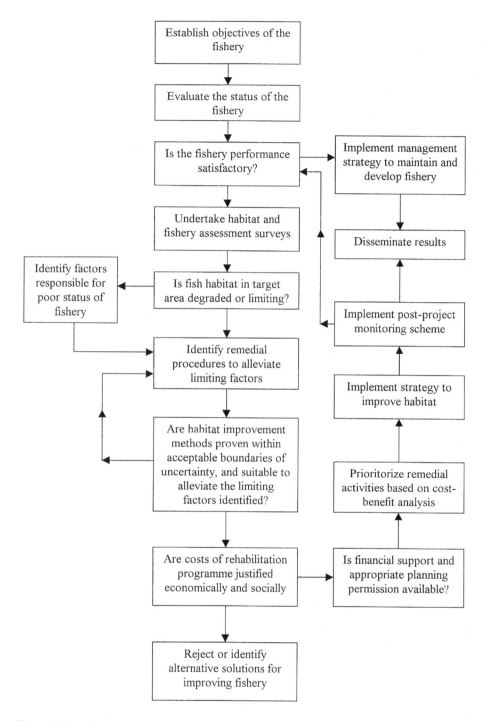

Figure 21.3 Flow chart to illustrate planning, implementation and feedback mechanisms for rehabilitating inland fisheries

pre-improvement stream survey (van Zyll de Jong 1991) provided baseline data for the analysis of limiting factors. This inventory was analysed for each species and those areas most likely to limit salmonid production were identified. A diagnostic dichotomous key, similar to that described by Reeves *et al.* (1991), was used to identify limiting factors from stream inventory information. On the basis of this analysis, sites were identified for improvement techniques.

The present planning process did not give the output expected, but it helped. It permitted a review of the status and opportunities for improvement, based on the best available information. A lack of quantification of relevant biological information at the site specific, local or regional level used in the planning phase contributed to the unexpected result. In addition, only a gross indication of habitat conditions was qualified. This was not sufficient for planning and design purposes. To overcome this information gap, the diversity of information on mechanisms regulating aquatic ecosystems must be fully documented. The controlling factors have to be elucidated and quantified, before it is possible to model the habitat needs and predict the outcome of rehabilitation work more accurately (Cowx 1994). This must be accomplished before any confidence can be attributed to likely improvement works, and rehabilitation techniques will achieve the desired objective.

A new planning approach must be delineated which will provide the necessary information and the ability to design effective improvement programmes. Critical in the development and organisation of this planning approach is the application of a classification system. One way to organise and representatively sample a variable environment is through a classification system. The general intent of classification is to arrange units into meaningful groups based on different spatial scales. A hierarchical landscape or catchment classification scheme that will allow several scales of resolution is most appropriate. Landscape classification schemes that examine macro (ecoregion and watershed), meso (stream) and microhabitat parameters can be used for assessing aquatic ecosystems if relationships among them exist. The character and succession of aquatic ecosystems depends to a large extent on the character of the catchment it drains, and, as such, spatial patterns in aquatic ecosystems should correspond to patterns in landscape. Such a system gives an organised view of spatial and temporal variation among and within different systems. The hierarchical framework is useful because it provides integration of data from diverse sources and at different levels of resolution, and it allows the scientist or manager to select the level of resolution most appropriate to their objectives. It is critical to look for patterns across these scales of resolution to delineate what resolution is necessary for management or scientific enquiry. The classification system and appropriate database will enable the identification of gaps in the data and limitations in the current data collection practices. By understanding the current data limitation it should be possible to streamline and standardise the methods of collection, archival and analysis.

Classification schemes should also incorporate a comparative approach to the study of stream ecosystems. This approach is based on two major premises: (1) that ecological structures and processes observed in one stream will be governed by the same processes in streams with a similar character; and (2) that comparisons of lentic and lotic areas and their biotic communities over space and time, and over a wide range of physical, chemical and climatic conditions, will provide a much broader insight than studies at a single point in time (Evans *et al.* 1987). Methods include monitoring of long-term trends (natural variation), experimental manipulation of habitats and fish communities, and adaptive management approaches. The synthesis of results should provide data on the structural and functional responses of fish communities and populations over a wide range of temporal and spatial scales, which can be used to predict the outcome of the rehabilitation scheme.

Standardised fish and habitat assessment and improvement techniques should be used so that all project results are comparable. The standardisation of methods and the removal of the subjective element in habitat data collection are needed to allow comparison of applications in varying habitats between rivers and within the same river. These steps should increase the predictive ability of habitat-based models. A standardised set of data collection procedures and a classification system for water bodies must be developed for each region. The output of each system should provide the ability to develop regional models at different spatial and temporal scales (e.g. microhabitat (pool-riffle-flat), mesohabitat (stream system scale), and macrohabitat (regional scale–ecoregional/sub-ecoregions). This would allow the formulation of management options that are directed at an appropriate spatial and temporal scale.

The third element of an improved scheme for planning habitat rehabilitation must be based on good study designs which attempt to limit the effect of the following problems: (1) fish may simply redistribute themselves in response to habitat improvement works, moving from unimproved areas to improved areas with no increase in total population; (2) numbers may increase in the modified areas, but it docs not necessarily follow that natural recruitment in the population will be enhanced or any bottleneck to recruitment removed; and (3) time lags or time delays in achieving the desired improvements suggest the need for long-term monitoring of rehabilitation projects. Simply, study designs should be placed in an appropriate spatial and temporal context. The classification system and comparative approach to habitat improvement and rehabilitation projects should allow for good study design. It should also provide the opportunity for replication and control.

21.4 Conclusion

The results of the case study had a positive effect on juvenile Atlantic salmon densities. The techniques, however, had little effect on the target species, brook trout. This result is not unexpected, considering the lack of information on pre-perturbation habitat and fish population/community dynamics. To restore habitat to its original condition, a quantifiable picture or understanding of what factors interacted to create it is needed. The results of the case study were unable to identify whether the habitat created or the fish response was positive in terms of its objective 'to rehabilitate degraded habitat to its pre-perturbation state'.

To achieve success in improvement and rehabilitation projects, quantifiable information on the biological and physical factors that created the habitat condition to be rehabilitated or improve must be provided. This information is a necessary antecedent to planning and experimental design. A first step in the process would be to develop a clear understanding of the diversity of aquatic systems on varying temporal and spatial scales and classify them into distinct groups. The creation of a classification should enable comparison of unaffected areas with pristine habitats in the same classification group. It will be possible to develop models that suggest the degree and magnitude to which factors need to be altered to create the 'pristine' state, and over what spatial and temporal scales the effect of manipulation would need to be monitored. It is also suggested that the collection of data should be comparable through standarisation of experimental methods and techniques. These steps should provide scientifically defensible conclusions, whether or not the project achieves its objectives.

Acknowledgements

The authors would like to thank the following individuals for their assistance in this project: Dr R. John Gibson, Mr Tim A. Anderson, Mr Lloyd J. Cole, Mr Keith Clarke and the White Bay Central Development Association. I would also like to thank the Canada Newfoundland Agreement for Salmonid Enhancement and Conservation and Environmental Partners for funding various stages of this project. Finally, we would like to give a special thanks to Ms Barbara Genge for her commitment and continuing support for the health and conservation of the Salmon River.

References

Buchanan R.A., Scruton D.A. & Anderson T.C. (1989) A technical manual for small stream enhancement in Newfoundland and Labrador. *Inshore Fisheries Development Agreement*, 108 pp.

Cowx I.G. (1994) Strategic approach to fisheries rehabilitation. In I.G. Cowx (ed.) *Rehabilitation of Freshwater Fisheries*. Oxford: Fishing News Books, Blackwell Science, pp. 3–11.

Cowx I.G. & Welcomme R.L. (1998) *Rehabilitation of Rivers for Fish*. Oxford: Fishing News Books, Blackwell Science, 260 pp.

Evans D.O., Henderson B.A., Bax N.J., Marshall T.R., Oglesby R.T. & Christie W.J. (1987) Concepts and methods of community ecology applied to freshwater fisheries management. *Canadian Journal of Fisheries and Aquatic Sciences* **44** (Suppl. 2), 448–470.

Everest F.H., Sedell J.R., Reeves G.H. & Bryant M.D. (1991) Planning and evaluating habitat projects for anadramous salmonids. *North American Journal of Fisheries Management* **10**, 68–77.

Gibson R.J. (1973) Interactions of juvenile Atlantic salmon (*Salmo salar* L.) and brook trout (*Salvelinus fontinalis* Mitchill). *International Atlantic Salmon Foundation Special Publication Series* **4**, 181–202.

Gibson R.J. (1993) The Atlantic salmon in fresh water: spawning rearing and production. *Reviews in Fish Biology and Fisheries* **3**, 39–73.

Gibson R.J., Hillier K.G., Dooley B.L. & Stanbury D.E. (1993) Relative habitat use, and inter-specific and intra-specific competition of brook trout (*Salvelinus fontinalis*) and juvenile Atlantic salmon (*Salmo salar*) in some Newfoundland rivers. In R.J. Gibson & R.E. Cutting (eds) Production of juvenile Atlantic salmon (*Salmo salar*) in natural waters. *Canadian Special Publication in Fisheries and Aquatic Sciences* **118**, 53–69.

Hamilton K. & Bergersen E.P. (1990) *Methods of Estimating Aquatic Habitat Variables*. Boulder, CO: Colorado State University, Cooperative Fisheries Research Unit, 94 pp.

Heggenes J. & Saltveit S.J. (1990) Seasonal and spatial microhabitat selection and segregation in young Atlantic salmon, S*almo salar*, and brown trout *Salmo trutta* in a Norwegian River. *Journal of Fish Biology* **36**, 707–720.

Oglesby R.T. & Christie W.J. (1987) Concepts and methods of community ecology applied to freshwater fisheries management. *Canadian Journal of Fisheries and Aquatic Sciences* **44** (Suppl. 2), 448–470.

Reeves G.H., Everest F.H. & Sedell J.R. (1991) Responses of anadromous salmonids to habitat modification: How do we measure them? *American Fisheries Society Symposium* **10**, 62–67.

Van Deventer J.S & Platts W.S. (1985) *Microcomputer Software System for Generating Population Estimates From Electrofishing Data – User Guide for Microfish 3.0, General Technical Report INT-254*. USDA Forest Service, Intermountain Research Station, 29 pp.

van Zyll de Jong M.C. (1991) *Stream and Pond Inventory: Analysis of Limiting Factor to Salmonid Production on the Salmon River Main Brook Newfoundland*. White Bay Central Development Association, 84 pp.

van Zyll de Jong M.C., Cowx I.G. & Scruton D.A. (1997) An evaluation of instream habitat restoration techniques on salmonid populations in a Newfoundland stream. *Regulated Rivers: Research and Management* **13**, 603–614.

Zippin C. (1958) The removal method of population estimation. *Journal of Wildlife Management* **22**, 82–90.

Chapter 22
Provision for the juvenile stages of coarse fish in river rehabilitation projects

B.P. HODGSON

Aquatic Management Services, 3 Selkirk Drive, Curzon Park, Chester CH4 8AQ, UK

J.W. EATON

School of Biological Sciences, Jones Building, University of Liverpool, PO Box 147, Liverpool L69 3BX, UK

Abstract

Options for juvenile fish habitat enhancement on the lowland reaches of the Welsh Dee, where coarse fish populations have been impacted by flow regulation and anthropogenic activity within the floodplain, are recommended. A rehabilitation technique, applicable to the main river corridor, offers aquatic and bankside ecological benefits. A scheme to supply the river with juvenile coarse fish from a floodplain lake could complement the in-channel initiative. Attention is drawn to the relevant risks associated with such schemes and the need to advance rehabilitation in a structured way, but recognising that success is not guaranteed.

22.1 Introduction

Anglers often regard fisheries science with some reservations because, although they recognise the need for research to provide a sound basis for the management of their sport, they also know that the time taken to achieve tangible results will be measured in years. This contrasts uncomfortably with their understandable desire for a more immediately productive fishery. For fisheries managers, this can present a dilemma with respect to river rehabilitation projects. This is because on numerous occasions actions are directed more to appease anglers' frustration by providing short-term solutions, through schemes that are not soundly based and are therefore prone to failure and wastage of money, than to long-term, sustainable solutions to fishery problems. This could explain why so few coarse fish rehabilitation schemes have been fully evaluated to establish whether they have achieved their initial aims and objectives, however loosely these may have been defined originally (Cowx 1999).

In natural river systems, adult coarse fish recruitment is linked to a number of factors, one of which is the success or otherwise of the juvenile phase. Strong initial production of a juvenile year class is no guarantee of strong representation in the adult life stage; mortality can be high without the right conditions for advancement of the juvenile fishes.

This chapter illustrates the importance of habitat diversity to provide territory suited to the specific requirements of juvenile coarse fish. A case study from the River Dee in North Wales, which was initiated to establish the reasons for the decline in roach *Rutilus rutilus* (L.), a key native species for match angling on the river, is presented.

The Welsh Dee has been recognised historically as a major coarse fishing venue in its lower reaches and much match angling has been pursued in the past for roach, perch, *Perca fluviatilis* L., bream, *Abramis brama* (L.) and pike, *Esox lucius* L. During the 1960s and early 1970s there was a significant decline in the roach populations, as they were replaced by a dominance of dace, *Leuciscus leuciscus* (L.), (Pearce 1983). Since 1953 the Dee has had its natural flow pattern progressively regulated, with the main change taking place in 1965 when Llyn Celyn reservoir (325 ha) was completed (Lambert 1988). From then on summer flows have been enhanced on average by 2–3 times over the natural state. The subsequent construction of Llyn Brenig reservoir (370 ha) on the Alwen catchment of the Dee in 1979 completed control measures for a river which is now considered one of the most heavily regulated in Europe.

At the time when Llyn Brenig was built, there was little consideration of the implications of flow regulation on the coarse fish populations of the Lower Dee, whereas there was much emphasis on the potential impact on the salmon, *Salmo salar* L., resource. There was a large baseline research programme by the former Water Resources Board, to study the potential effects of the regulation system, but relatively few of the findings were ever published. Following the construction of Llyn Celyn, Blezard *et al.* (1970) suggested that regulation would be largely beneficial to coarse fish populations because of the greater abundance of cleaner water, especially in summer. Indeed, it was considered that, if any change did take place, it would be an increase in abundance rather than the reverse.

In reality, there was considerable detrimental change to the native coarse fish population. Hodgson and O'Hara (1994) linked the coarse fish population shifts in the 1970s and 1980s to the effects of changes in the river regime caused by regulation. However, other anthropogenic factors were involved, together with natural progressive influences which changed the character of the river and compounded the level of impact on the fish population (Fig. 22.1). Particularly vulnerable were roach, bream and perch, which had limited home ranges and were largely dependent upon habitats within the main river channel, especially the habitat upon which the juvenile phases depended.

To counter this problem, a number of trials were pursued within the main river corridor to find the most effective way of creating more stable and beneficial types of marginal habitat that could be utilised by juvenile fish.

As well as the impact of regulation on the ecology of the river, wave wash from recreational boats was a particular problem on the Lower Dee, for example as a major obstacle to the establishment of shallow water and marginal vegetation. Early trials using wooden pallet barriers were found to be effective in reducing on-shore energy from wave wash and achieved partial stability of sediments, but such techniques were abandoned because colonisation by aquatic plants was limited. The

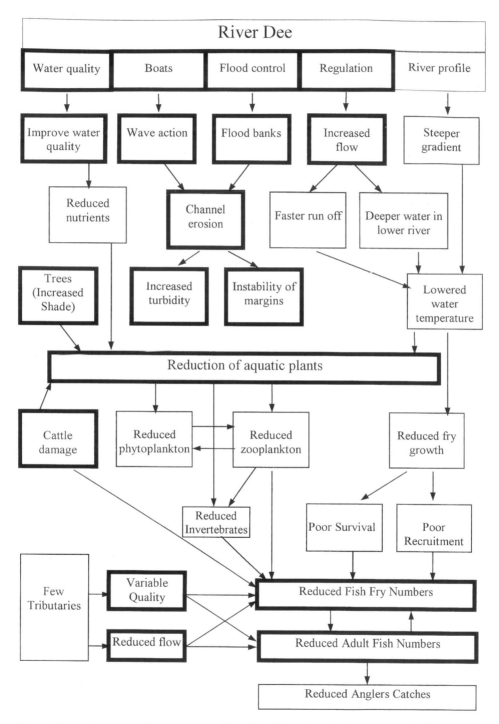

Figure 22.1   Pressures affecting coarse fish of the River Dee (main causes and effects in bold)

benefits derived did not warrant the effort or expense involved in installing them, nor were the structures visually enhancing to the environment.

22.2   Materials and methods

An assessment was first made of the suitability and types of channel substrate that were available within the coarse fish zone of the Lower Dee. This was achieved by boat surveys using an Ekman grab and sieving of the respective components. It was expected that fine silts would be the main constituents of sediments in the margins of the river, because fine suspensions are normal in the water column under summer flow conditions.

The site selected, at Heronbridge, near Chester, was a representative, largely unvegetated section of river banking on the outside of a bend, which measured 120 m long and 10 m wide, with a uniform bank slope of 1:10. Sparse clumps of vegetation were first removed by means of a mechanical excavator and top soil was spread evenly over the entire area to a depth of 10 cm. Work was advanced during early summer when the river level was close to its lowest, to ensure that the area planted would extend into the water under normal regulated summer flows. Along randomly selected sections, 4.5-m rolls of Netlon Tensar matting were laid down the slope and into the water and bedded in below the water line with the aid of deposited stone (20–30 cm diameter). At the top of the slope the material was embedded into the banking. Adjacent rolls were overlapped, taking account of the direction of flow of the river, and all sheets were fixed securely with 15-cm metal pegs at regular intervals. Gravel chippings (2.5 mm) were brushed into the double weave of the matting to hold it firmly in place. As well as the netloned areas there was an equivalent section where no material was laid.

The trial area was then split into plots measuring 4 m^2 and the meshed and unmeshed areas were systematically planted with *Typha latifolia* L. plants at 1-m intervals. In the case of the Tensar matting plantings, this was achieved by cutting the mesh. *Nuphar lutea* L. was planted along the margins where the large tipped stone was located. The whole area was then fenced along the landward edge to prevent cattle from invading and damaging the trial plots. Over a 4-year period the extent of spread of *T. latifolia* was recorded, together with the calculated number of stems m^{-2} at the end of the trial period, established by counts from a 1 m^2 quadrat.

22.3   Results

It was found that the bed substrates in the central parts of the channel in mid-summer become progressively finer down the catchment (see Fig. 22.2 for the lower 35 km of the River Dee). It should be noted that the silt content of the substratum at the side of the river increased to between 60 and 80% during the same period. In winter, marginal silt decreased to a level comparable with that found in mid-channel.

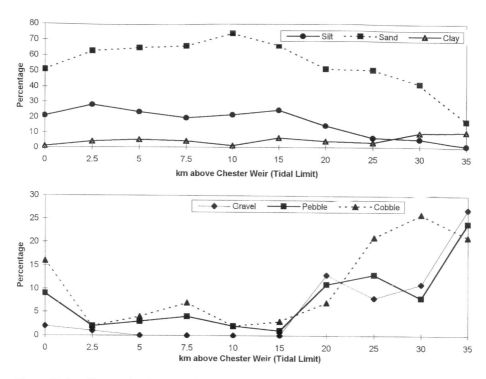

Figure 22.2   Changes in channel substratum composition in the lower 35 km of the River Dee

This would suggest that higher flows in winter progressively flushed out the finer components. It was also found that marginal substrates were highly mobile and this also restricted their colonisation by rooted vegetation, particularly in the more exposed areas. In the few locations where natural or man-made features were to be found (i.e. inlets, fallen trees, jetties) the marginal substrates were more stable and vegetation did establish.

Tensar mat was found to be effective in consolidating the margin of the trial area, but not in promoting the growth of a large reed like *T. latifolia* (Table 22.1). Although the matting made the sediments more stable, the mesh constricted the expansion of the larger, emerging shoots. Constrictions occurred around the bases of the stems and at times of higher river flows these made the plants more susceptible to breakage; consequently, there were losses from the meshed areas in successive winters. Plant expansion within meshed plots A, B and E was therefore restricted by the mesh but it was found that lateral expansion into the water, outside the mesh, did occur and once achieved, expanded at a comparable rate to that found in the wholly unmeshed areas (Table 22.1).

In the unmeshed plots C and D, the area of growth achieved was three times that for meshed plots A, B and E after the 4-year period. This expansion created a greater than three-fold increase in shoot density from an estimated 880 to 2800 per plot between meshed and unmeshed areas. Slower development was, however, recorded

Table 22.1   The colonisation of *Typha latifolia* in the Heronbridge trial

	A	B	C	D	E	F
Plots	Mesh	Mesh	No mesh	No mesh	Mesh	No mesh
Total area (m^2)						
Start	16	16	16	16	16	16
End of 1st year	22	20	25	30	28	24
End of 2nd year	28	20	40	54	32	31.5
End of 3rd year	31.5	31.5	78	78	38	44
End of 4th year	34	34	104	104	44	60
Area in water (m^2) after 4th year	14	15	42.5	42.5	16.5	22.5
% in water after 4th year	41	44	41	41	38	38
Plant numbers						
Stems m^{-2} at start	1	1	1	1	1	1
Stems m^{-2} in 4th year	26	26	28	26	20	14
Estimated stems within plots	884	884	2912	2704	880	840

for unmeshed plot F. This was because rapid advancement of naturally seeded willow *Salix* species had taken place at the rear of the plot and this progressively excluded other plants by a combination of vigorous growth and increased shading.

At all sites, whether meshed or unmeshed, around 40% of the growth achieved was within the water itself. This indicates that once the plants were established, growth could continue into the water space and provide the types of habitat that juvenile fish colonise and benefit from. Tables 22.1 and 22.2 show the extent of enhancement achieved from this trial after only four years.

It should also be recognised that a greater range of additional plant species colonised the area where Tensar mat was present than where it was absent (Table 22.2). Smaller marsh plants did particularly well in the meshed areas because their root systems were able to penetrate the meshes and use it to increase stability. Lateral growth of the species was also easier to accomplish and this probably accounted for the smaller macrophytes extending further into the water in such areas.

By the end of the third year, willow was starting to develop and, although not dominating at this stage, was beginning to encroach and displace other plants. The extent of plant growth by the end of the fourth year, together with the initial development of willow, can be seen in Table 22.2.

22.4   Discussion

Establishing the habitat characteristics on a river system that are influential for juvenile coarse fish growth and development is a complex process (Cowx 1999). Mann (1995) indicated that there were three key environmental attributes essential for recruitment success:

Table 22.2 List of plant species colonising the experimental length of the River Dee at Heronbridge, near Chester

Plant species	Year 1 Tensar mat	Year 1 No mat	Year 2 Tensar mat	Year 2 No mat	Year 3 Tensar mat	Year 3 No mat	Year 4 Tensar mat	Year 4 No mat
Emergent species								
Agrostis stolonifera L.							×	×
Alnus glutinosa L. Gaertner							×	
Apium nodiflorum (L.) Lag.	×		×		×	×	×	
Aster tripolium L.							×	
Bidens cernua L.	×		×		×		×	
Bidens tripartita L.				×	×	×		
Bolboschoenus maritimus (Asch.) Palla.			×		×		×	
Callitriche platycarpa Kuetz								×
Carex remota L.						×		
Carex hirta L.								×
Epilobium hirsutum L.	×		×		×	×	×	×
Galium palustre L.							×	
Glyceria maxima (Hartman) O. Holmb.	×		×		×		×	
Iris pseudacorus L.							×	
Juncus articulatus L.	×	×	×	×	×	×		×
Juncus bufonius L.			×	×	×	×		
Juncus effusus L.	×		×		×		×	×
Lycopus europaeus L.		×		×		×	×	×
Lysimachia nummularia L.					×			
Lythrum salicaria L.			×		×	×	×	×
Myosoton aquaticum (L.) Moench							×	
Oenanthe crocata L.						×	×	
Persicaria hydropiper (L.) Spach	×		×		×	×	×	×
Phalaris arundinacea L.	×		×		×		×	×
Phragmites australis (Cav) Trin.ex Steudel					×		×	
Plantago major L.						×	×	×
Rorippa nasturtium-aquaticum (L.) Hayek	×		×					
Rorippa sylvestris (L.) Besser	×		×		×			
Salix spp.					×	×	×	×
Scrophularia auriculata L.							×	×
Solanum dulcamara L.							×	×
Sparganium erectum L.			×		×	×	×	
Tanacetum vulgare L.			×				×	×
Typha latifolia L. (P)	×	×	×	×	×	×	×	×
Veronica anagallis-aquatica L.					×			
Veronica beccabunga L.							×	×
Floating leaved species								
Lemna minor L.								×
Nuphar lutea L. (Smith) (P)	×	×	×	×	×	×	×	×
Nymphaea alba L.							×	
Ranunculus flammula L.			×					
Sparganium emersum Rehmann.		×		×	×	×		
Submersed species								
Elodea nuttallii (Planchon) H. St. John	×		×				×	
Total species	13	5	19	7	21	16	29	19

(P) – Planted at the start of the experiment.

- availability of spawning medium close to refuges for newly hatched larvae;
- sufficient microhabitat diversity to satisfy changing needs of developing larvae;
- backwaters that give sanctuary during floods and provide specialised habitats for phytophilous species.

Copp (1990, 1992) showed that roach larvae became positively associated with microhabitats having shallow depths, aquatic vegetation and warmer water temperatures. From micromesh seine netting it was established that 0 + roach and dace populated the shallow margins of the Lower Dee and on examination of gut contents of the fry it was found that roach, in particular, were active feeders at these locations.

Mann (1973) determined that growth in the first year was important for the successful survival of juveniles over the first winter. Consequently, because the marginal areas are important to the juvenile fish on the Dee, steps were taken to improve these habitats and thereby enhance juvenile development during their first year of growth. Unfortunately such areas were few because three influences combined to eliminate vegetation in most shallow margins:

(i) encroachment and shading by bankside trees which created zones uncolonisable by herbaceous plants;
(ii) extensive cattle damage of marginal habitat;
(iii) wave wash from recreational boat traffic which made substrates unstable and actively displaced vegetation.

The trials with Tensar matting at Heronbridge demonstrated that bankside margins can be consolidated by the installation of mesh. This enhances the development of smaller plants, whether marsh type or terrestrial, and demonstrates that following establishment of *T. latifolia*, aquatic microhabitats are created which can be colonised by juvenile cyprinids. The outer fringe of *N. lutea* also established well by stabilising the substrate with rock for rooting, and this providing additional habitat for juvenile fish. Weatherley (1985) previously demonstrated the value of *N. lutea* to juvenile roach on the River Dee. The retention of this desirable system in a sustainable form did, however, present difficulties, because advancement of willow *Salix* sp, was found to be displacing desirable plant species.

The trial revealed that consolidatory mesh was not required to successfully establish *T. latifolia* at the margins. This could be adequately achieved by digging the plants into the substrate and, once introduced, they were not easily displaced. The extent of expansion into the channel was found to depend upon the intensity of erosive forces from the river and the extent of the shallow margin. For many of the larger plant species, establishment by seedlings is rare, either because seed production is not a major strategy, or because the seedlings are vulnerable to destruction by waves, drying or other influences. This suggests that the problem phase of colonisation of the margins by plants is the initial one of becoming firmly rooted. The unstable nature of the substratum in river systems like the Dee probably prevents this process occurring naturally, making stable rooting difficult for seedlings and larger vegetative fragments alike. The limitations imposed on larger emergent aquatic plants through shoot restriction could be used to actively curb this vegetation component and thereby encourage a flora of broad-leaved species, grasses and rushes to benefit biodiversity, conservation and landscape quality along the bankside. Netlon mesh is unlikely to slow the ingress of willow, as this is initially narrow stemmed, very strong and therefore able to overcome such resistance imposed by this material.

Habitat creation of the kind outlined can prove to be effective in producing more localised, seasonal sanctuary and feeding areas, but the scope for providing sufficient areas on a large lowland river system like the Dee will be dependant upon the availability of suitable sites that can be modified. More areas can be created by the selective removal of trees that totally shade the shallow bankside margins. Along the 20 km of the lower River Dee, about 60% is shaded by trees, and the extent of cover has increased in recent years. With large river systems, in particular, habitat degradation is a common feature and alternative proposals may have to be considered if adequate areas of spawning habitat are to be created. These could take the form of artificial backwaters off the main river channel (Linfield 1985), enhancing tributary spawning habitats, or possibly the creation of floodplain recruitment lakes.

The fecundity of coarse fish is generally high and, in the case of roach, a 250 g female can produce over 20 000 eggs (Mann 1973). In the wild, survival from egg to fry at the end of the first summer can be as little as 3.5% (Easton & Dolben 1980; Diamond 1985). Dangers of wash-out of fry and fish habitat degradation are also more prevalent in the main channel of highly managed river catchments such as the Dee, because of greater or unusual flow characteristics (O'Hara 1976). Therefore, off-channel solutions may be worth consideration.

It was estimated that a 2.2-ha fry-rearing lake in the floodplain of the Dee could produce 400 000 roach fry at the end of each summer season, which would achieve a stock density target for the system of 8–10 g m^{-2} roach biomass (APEM 1997). Initial capital cost of construction is high (£350K), although production is calculated to be cost effective at £0.11 per roach fry, compared to a hatchery-reared fish at £0.32.

Although such a system for supplying fish to a large river system is only at the development stage, it may offer a supplement or alternative to creating smaller areas for juvenile recruitment in the main river channel. Although an off-line, artificial spawning facility will produce large quantities of juvenile fish, it is a form of extensive fish farming that will require long-term financial commitment to maintain efficiency of operation. In-channel improvements will, however, be more modest investments and could be pursued by angling clubs and riparian owners. They also:

- restore a more natural river state;
- improve the aesthetics of the channel, especially in tourist areas where landscape value is important;
- produce 'wildlife gain' from aquatic through to avian species, which supports the Agenda for Sustainable Management promoted by English Nature (1997).

Many river systems have been highly modified by man over time and some aspects of these changes are irreversible. The options presented seek to create an environment which is neither the historical pristine state nor the situation that exists at the present time. Such an environment is still subject to the artificial, dynamic forces of a managed river corridor and therefore, there is no guarantee of success. Greater stringency on public funding of rehabilitation projects will inevitably lead to the requirement for strong justification to develop ambitious fisheries schemes in the future. Questions will also be asked about whether projects on rivers to benefit angling are justified in the light of the anglers census (NOP 1994) which reveals that there is a movement away from river-based angling to stillwater fishing. River-based fishing clubs may wish to continue to pursue schemes that have few tangible benefits, but in this highly emotive arena where action is sometimes considered more desirable than appraisal, small-scale efforts should still not be discouraged. Instead, advice on projects should be presented in a way that follows best working practice with the available evidence (Cowx & Welcomme 1998). This would structure enhancement schemes more effectively and allow progress still to be made on problematical river systems.

References

APEM (1997) *Floodplain Lake Feasibility Study – Lower River Dee.* Mold: Environment Agency, 55 pp.

Blezard N., Crann H.H., Iremonger D.J. & Jackson E. (1970) Conservation of the environment by river regulation, *Association of River Authorities, Annual Conference, Chester, 1970*, 70–115 pp.

Copp G.H. (1990) Shifts in the microhabitat of larval and juvenile roach, (*Rutilus rutilus* L.), in a floodplain channel. *Journal of Fish Biology* **36**, 683–692.

Copp G.H. (1992) Comparative microhabitat use of cyprinid larvae and juveniles in a lotic floodplain channel. *Environmental Biology of Fishes* **33**, 181–193.

Cowx, I.G. (1999). Factors influencing coarse fish populations in rivers, *R&D Note 460*. Bristol: Environment Agency, 184 pp.

Cowx I.G. & Welcomme R.L. (1998) *Rehabilitation of Rivers for Fish.* Oxford: Fishing News Books, Blackwell Science, 260 pp.

Diamond M. (1985) Some observations of spawning by roach, (*Rutilus rutilus* L.) and bream (*Abramis brama* L.), and their implications for management. *Fisheries Management* **16**, 359–368.

Easton K.W. & Dolben I.P. (1980) The induced spawning and subsequent survival and growth of roach (*Rutilus rutilus* L.). *Fisheries Management* **11**, 59–66.

English Nature (1997) *Wildlife and Freshwater – An Agenda for Sustainable Management.* Peterborough: English Nature, 56 pp.

Hodgson B.P. & O'Hara K (1994) Fisheries management of the Welsh Dee – A regulated river. *Polskie Archiwum Hydrobiologii* **4**, 331–345.

Lambert A. (1988) Regulation of the River Dee. *Regulated Rivers: Research and Management* **2**, 293–308.

Linfield R.S.J. (1985) The effects of habitat modification on freshwater fisheries in lowland areas of Eastern England. In J.S. Alabaster (ed.) *Habitat Modification and Freshwater Fisheries.* London: Butterworth, pp. 147–156.

Mann R.H.K. (1973) Observations on the age, growth, reproduction and food of the roach (*Rutilus rutilus* L.), in two rivers in southern England. *Journal of Fish Biology* **5**, 707–736.

Mann R.H.K. (1995) Natural factors influencing recruitment success in coarse fish populations. In D.M. Harper & A.J.D. Ferguson (eds) *The Ecological Basis for River Management*. London: Wiley, pp. 339–348.

National Opinion Poll (NOP) (1994) NOP, National angling survey, *Fisheries Technical Report. No 5*. Bristol: National Rivers Authority, 29 pp.

O'Hara K. (1976) An ecological study of fishes, with particular reference to freshwater species, in the Welsh Dee at Chester. PhD thesis, Department of Zoology, University of Liverpool, 185 pp.

Pearce H.G. (1983) Coarse fish stocking to British rivers. *Proceedings of the 14th Institute of Fisheries Management Study Course, West Bridgford*. West Bridgeford: Institute of Fisheries Management, pp. 209–220.

Weatherley N.S. (1985) The feeding ecology of juvenile fish in a lowland river. PhD thesis, Department of Zoology, University of Liverpool, 205 pp.

Section VI
Management

Chapter 23
Principles and approaches for river fisheries management

R.L. WELCOMME

Renewable Resources Assessment Group, Imperial College, London UK (e-mail: gbz01@dial.pipex.com)

Abstract

The current state of knowledge on the fisheries ecology of large rivers is summarised and used as a basis for defining major approaches to the management of riverine fish and fisheries. There are many users of rivers, each with their own perception, agenda, pressure groups and financial interest. No one group should be allowed to dominate, nor should it act without reference to other groups. This implies collaboration for management among all interested parties and agencies.

Management and rehabilitation of river fisheries can be carried out at three levels: the fish, the fishery, and the ecosystem. These can be managed by a variety of techniques, depending on the priorities and goals of the society. Techniques for management of the fish range from introductions and stocking to biological control of unwanted species. The fishery can be managed by controlling access, by conventional controls on gear and by giving the responsibility to the fishing communities through co-management systems. The ecosystem can be managed to conserve, to mitigate or to rehabilitate resources depending on the social, economic and policy dimensions of the external stresses. Each of the approaches to management has social, economic and policy implications that have to be taken into account when considering any particular strategy.

Keywords: River fisheries, management, enhancement, rehabilitation.

23.1   Introduction

Until comparatively recently, large rivers have been largely ignored by limnologists, mainly because of the difficulties in sampling, although pioneering work was done in the Danube (Antipa 1910) and the Rhine (Lauterborn 1918; Richardson 1921). As a result, understanding of river ecology was drawn mainly from small temperate streams, whose processes are very different from the larger, potamonic lower reaches of the system. Some understanding of the processes regulating large rivers was gained through studies on tropical systems in the late 1970s. This knowledge began to be transferred to rivers of the temperate zones, resulting in reviews such as that of the Large Rivers Symposium in 1986 (Dodge 1989). Subsequently, there has been a rapid increase in the literature on river biology from about 25 papers in 1975 to about 2500 a year over the last 5 years. The annual volume of publications has made it extremely difficult to keep the whole discipline under review.

Whilst intellectual curiosity may be a prime mover in this explosion of knowledge, there have been recent trends at a global level which have provided political motivation to promote these efforts. Concerns over the environment in general led to the formulation and international adoption of the Convention on Biological Diversity, whereby adhering states are legally obliged to take steps to protect the diversity of their living resource. The rapid decline of many of the marine fisheries resources due to overfishing (Grainger & Garcia 1996) led to the elaboration and adoption of the Code of Conduct for Responsible Fisheries by member countries of the Food and Agriculture Organisation (FAO). This Code (FAO 1995) is oriented towards sustainability and the protection of stocks. Water is increasingly perceived as a major limitation to sustainability or further expansion of human activities in many areas of the world and ways are being sought for its more efficient use. These include proposals for further modifications of unregulated and already regulated systems, such as the Hidrovia Project for the Parana/Paraguay system. As a counterweight, there is growing concern at the impacts of human interventions on aquatic ecosystems and a corresponding desire to mitigate these impacts and to restore damaged systems. There is a widespread perception of the diminished role of centralised government in the management process, with a consequent devolution of power through systems of co-management or assignment of rights to fish. All these factors now shape attitudes towards development and management of inland fisheries, against which current perceptions of the functioning of these systems should be viewed.

23.2   Principles

Early synthesis on river biology (Lowe-McConnell 1975; Welcomme 1979; Vannote *et al.* 1980; Junk *et al.* 1985) advanced a series of hypotheses on the functioning of riverine ecosystems which have not been contradicted by the later literature. These advances are expressed in two major integrating concepts. The river continuum concept (Vannote *et al.* 1980) describes the longitudinal evolution of trophic relationships from headwaters to the mouth of a river. It has generated a considerable amount of literature, qualifying its precepts (e.g. Cushing *et al.* 1983; Statzner & Higler 1985), although it remains perhaps the best generalised description of the overall shifts occurring along streams. The flood pulse concept (Junk *et al.* 1989) describes the lateral relationships between the floodplain and the river. It has proved less controversial and has been applied in several particular situations for assessment of impacts of projects that involve encroachment on the floodplain (e.g. Bonetto & Wais 1990).

The major advances in understanding of major river systems have principally concerned the relationship between the main channel and the riparian flood zone. As a consequence, there has been a tendency to regard the channel itself mainly as a conduit for migration and a refuge in adversity. It is clear that this situation is overly simplistic. Most large temperate rivers have lost all or most of their floodplains, and fish faunas, originally evolved for a migratory habit, have had to adapt to life in the main channel and to a restricted number of remaining backwaters. Many tropical rivers have elements of their fauna that are confined to the main channel and which apparently assume dominance in years of low flow. Some recent advances in deep water sampling in the Amazon and the Mekong now indicate that complex specialised faunas exist within the deeper waters of the main channel whose ecology and relationship to the rest of the river fauna is not clear (see, for example, http://eebweb.arizona.edu/fish/calhamaz.html). However, these species have only recently come to the attention of taxonomists and have played no role in fisheries so far.

In summary, the current state of knowledge on the fisheries ecology of large rivers is as follows. The morphology of any river is a function of the form of the landscape and the amount of precipitation in its basin. Under natural regimes there is an integral relationship between the river channel and its floodplain, the primary attributes of which are longitudinal connectivity within the river channel and lateral connectivity between river and floodplain. The progressive breakdown of allochthonous material forms the trophic base in low order streams and is progressively transmitted downstream to the larger channels. The generation of material on the floodplain through the release of stored nutrients by the flood pulse is the major driving factor in higher order river ecosystems. River fish faunas are highly complex, usually consisting of large numbers of species with a range of feeding and breeding strategies that aim to fully exploit the rich variety of feeding, breeding and shelter habitats in the system. Migration is a particularly common strategy, aimed at avoidance of stressful conditions at low water. In some species it also increases breeding success by timing the appearance of young fish so as to place them on the food-rich floodplains at the beginning of the flood. River fish assemblages are therefore highly adapted to the variable hydrological regimes caused by flooding. In regulated rivers, the natural form of the river is impacted by the totality of land and water use patterns in the basin, which produce breakdowns in longitudinal and lateral connectivity, as well as a simplification in form and function. River faunas in such systems may be stressed by: (a) modifications to the system through interruptions in longitudinal and lateral connectivity; (b) simplification, in the form of the system reducing diversity of habitats; (c) modifications to the quantity, quality and timing of water in the system; and (d) inappropriate fishing practices, including destructive methods, excess effort and bad timing.

River faunas respond to fishing and general environmental stress in a similar manner, with the successive elimination of the larger and longer-lived individuals and species from the assemblage. Dams and reservoirs that introduce discontinuities into the river continuum (Ward and Stanford 1995) and reduce flooding on the floodplains downstream, produce the biggest stresses on rivers. Detachment of the floodplain from the channel through levees and bunds also produces severe effects. Interruptions of longitudinal connectivity through dams suppress movements of migratory species. Interruptions of lateral connectivity through levee construction and floodplain reclamation suppress floodplain-spawning species. Rivers and their faunas appear very resilient to normal levels of natural stress and measures to improve or rehabilitate them through management can produce rapid, positive results within the system (Cowx & Welcomme 1998).

23.3   Approaches to management

There are many users of rivers, each with their own perception, agenda and financial interest. As a principle of river management no one group should be allowed to dominate, nor should it act without reference to other groups. In reality some uses such as navigation, power generation, irrigation and domestic consumption are currently seen by society as being of overriding economic and social significance. As such they dominate planning for allocation of water resources and the living aquatic components of the system are all too frequently neglected or omitted entirely. For example, until recently there was no explicit section dealing with freshwater living resources in the Convention on Biological Diversity (UNEP 1994) and it was only in September 1997 that the Subsidiary Body on Scientific, Technical and Technological Advice (SBSTTA) of the Convention discussed this issue. Equally the Code for Responsible Fisheries only addressed the special problems on inland fisheries when the technical guidelines for inland fisheries were published (FAO 1997a–c). Furthermore, most major international congresses on water tend to treat water as a commodity only, and exclude consideration of the living components of the system. Water is now seen as the major limiting resource in many parts of the world. As pressures on water for public supply, industry and agriculture grow, the living aquatic resource and fishery will find increasing difficulty in justifying their place in the multi-use system. Against this background, it is clear that inland fishery authorities are not fully in control of the resource they manage and that decisions taken elsewhere tend to determine the evolution of the fishery. Management of the river for fisheries, therefore, frequently represents a series of compromises with external constraints and becomes as much a question of negotiating space for the conservation of a viable ecosystem as the conventional management of the fishery itself. This means that the fishery interests have to be clearly represented within appropriate basin-oriented mechanisms for allocating river water and its physical setting.

There are a variety of purposes for management within the fishery. Originally rivers were fished mainly to provide a source of food. This still remains the primary function in many areas of the world, although food fisheries themselves can appear under several guises, ranging from commercial fisheries for profit and urban supply to small-scale operations for subsistence. More recently, recreational fisheries have appeared as a major purpose, although here too there is a range of use from pure sport fisheries to fisheries for domestic consumption. More recently, conservation has emerged as a motivation in its own right. This variety of purposes is strongly segregated into two main strategies for inland water management, one production oriented and one conservation oriented (Table 23.1). Originally this dichotomy was strongly geographically linked with the affluent, temperate nations opting for conservation and the poorer, tropical nations preferring production. In the last few

Table 23.1 Differing strategies for management of inland waters for fisheries in developed (mainly conservation orientated) and developing countries (mainly production orientated)

	Conservation (developed)	Production (developing)
Objectives	● Conservation	● Provision of food
	● Recreation	● Income
Mechanisms	● Sport fisheries	● Food fisheries
	● Habitat restoration	● Habitat modification
	● Environmentally sound stocking	● Enhancement through intensive stocking and management of ecosystem
	● Intensive, discrete, industrialised aquaculture	● Extensive, integrated, rural aquaculture
Economic	● Capital intensive	● Labour intensive

years there are signs that the geographical basis for differentiation is breaking down as an increasing number of tropical countries choose to manage their inland resources for recreation and conservation.

The overall impression is that, on one hand, there are patterns of intensive use whose main objectives are to manipulate the population structure and productivity of inland waters in the interests of the goals defined by society for food or recreation. On the other hand, there is a trend to emphasise the conservation values of the resource and to strictly limit the scale of management intervention, if not completely prohibit exploitation.

Management of rivers for fisheries can be carried out at three levels – the fish, the fishery and the ecosystem.

23.3.1 *Management of the fish*

Fish assemblages can be managed to improve production of species favoured by commercial or recreational interests, to make up for shortfalls in production arising from overfishing or environmental change, and for conservation of threatened species and stocks. A variety of techniques are available and have been reviewed in Welcomme and Bartley (1998) and by another conference of this series (Cowx 1998). These include:

● stocking natural waters to improve recruitment, bias fish assemblage structure towards favoured species, or to maintain productive species that would not breed naturally in the system;
● introduction of new species to exploit under-utilised parts of the food chain or habitats not colonised by the resident fauna;
● elimination of unwanted species;
● construction of biased and selected faunas;

- genetic modification to increase growth, production, disease resistance and thermal tolerance of the stocked material.

Stocking and introductions of fish are old practices and in the past the main motivations for such management measures were political and cosmetic. Few questions were asked as to their biological efficiency and cost effectiveness. As a result, attempts to supplement populations artificially by centralised governments have not always been successful when economies were privatised or when the fishermen were asked to assume the burden of the cost of stocking. It now appears that the degree to which such measures can be adopted depends largely on the priorities and goals of the society. For instance, in temperate countries the high priority currently being assigned to conservation of resources seen as natural, is limiting the degree to which manipulations of the population in the interest of exploitation for food or sport is permitted. In tropical and food deficit countries, however, the growing emphasis on intensification is leading to some of these strategies being explored and adopted. In these areas, the long-term success of enhancement policies will depend on a systematic evaluation of their cost effectiveness, and their social and ecological impact. Stocking of anadromous species such as the Atlantic salmon *Salmo salar* L., the Pacific salmons *Onchorhynchus* and the sturgeons *Huso huso* (L.) and *Acipenser* spp have undoubtedly contributed to the survival of these species in the face of heavy exploitation and environmental threat. Programmes to increase the production of floodplain oxbow lakes through stocking in Bangladesh have equally been successful in that they have produced notable increased in production (PIU/BRAC/DTA 1995). The success of the prolonged stocking programmes for coarse fishes in European inland waters is less certain, however, and the usefulness of a more general application of stocking technologies to large river systems remains to be explored.

23.3.2   *Management of the fishery*

Fisheries in regulated rivers, lakes and reservoirs are commonly managed according to conventional approaches developed for the management of marine fisheries. Special considerations in floodplain rivers and in some larger regulated systems make such approaches to fishery management difficult to apply and largely inappropriate.

   1. *Fish assemblages in rivers are highly complex. The number of species in a river is strongly correlated with its basin area* (Oberdorff *et al.* 1995). Even though not all species are necessarily present in any one reach, even small streams have numerous species of differing total lengths, as well as in trophic and habitat preferences. Examination of the length structure of fish assemblages in rivers (McDowall 1994) shows that the majority of species in any river are small and as there is a gradual shift downwards in mean length with increases in fishing effort, the number of species in the catch will increase. The species present, however, will be smaller and faster

growing and the length-related relationships identified by Allen (1971) and Pauly (1980) predict that standing stocks will decrease and total production rise. These relationships also predict that while individual species in the assemblage may conform to standard surplus yield models, the overall catch curve rises initially with effort to reach an asymptote which may be sustained over a considerable range of increasing effort. Lae (1997), for instance, described the relationship of catch and effort in a multi-species fishery of this type in the coastal lagoons of West Africa by the formula:

$$\text{Log}_e(Y+1) = \text{Log}_e(Y_{\max})(1 \cdot \exp - (af))$$

where Y is the yield in kg ha^{-1} yr^{-1}; f is the effort (number of fishermen km^{-2}) and $Y_{\max}$ is the asymptotic level of yield.

Unfortunately, there appears to be no system available to predict the level at which the asymptote ($Y_{\max}$) will occur, other than the use of generalised predictors such as Morpho Edaphic Index in lakes and yield/area equations in rivers (MRAG 1994a, b).

Eventually, when effort reaches sufficiently high levels, the assemblage may become sufficiently impoverished as to become destabilised and collapse. This, however, appears to be extremely rare as more frequently economic factors limit the rise in effort and prevent this level of overfishing. Very high effort fisheries are usually the result of high population densities brought about by local economic expansions. These in themselves tend to place pressure on the resource through pollution and environmental modification. Changes in fish assemblages subject to such stresses parallel those produced by fishing, and the combined effects of fishing and environmental degradation may well be synergistic (see Halls (1998) for example). In these cases the manager has to distinguish between fishing and environmental impacts in order to address the two issues through the appropriate mechanisms.

There are several implications for management of the fishery in this process. First, classical terms such as 'overfishing' are difficult to apply. Individual species may be overfished and disappear from the fishery but the assemblage as a whole continues to produce at a high level, albeit of fish which may not have the same value as those that have disappeared. In this context overfishing can only be deemed to occur with reference to some defined value, such as a particular group of species, quality, size etc. Second, the fishery can absorb increased amounts of effort, either as labour or as improved technology, than would be supported by a fishery concentrating on only the larger species in the assemblage.

2. *Fisheries in large rivers are often extremely complex, both in the gear and strategies used for fishing, and in the social and economic context of the fishery.* Broadly speaking there are two main types of river fishery: commercial and artisanal/subsistence, which have differing social and economic regulating mechanisms. Furthermore, there are differences in the ways in which the different types of gear affect the fish stock and respond to regulation. Commercial fisheries are usually

capital intensive operations that are based on only a few types of gear, usually gill nets or seines that select for a narrow band of species and sizes. Artisanal and subsistence fisheries tend to be more labour intensive. They target separate elements of the large number of ecologically diverse fish species and their different life stages by a diversity of static and active gear types, which are often deployed seasonally.

3. *River fish populations fluctuate widely in abundance in response to year-to-year fluctuations in precipitation and flood strength.* Similar variations occur in the relative abundance of individual species (see Welcomme (1995) for review). Furthermore, the time taken for hydrological variation to be reflected in the fishery varies. In lightly fished, temperate systems it may take several years for the fish to recruit to the fishery and best correlations are with $t = 4$, 5 or 6 (Kryhktin 1975; Holcik & Kmet 1986). In these cases the mean size of fish in the catch is relatively large. In heavily fished tropical systems lag times are much shorter (t or $t + 1$) and the fish in the catch are much smaller. In systems where fishing pressure has increased over the time series a shift in lag time is likely to have occurred. Such data sets as the River Niger (Welcomme 1985; Lae 1994) confirm that increasing fishing pressure leads to a reduction in t which then becomes a useful primary indicator of the overall exploitation status of the fishery.

Strategies for management of the fishery

The sustained plateau of yield and the succession from large to small fish mean that those responsible for managing production fisheries can select either explicitly or implicitly from a range of options. These may be simplified as four critical points in the evolution of the fishing-up process:

(i) aiming the fishery at only the most valuable larger species;
(ii) maximising economic yield;
(iii) maximising ponderal yield but retaining a reasonable quality of product;
(iv) maximising the employment (or distribution of the benefits of the fishery) by allowing the effort to rise.

In reality it is not uncommon to see unsuccessful attempts at managing fisheries for a combination of these objectives and many rivers combine these two types of fishery, producing complex social and economic interactions (Hoggarth & Kirkwood 1996).

Different types of fishery need different approaches to management. Commercial fisheries using mesh-selective gears such as gill nets or seines, for example the fisheries of the lower Orinoco (Novoa 1986), are generally oriented to production of large fish for urban markets. In such fisheries the mesh-selective gear tends to successively remove the larger individuals and species, producing a progressive drift from large to smaller mesh nets and a correlated drift downwards in mean length of fish caught. As the trend to adopt smaller and smaller meshes occurs in response to the disappearance of the larger elements from the community, so increased numbers of species become vulnerable to capture. In these cases, the best management strategy may be to limit the mesh sizes to those corresponding to the size ranges and species defined by the objectives of the fishery and rely on the cost–benefit relationship to regulate effort (input controls *sensu* FAO 1997a–c). The multiplicity of gears preferred by many artisanal and subsistence fisheries simultaneously exploit the complete range of species, size groups and ages of the assemblage. In this case imposition of mesh limits or the selective prohibition of gear may be impossible to implement without draconian legislation and the correct management approach appears to be to impose limits on effort, usually by limiting access to the fishery (output controls *sensu* FAO 1997a–c).

Measures such as creation of reserve areas where breeding stocks may be maintained for sustainable fishing, or limitations on the size and number of barrage traps are also used locally. Other solutions developed in marine fisheries, such as individually transferable catch quotas (ITQs) associated with a total allowable catch (TAC), could readily be applied to certain types of gear such as fixed engines, or even to individual commercial boats. However, they have not been widely used in inland fisheries to date except in some recreational fisheries that already apply bag limits.

So difficult are these problems in both marine and inland environments that there is a crisis of management throughout much of the fisheries world. In inland waters it has proved difficult to legislate for different approaches to different fisheries within the same basin. More often centralised authorities attempt to impose a single legislation, usually aimed at controlling the most important fishery economically, which is inappropriate to other fisheries. This results in social inequities and difficulties of enforcement. The inadequacy of such solutions has led to the idea of involving fishermen's communities in the regulation process, making possible a more flexible and localised approach to management.

23.3.3 *Management of the environment*

Environmental change produces impacts on fish assemblages that resemble those produced by fishing, whereby many of the larger species tend to disappear and smaller more opportunistic species become more abundant. However, certain human interventions can target specific elements of the assemblage. Such changes usually follow from alterations in connectivity. Thus cross-channel dams will impede longitudinal migration, resulting in the disappearance of obligate migratory species. Lateral dikes and levees will prevent lateral migration, reducing the abundance of phytophylic species. Where the floodplain is reclaimed for human occupation, permanent floodplain-resident species will disappear and, as much of the productivity in large river systems originates in the floodplain, there will also be a reduction in overall yield from the system (Welcomme 1995).

These changes result mainly from the impact of human activities other than fisheries on the aquatic environment. Gross changes such as those resulting from major hydraulic engineering projects for energy production or irrigation can be readily addressed through impact studies, although the apparent low value of fisheries relative to such activities may not influence the development of the project.

More insidious in some ways is the general encroachment on the floodplain of agriculture, human habitation, urbanisation etc., all of which gradually erode the area and the morphology of the resource base. Management of the aquatic environment in the context of multi-purpose use is therefore a prerequisite for the survival, if not the improvement, of fisheries in these regions. Five main responses to environmental degradation are possible:

 (i) *do nothing* where pressures from competing uses are excessively strong;
 (ii) *protect* the system from further decline where there is sufficient economic and social slack in the system;
(iii) *mitigate* for impacts that arise from continuing competing interventions;
(iv) *rehabilitate* where social and economic pressures have eased;
 (v) *enhance* where the value of the aquatic component is sufficiently important.
(vi) *Do nothing*.

There are many situations where the economic and social importance of the competing activities make it difficult for fisheries interests. In such circumstances, and where there are major impacts on the aquatic system, interventions for the improvement of fish habitat and stocks are usually unsuccessful and it is better to wait until conditions improve before attempting remedial action.

Protection

Where natural or acceptable environmental conditions persist in systems, consideration should be given to protecting all or part of these through legislation setting up reserves or sanctuaries. It is difficult to protect entire systems in this way in the face of growing population pressure, but similar arguments apply here to those used in rehabilitation. The dynamics of fish populations in floodplain rivers probably create enough surplus so that a smaller area of floodplain than exists under pristine conditions is needed to maintain a reasonable level of fish stocks. Thus the concept of reserves or rehabilitation areas strung along the river course rather in the form of a string of beads has been adopted in several areas. The ratio of reserve to modified area would depend on the nature of the individual system. However, such schemes should attempt to maximise lateral connectivity, and channel and floodplain diversity. Longitudinal connectivity is less easy to guarantee, especially where long-distance migratory species are involved and cross-channel dams occur.

Mitigation

Where competing users are economically important, but where the aquatic environment can benefit from interventions, a number of mitigating actions are possible. These include the alleviation of pollution by better treatment plants, creation of spawning beds where these have been destroyed, stocking to make up for spawning failure and poor recruitment, negotiating mitigating flows downstream of dams, weed cutting, dredging and desilting channels. Because the major causes of the disturbance remain in the system, mitigating activities of this kind usually require continuous intervention and considerable financial outlay. Increasingly the user causing the problem is required to bear the costs of mitigation and provisions for mitigating activities should be built into any new projects which are likely to create long-standing environmental hazards.

Rehabilitation

Where pressures from other users have eased there may be a possibility of restoring natural or quasi-natural features to the river. In contrast to mitigation, which generally requires sustained inputs, rehabilitation requires a one-off investment, after which natural processes should maintain the system. Of course, in most rivers regulation and modification will persist elsewhere in the system and thus fully natural regimes will not re-establish. This is particularly true of erosion–deposition processes and it may always be necessary to intervene with such operations as dredging to dispose of material which the modified flow regimes cannot handle. Rehabilitation aims mainly to restore the system to as near pristine conditions as possible through:

 (i) restoration of channel diversity;
 (ii) restoration of longitudinal connectivity; and
(iii) restoration of lateral connectivity.

Techniques for these have been explored in an increasing number of recent publications, such as Cowx and Welcomme (1998).

Intensification

Where the priorities of society are towards increased food production, a variety of physical approaches may supplement biotic ones, such as introductions and stocking. Many such techniques have been used in lakes but are more difficult to apply in floodplain rivers. Development of technologies to manage rivers in this way is still in the early stages, but the following are common in various parts of the world:

 (i) fertilisation of enclosed floodplain waterbodies;
 (ii) installation of brush parks in the main channel and floodplain water bodies;
(iii) construction of new floodplain water bodies in the form of drain-in ponds;
(iv) cutting of parts of river channels and bunding of parts of floodplain water
       bodies to serve as fish ponds.

23.4   Social and economic implications of management

Management is as much a political, social and economic activity as it is a technical one. While the present chapter deals primarily with technical aspects, some mention of the societal implications of the various measures is needed to retain a sense of the place of each in the development process. The following section is, therefore, far from exhaustive. Most of the approaches described imply a degree of social and economic change that in some cases can be severe. Current views on fisheries management emphasise the need to limit effort in the fishery through limitation of access. This automatically creates social inequities by denying access to some existing or potential users and by the assignation of rights to a limited group. It has also become obvious that artificial limitations on effort by a centralised authority are ineffective because of its incapacity to respond to the fluctuating nature of the river fish resource. The required flexibility can only be attained if the local fishermen are involved with at least part of the responsibility for management. There is thus a worldwide trend to charge fishing communities with the management and improvement of their resource, either directly through assignment of rights by governments, or less directly by extending the period of leases and licences.

Certain approaches to the management of the fishery intensify inequities in that limitations on gear or mesh size almost always affect the poorest and worst equipped of the fishermen. Societies, therefore, should carefully consider the degree to which they wish to compromise their poorer elements in the interests of efficient management of the resource.

The limitation of access and the creation of a concept of ownership among fishermen is essential if the fishery is to be well managed and particularly where investments are to be made in enhancements. Essentially this involves the passage from public to private ownership, with the fishermen having the privilege of access but being charged with the cost of developing the resource. In return, the state should pass legislation to guarantee that the developer of the fishery reaps the benefit from his investment, is empowered to negotiate with other users of the water for adequate environmental quality for his fishery, and can be compensated for damage to his fishery from other sources.

Rehabilitation of watercourses also involves some of the above issues, especially where fishermen's associations carry out restoration. More usually, however, rehabilitation of larger watercourses is an activity of central or local government, as it involves considerable costs as well as trade-offs with riparian landowners. It is rare that those holding title to riparian lands can be dispossessed and buy-back schemes whereby governments acquire title to tracts of land for wetland restoration are now common. Floodplain occupants can also be encouraged to leave their land if the government refuses to subsidise floodplain occupation through insurance or through flood protection schemes.

The move from simple, open access capture fisheries to more complex systems and the assumption of increased responsibility for the management of the resource implies the setting up of infrastructures to handle decision making within the fishery, and negotiation between the fishery and other users of the resource. This in turn requires increased knowledge on the part of the fishermen and their advisors, as well as better financial institutions to handle the increased capitalisation required. The role of government in this process should be to create a facilitating environment by setting up adequate institutions for finance, decision making, negotiation and setting of safeguards, and for extension and training.

23.5 Conclusion

The range of techniques available to the managers of inland waters, and particularly rivers has changed little since the end of the nineteenth century. What has changed, however, is the relative importance of the various strategies used. Initially, interventions on the fish through stocking or on the environment through provision of spawning grounds and shelter were local and confined to valuable fisheries such as the recreational fisheries for salmonids in the North Temperate Zone. Most inland fisheries elsewhere were managed by conventional systems such as licensing, closed seasons and areas, and mesh size limitations imposed by some central authority. These conventional approaches obviously continue to have a place in the range of options for management, although it is now clear that they are generally ineffective on their own and need to be combined with some devolution of management to the fishing communities.

The evolution of society, with greater pressures of human population and consequently greater demands for water and its associated products, has conditioned inland fisheries in two ways. First, the growing demand for food in many areas of the world has led to substantial overfishing of the available resources. This means that new methods of sustaining production are being sought, enhancement of inland fisheries amongst them. Second, the major pressures on inland fisheries are now no longer the fisheries themselves but the deformation of the aquatic environment in response to a range of other human needs. This means that the management of the resource is no longer the domain of the fisheries manager alone but should be the responsibility of a collective decision-making process which integrates all the interest in the river or lake basin. New technologies to enhance fisheries on the one hand and to mitigate or rehabilitate aquatic systems in the face of alternative users on the other are emerging in response to these changes. The emergence of such technologies is coupled with a change in social perceptions and economic climate which favour the swing from conventional low input, open access, reactive management of the fishery to high input, privatised, proactive management of the fish and the environment.

Scientific knowledge is a major guiding principle in this process. More efficient management, be it for conservation or for intensification of production, depends on an understanding of the systems involved. Certainly recent insights into the functioning of rivers have brought about far-reaching changes in policy, as a result of which approaches to management and conservation are very different today than they were 30 years ago. Furthermore, a variety of indicators have been developed for the evaluation of environmental quality, and water quality and quantity. These are used for formulation of management strategies, negotiation with other users and for monitoring the effects of management. Present models are, however, incomplete and several questions need to be answered so that management can be further improved. Obviously the complexity of the faunas and the aquatic ecosystems themselves will provide rich grounds for basic ecological research for many years, particularly with regard to evolutionary theory, energy flow dynamics and equilibrium/variability states. More urgent perhaps, are such basic questions as: Are tropical rivers different from temperate ones or can the same generalised models be used for both? How do complex species assemblages react to stress and species invasions? What is the natural resilience and recuperative power of river fish assemblages to various types of insult? How much does the river need to be maintained/restored to conserve a reasonable species balance? What are the dynamics and cost–benefit ratios of stocking with selected species? The answers to these and other similar questions will doubtless emerge from future research and form the basis for new models of management.

References

Allen F.R. (1971) Relation between production and biomass. *Journal of the Fisheries Research Board of Canada* **28**, 1573–1581.

Antipa G. (1910) Reguinea inundaibila a Dunarii. Starea ei actuala si mijloacele de a o pue valore. *C.Gobl Institutul di Arti Grafice, Bucharest*, 318 pp.

Bonetto A.A. & Wais I. (1990) El concepto de 'pulso de inundacion' en relacion a las planicies aluviales del systems fluvial Parana-Paraguay. *ECOSUR* 16, 85–98.

Cowx I.G. (1994) Stocking strategies. *Fisheries Management and Ecology* 1, 15–30.

Cowx I.G. (ed.) (1997) *Stocking and Introduction of Fish*. Oxford: Fishing News Books, Blackwell Science, 456 pp.

Cowx I.G. & Welcomme R.L. (1998) *Rehabilitation of Rivers for Fish*. Oxford: Fishing News Books, Blackwell Scientific, 260 pp.

Cushing C.E. McIntire C.D., Cummins K.W., Minshall G.W., Petersen R.C., Sedell J.R. & Vannote R.L. (1983) Relationships among chemical, physical, and biological indices along river continua based on multivariate analyses. *Archiv für Hydrobiologie* **98**, 317–326.

Dodge D.P. (ed.) (1989) Proceedings of the International Large Rivers Symposium (LARS). *Canadian Special Publication on Fisheries and Aquatic Science* **106**, 629 pp.

Food and Agriculture Organization (FAO) (1995) *Code of Conduct for Responsible Fisheries*. Rome: FAO, 41 pp.

FAO (1997a) Inland fisheries. *FAO Technical Guidelines for Responsible Fisheries* **6**, 36 pp.

FAO (1997b) Draft Framework for the responsible use of introduced species. *FAO Fisheries Technical Report* **541**, 123–136.

FAO (1997c) Fisheries Management. *FAO Technical Guidelines for Responsible Fisheries* **4**, 82 pp.

Grainger R.J.R. & Garcia. S.M. (1996) Chronicles of marine fishery landings (1950–1994): Trend analysis and fisheries potential. *FAO FisheriesTechnical Paper* **359**, 51 pp.

Halls A.S., Hoggarth, D.D. & Debnath K. (1999) Impacts of hydraulic engineering on the dynamics and production potential of floodplain fish population in Bangladesh. *Fisheries Management and Ecology* **6**, 261–286.

Hoggarth D.D. & Kirkwood G.P. (1996) Technical interactions in tropical floodplain fisheries of South and South-east Asia. In I.G. Cowx (ed.) *Stock Assessment in Inland Fisheries*. Oxford: Fishing News Books, Blackwell Science, pp. 280–292.

Holcik J. & Kmet T. (1986) Simple models of the population dynamics of some fish species from the lower reaches of the Danube. *Folia Zoologica* **35**, 183–91.

Junk W.J., Bayley P. & Sparks R.E. (1985) The flood pulse concept in river floodplain systems. In D.P. Dodge (ed.) *Proceedings of the International Large Rivers Symposium. Canadian Special Publication on Fisheries and Aquatic Science* **106**, 110–127.

Krykhtin K.L. (1975) Causes of periodic fluctuations in the abundance of the non-andromous fishes of the Amur River. *Journal of Ichthyology* **15**, 826–829.

Lae R, (1995) Modificatio des apports en eau et impact sur les captures de poisson. In J. Quensiere (ed.) *La Peche dans le Delta Central du Niger*. Paris: ORSTOM, pp. 255–265.

Lac R. (1997) Does overfishing lead to a decrease in catches and yields? An example of two West African coastal lagoons. *Fisheries Management and Ecology* **4**, 149–164.

Lauterborn R. (1918) Die geographische und biologische Gliederung des Rheinstromes – *Sber. Heidelberg Akademie Wissenschaft Mathematika-Naturwissenschaft* Kl. B.: 87 S.; Heidelberg (Carl Winters Universitätsbuchhandel).

Lowe-McConnell R.H. (1975) *Fish Communities in Tropical Freshwaters*. London: Longman, 337 pp.

Marine Resources Assessment Group (MRAG) (1994a) *A Synthesis of Simple Empirical Models to Predict Fish Yield in Tropical River Fisheries*. London: MRAG, Imperial College, 29 pp.

MRAG (1994b) *Floodplain Fisheries Projects; Biological Assessment of the Fisheries*. London: MRAG, Imperial College, 186 pp.

McDowall R.M. (1994) On the size and growth of freshwater fish. *Ecology of Freshwater Fish* **3**, 69–79.

Novoa D. (1986) Una revision de la situacion, actual de las pesquerias multiespecificas del Rio Orinoco y una propuesta del ordanamiento pesquero. *Memoria del Sociedad de Ciencias Naturales La Salle* **46**, 167–91.

Oberdorff T., Guegan J-F. & Hugeueny B. (1995) Global scale patterns of fish species richness in rivers. *Ecography* **18**, 345–352.

Pauly D. (1979) Theory and management of tropical multi-species stocks. *ICLARM Studies and Reviews* **1**, 35 pp.

Pauly D. (1980) On the interrelationships between natural mortality, growth parameters, and mean environmental temperature in 175 fish stocks. *Journal du Conseil International pour l'Exploration de la Mer* **39**, 175–192.

PIU/BRAC/DTA (1995) *Quarterly Statistical Reports* **13**, 66 pp.

Richardson R.E. (1921) The small bottom and shore fauna of the middle and lower Illinois River and its connecting lakes, Chillicothe to Grafton: its valuation: its sources of food supply; and its relation to the fishery. *Illinois State Natural History Survey Bulletin* **13**, 363–522.

Statzner B. & Higler B. (1985) Questions and comments on the river continuum concept. *Canadian Journal of Fisheries and Aquatic Sciences* **42**, 1038–1044.

United Nations Environment Programme (UNEP) (1994) *Convention on Biological Diversity*, UNEP/CBD/94/1, Switzerland, 34 pp.

Vannote R.L., Minshall G.W., Cummins K.W., Sedell J.R. & Cushing C.E. (1980) The river continuum concept. *Canadian Journal of Fisheries and Aquatic Sciences* **37**, 130–137.

Ward J.V. & Stanford J (1995) The serial discontinuity concept. Extending the model to floodplain rivers. *Regulated Rivers: Research and Management* **10**, 159–168.

Welcomme R.L. (1979) *Fisheries Ecology of Floodplain Rivers*, London: Longman, 317 pp.

Welcomme R.L. (1985) River Fisheries, *FAO Fisheries Technical Paper* **262**, 330 pp.

Welcomme R.L. (1995) Relationships between fisheries and the integrity of river systems. *Regulated Rivers: Research and Management* **11**, 121–136.

Welcomme R.L. & Bartley D.M. (1998) An evaluation of present techniques for the enhancement of fisheries. *Fisheries Ecology and Management* **5**, 351–382.

Chapter 24
Fisheries science and the managerial imperative

P. HICKLEY and M. APRAHAMIAN

Environment Agency, National Coarse Fisheries Centre, Arthur Drive, Hoo Farm Industrial Estate, Worcester Road, Kidderminster DY11 7RA, UK (e-mail: Phil.hickley@environment-agency.gov.uk)

Abstract

The successful management of river fisheries depends upon effective relationships between scientists and decision makers. It is considered, however, that there is scope for improvement in the partnership. Freshwater fisheries science needs to become less of a self-motivated discipline and more aligned to the management questions being asked. Managers, on the other hand, must better appreciate what science can provide and improve the quality of the information they request. This chapter examines the role of fisheries science, gives details of the perceived priorities for future work, identifies the pressures faced by present-day fisheries managers and describes some of the coping mechanisms. A better mutual understanding of the social, technical, economic and political matrix by scientists, managers and user groups should lead to more robust and defensible decision making.

24.1 Introduction

Field survey work, data processing and analysis of results play an essential role in any fisheries management activity, but one of the most critical stages is the translation of scientific findings into hard management decisions. Hickley (1996) commented that scientists and managers need to be encouraged to develop mechanisms for interpretation of results with the same enthusiasm that has previously been shown towards fish capture techniques. The need for better communication between the scientific community and managers or administrators has been raised many times (Balon & Bruton 1986) and, in this context, several authors have recently expressed consternation about the relationship between the two sides:

- Rose (1997): 'Quantitative methods and population models are cornerstones of fisheries science and management. They should remain so . . . Data are sometimes bountiful. However, wisdom to interpret those data is often scarce . . . Science has lost much credibility because of its head-in-the-sand approach, and modelling with hindsight has helped little to conserve fisheries.'
- Pauly (1994): 'One begins to ask why so many of the things scientists do remain unconnected, left for the general public to interpret . . . Either the environ-

mental scientists help reconstruct the systems of which they have studied parts, explain how these parts interact and how the whole fits, or others will do it for them.'

- Noss (1996): 'Scientific abstractions and fancy technologies are no substitute for the wisdom that springs from knowing the world and its creatures.'

Thus, the problem is one of fisheries science needing to become less of a self-motivated discipline and more aligned to the management questions being asked. Managers, on the other hand, must better appreciate what science can provide and improve the quality of the information they request. What the scientist thinks the manager has requested must be the same as what the manager thinks the scientist is going to provide. However, Lévêque (1997) questioned whether decision makers are willing to accept advice that does not conform to their schemes and whether scientists have sufficient knowledge to give clear and unequivocal advice. For example, serious conflicts between scientists and decision makers often occur in the context of stocking and introductions, e.g. the transfer of Nile perch, *Lates niloticus* (L.) into Lake Victoria (Reynolds & Greboval 1988) and the proposal to introduce Tanganyika sardines, *Limnothrissa miodon* (L.) and *Stolothrissa tanganicae* (Bloch) into Lake Malawi (Turner 1982; Eccles 1985).

This chapter examines the role of fisheries science, gives details of the perceived priorities for future work, identifies the pressures faced by present-day fisheries managers and describes some of the coping mechanisms.

24.2   Role of fisheries science

Under UK legislation (Environment Act 1995) the Environment Agency has a general duty to maintain, improve and develop 'salmon, trout, freshwater fish and eel fisheries' in England and Wales. Within these groups the types of fisheries can be wide-ranging, e.g. recreational or commercial, river or lake, casual or competition, natural or artificial. This basic fisheries resource needs to be managed so as to optimise the social and economic benefits from its sustainable exploitation. Fisheries science should support the interface between facts and perceptions in understanding and managing overall fishery performance, i.e. the total package of fish stocks, fish habitat, fishing activity and angler environment (Fig. 24.1). In doing so, fisheries science output can target most categories of environmental management, e.g.:

- fisheries – protection and enhancement of stocks;
- water resources – minimum acceptable flows, impact of drought;
- water quality – state of the environment, quality objectives;
- flood defence – impact of schemes;
- conservation – fish habitat, threatened species;
- sustainable development.

In a typical situation where management activity moves from the assessment of status through to implementation of decisions, fisheries science has a key role to

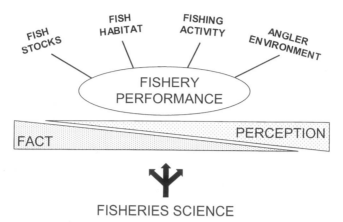

Figure 24.1   The components of overall fishery performance and the place of fisheries science in supporting the management of facts and perceptions

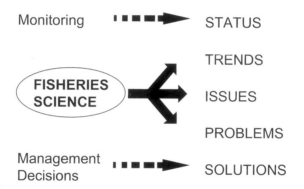

Figure 24.2   The role of fisheries science in fisheries management activity

fulfil in aiding the determination of trends, issues and the associated problems that require resolution (Fig. 24.2). For example, in the guidance given to the Environment Agency by the UK government on contributing to sustainable development, it is stated that decisions should be based on the best available scientific information and analysis of risks.

24.3   Priorities

With regard to improving the interface between science and management, common areas of need can be recognised at all levels. What may be perceived initially as a local information requirement will often, when upwardly consolidated, be reflected as underpinning some international concern.

24.3.1 *Local priorities*

A questionnaire was used to assess the priority given by local fisheries staff to obtaining information in 12 key subject areas of importance to the Environment Agency (1999). Funding opportunities were indicated as being high priority by 81% of respondents, habitat improvements by 73% and customer liaison by 65%. In decreasing order of rating by respondents, the remaining topics deemed to be of high priority information need were: minimum flows (62%); cormorant predation (58%); fish recruitment (58%); mitigation (58%); fish migration (54%); new survey methods (50%); disease (42%); licence sales (42%); pollution (38%); restocking (35%); enforcement (31%); access for angling (23%); and closed seasons (12%). Apart from the purely administrative topics, such as funding and licence sales, most of these information needs are dependent to some degree upon fisheries science.

24.3.2 *National priorities*

In its Environmental Strategy, the Environment Agency (1997) has identified nine key environmental themes under which its duties for regulation and priorities for management can be addressed. All nine themes can be deemed to have links to fisheries.

(1) *Addressing climate change* – resource status assessment, ecological and population process modelling in marine and freshwater environments.
(2) *Improving air quality* – quantifying impacts of air-borne contaminants on fisheries, e.g. acidification.
(3) *Managing water resources* – defining flow needs for fish and fisheries.
(4) *Enhancing biodiversity* – establishing fish community structure and size; stock management in protected habitats.
(5) *Managing freshwater fisheries* – technical basis of sustainable policies and strategies; understanding, guidance and techniques.
(6) *Delivering integrated river basin management* – coordination, planning and technical guidance for fisheries action plans and strategies.
(7) *Conserving the land* – measuring the impacts of land use on fish, e.g. erosion and siltation, contributing to policy development.
(8) *Managing waste* – advice on means to detect impacts of waste.
(9) *Regulating major industry* – effective standards for environmental protection.

24.3.3 *International priorities*

Accounting for local and national requirements can produce key worldwide recommendations for the future. With regard to fisheries science and management issues, such recommendations have been forthcoming from a series of international workshops, namely, EIFAC Symposium on Management of Freshwater Fisheries (van Densen *et al.* 1990), International Symposium and Workshop on Rehabilitation of Freshwater Fisheries (Cowx 1994a), Eighteenth Session of EIFAC (FAO 1994) and EIFAC Symposium on Recreational Fisheries (Hickley & Tompkins 1998). By consolidating these recommendations into common activity areas the following international priorities can be identified.

State of the environment:

- clarification of objectives for collection, analysis and documentation of data;
- adoption of a more consistent approach, so as to provide comparable baseline data;
- a pan-European survey of recreational fisheries, e.g. in the year 2000;
- monitoring of the availability and quality of inland fisheries.

Research and development:

- fisheries rehabilitation methodology;
- optimal habitat requirements for different species;
- socio-economic value estimates of fisheries, especially recreational fisheries.

Education and liaison:

- education of decision makers, user groups and general public to reduce conflicts in multiple-use situations;
- improved understanding and communication between managers, scientists and fishermen.

Best practice:

- guidelines for decision making and planning;
- establishment of a code of good practice for recreational fishing;
- the true economic value of fisheries to be included in decision-making processes with all projects related to environmental management.

Policy and strategy:

- integrated, sustainable, long-term approach to aquatic resource management to provide for the maintenance of biodiversity, the protection and enhancement of fish stocks, and optimal socio-economic benefits to society;
- catchment-based planning and administration of inland fisheries;
- periodic review of policies and strategies to ensure practices are appropriate to current problems.

24.4   Research and development

Appropriate research and development (R&D) can provide the basis for sound advice to decision makers. Although there are times when pure academic research can be justified, the general role of research and development in the context of fisheries management is to prevent decisions being compromised by lack of appropriate knowledge. For example, there has been no formal UK study on the capital or rental value of cyprinid fisheries (Postle & Moore 1998) and because the absence of such information was causing difficulties, the Environment Agency has recently commissioned an appropriate piece of research.

An environmental R&D programme should assist and develop environmental assessment and monitoring capability, support policy development for managing and protecting the environment, improve efficiency and effectiveness of the business, and increase strategic knowledge and understanding to inform possible future actions.

The need for R&D has been recognised in the European Union Fifth Framework Programme for Research and Technological Development (EU 1997). Agreement was reached in February 1998 by science ministers that the programme structure should comprise four themes:

  (i) improving the quality of life and living resources (2239 MECUs);
 (ii) creating a user-friendly information society (3363 MECUs);
(iii) promoting competitive and sustainable growth (2389 MECUs);
(iv) energy, environment and sustainable development (2088 MECUs).

Note that there was strong support for greater emphasis on socio-economic and ecological factors at the expense of technologies.

24.4.1   *Fisheries research and development*

Within the Environment Agency the business priorities driving the fisheries R&D programme are based on an overall action plan (Environment Agency 1998a) and include:

● effective, scientifically-sound methods to classify and assess fisheries resources;
● improved understanding of customer needs and fisheries economics to ensure fair financial strategies and better communications;
● formulation of rational policies and development of sustainable management strategies based on sound, high quality science;
● application of cost-effective methods for restoration and maintenance of fisheries.

It is important, also, to recognise that the resource comprises not just fish stocks, but includes their habitat and, crucially, the economic and social features of the fisheries that the stocks actually or potentially support. An understanding of the fishermen's environment is an essential and previously under-researched topic.

Three predominant research programme themes flow from the above rationale and these are given below with examples of current Environment Agency projects.

(1) *To describe the status of fisheries resources*: stock and habitat classification and inventories; evaluation of economic and social values; development of spawning target methodologies.

(2) *To understand the processes behind resource dynamics and response to pressures*: recruitment variation in coarse fish; impacts of endocrine disrupters; factors affecting salmon decline in chalk rivers.

(3) *To optimise decision-making and practical management procedures*: habitat restoration methods; swimming speeds in fish (for fish-pass design); life cycle simulation modelling to evaluate risks and options; development of lightweight electric fishing equipment.

Note that appropriate plans for the structured implementation of outputs should always be included in R&D programmes. It is essential that the management of outcomes is approached with the same skill and enthusiasm as the research itself. Ignoring the findings or the hobbyist mentality can both be equally wasteful of investment.

24.5   Pressures on fisheries and management

Not only are there the classic stresses and strains impacting upon the environment itself but there are now many additional pressures acting upon the fisheries manager. It is within these two scenarios that the scientist must operate.

24.5.1   *Pressures on fisheries*

Perturbations affecting the status of inland fisheries are many and varied. Although the primary pressures on fish stocks could at one time be considered to be intense fishing and pollution, more recent anthropogenic activities associated with land and water use developments have resulted in habitat loss and degradation. Principal impacts on fish stocks are summarised by Cowx (1994a):

● pollution, e.g. eutrophication, toxic waste, acid rain;
● river engineering, e.g. impoundments, land drainage, flood alleviation;
● habitat loss, e.g. land reclamation, removal of riparian vegetation;
● afforestation and other land use changes;
● introduction of and invasion by non-native species;
● over-harvesting;
● shifts in biotic communities stressed by environmental insults.

24.5.2 *Pressures on the fisheries manager*

The modern-day fisheries manager has to contend with many recently introduced concepts, albeit honourable, which can make progress less than straightforward. In the UK many of these new management pressures arise from government guidance, or similar.

● *Integrated catchment management.* Linking fisheries programmes with other water resource management regimes requires much consultation with stakeholders and compromise may be necessary for the resolution of conflicts. Of particular relevance is the duty placed on the Environment Agency to promote conservation of inland waters and their use for recreation.

● *Sustainable development.* The Environment Agency has to discharge its functions of protecting and enhancing the environment so as to make an appropriate contribution to sustainable development. Also, in this context, FAO (1995) urged users of living and aquatic resources to conserve aquatic ecosystems and stated that the right to fish carries with it the obligation to do so in a responsible manner.

● *Biodiversity.* The Convention on Biological Diversity, Rio, 1992, set the scene for addressing this issue. Conscious of the intrinsic value of biological diversity and of the ecological, genetic, social, economic, scientific, educational, cultural, recreational and aesthetic values of biological diversity and its components, the objectives of the Convention were the conservation of biological diversity, the sustainable use of its components, and the fair and equitable sharing of the benefits arising.

● *Social welfare.* The Environment Agency in carrying out certain of its functions, which include fisheries, has to have regard to any effect on social well being of local communities in rural areas.

● *Cost–benefit requirement.* In general, initiatives to generate environmental benefit are often constrained by financial considerations in terms of not imposing excessive costs (in relation to benefits gained) on either regulated organisations or society as a whole.

● *European directives.* EU directives which relate to overall water quality, and thus to fisheries, are implemented in the UK either under Acts and Statutory Instruments or directly through the policy and powers of the Environment Agency. The Fisheries Directive (78/639/EEC) sets physico-chemical limits for designated salmonid and cyprinid waters. The Habitats Directive (92/43/EEC) concerns the designation of Special Areas of Conservation for the protection of habitats, fauna and flora. The draft Water Framework Directive, currently being formulated, is designed to establish a framework for the protection of surface freshwater, estuaries, coastal waters and groundwater, and, as such, will protect and enhance the status of aquatic ecosystems.

● *Technical limitations.* The range of survey methods available means that some level of stock assessment can be carried out, whatever the nature of the water body being studied. The degree to which any appraisal is quantitative, however, often leaves much to be desired. This is especially so where the fishery is large and difficult to sample (Hickley 1996). Although, in recent years, improvements are evident in many methodologies, especially fish counters, hydro-acoustics and electric fishing, technical limitations remain an issue.

● *Public expectations.* Success is as much about management of perception as it is about reality. Appropriate levels of consultation with stakeholders are a present-day necessity. Government guidance to the Environment Agency requires the development of close and responsive relationships with the public and representatives of local communities. In addition, as public expectations rise, there is an increasing need for individual behaviours to become more customer focussed.

● *Political influences.* At the highest political level, government sets in motion the policies that influence the way in which those organisations that have statutory duties for fisheries, such as the Environment Agency, approach their activities. At the local level, there is scope for 'political' debate not only between regulators and stakeholders, but also as to what the role of the fisheries scientist should be in expressing a view. Many managers would expect scientific results to be accompanied by some degree of interpretation. However, Langford (1997) reported that at a meeting of the American Fisheries Society at least two presenters in one session stressed that their projects were to provide information for fish and habitat management but they were to provide no recommendations or opinions as to how it should be acted upon.

24.6   Mechanisms

Appropriate mechanisms are needed to control the various stages of data analysis, planning, prioritisation, decision making and outputs. Before any environmental management action is taken, a snapshot in time is required, followed by the identification of stresses and an appraisal of options, including both risks and opportunities as per the Environment Agency State, Pressure, Options and Response model (Environment Agency 1997). Guidance through the process of planning and managing an inland fishery, using a framework of sequential questions ranging from general issues to prescriptions for the application of specific techniques, is given by Welcomme (1998).

24.6.1   *Prioritisation*

When planning what needs to be done, especially in the all-too-common situation of limited resources, some degree of prioritisation becomes essential. Several basic techniques exist, some of which are more effective and suitable then others, according to the situation to be dealt with. As much background information as possible should be used, even though the final judgement on ranking will usually involve varying degrees of subjectivity. Nonetheless, if the process is properly structured, the appraisal of priorities is more likely to be successful. Two common approaches are: first, a priority-based budget approach using a simple scoring of the importance of the work to the business; and second, separate weighting of component parts against pre-set criteria and the summation of scores.

For the single score approach, a typical priority-planning rating scale might award 10 points for an activity perceived as 'essential' (e.g. a statutory duty) down to 1 point for 'unlikely ever to be funded'. The mid-range scores provide a decision fulcrum with 6 points for 'significant' and 5 points for 'desirable but to be dropped if funding is reduced'. The more complex process is exemplified by the Environment Australia Action Plan for Freshwater Fishes (Wager & Jackson 1993). To correctly prioritise the recommendations for different species within this conservation programme, each species was given a 'high', 'medium' or 'low' rating against each of five assessment criteria; degree of threat, recovery potential, genetic distinctiveness, ecosystem importance and social value.

24.6.2   *Risk analysis*

A key component of the management process that can be easily under-represented is risk analysis of potential actions. Some outcomes of proposed actions can depend on occurrences beyond any reasonable control and, thus, the absence of risk analysis in the decision-making process can be a major weakness of current fisheries management systems (Lane & Stephenson 1998). It is generally accepted that risk analysis comprises the two components of risk assessment and risk management (Balson *et al.* 1992).

The risk assessment element identifies possible consequences and their likelihood of occurrence, whereas risk management uses this information to evaluate and steer decision alternatives. The first stage in any risk analysis should be to define the risk scenario whereby the objectives and constraints of the fishery system are quantified and the variables, both controllable and uncontrollable, are described. Risk screening should be used to separate low-priority risks from high-priority risks. For complex, high-priority risks a tailored assessment is likely to be necessary, whereas for less complex risks a generic assessment may be adequate.

In general terms, the protocols surrounding decisions and their implementation should become more protective and specific as the risks increase, although the degree to which this occurs can be influenced by the acceptability of the particular risk involved (Fig. 24.3). Activities in this context could be such things as fish introductions, habitat modification or biomanipulation. For example, the stocking of an already resident species into a lake would not invoke the same level of protocol as introducing a non-native fish into a river system.

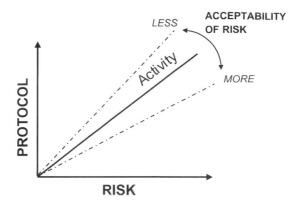

Figure 24.3   Schematic chart illustrating the principle of introducing more stringent protocols as the risk associated with any activity increases. The slope of the XY line can be varied according to the acceptability of the risks identified

24.6.3   *Options and decisions*

It is inevitable that at some stage managers have to undertake options appraisal and decision making. Clearly, there are advantages to following one or more of the established logical regimes such as SWOT (strengths, weaknesses, opportunities, threats) and STEP (social, technical, economic, political). It is most important that scientific information is fed into the management process at the right time and that it is of the right quality and quantity. The various input stages are indicated in the generic flow chart (Fig. 24.4). Existing facts should be questioned, information gaps identified and the appropriate data collected for interpretation. Specific examples of

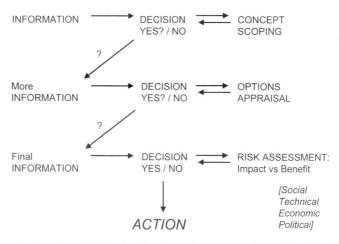

Figure 24.4   Generic flow chart indicating the dynamic process of applying fisheries science to a typical management decision making regime. The left-hand column represents provision of scientific information and the right hand column, consultation and debate

flow charts for decision making in practice are given by Cowx (1994b, 1999) for stocking and Langton *et al.* (1996) for habitat management.

24.6.4 *Provision of data*

Scientific outputs can be provided as information for managers in several forms.

- *Raw data.* This is rarely an appropriate form in which to supply information to decision makers although, often, it is asked for by way of appendices to reports. Such requests, usually justified under the guise of giving credibility to summary statements, can generate a suspicion that management is distrustful of its scientific advisors.
- *Processed data.* Results in this format are most appropriate where decision makers either have the expertise to make interpretive judgements or take care to involve the scientists in any debate. Data processing itself can range from analysis of real data, using classic statistical methods (Zar 1996) or complex modelling with 'black box' solutions (e.g. Mastrorillo *et al.* 1999), through to professional consensus, such as Delphi (Zuboy 1980). However, the more processed the output, the more care that is needed to ensure the correct issues are being addressed.
- *Interpretive reports.* Strategic overviews with recommendations are particularly useful in relation to the desirability of taking an integrated approach to fisheries management. In assessing current status and trends, the use of classification systems can be helpful. For example, the Environment Agency has, amongst others, both a fisheries classification scheme (National Rivers Authority 1994) and a habitat classification scheme (Raven *et al.* 1997). For the promotion of future activity, some form of action plan is desirable, which in draft form can support the consultation process and, in final form, becomes the agreed working document. For example, the Environment Agency produces Local Environment Agency Plans (Environment Agency 1997) to facilitate integrated catchment management and, in association with English Nature, Species Action Plans to address biodiversity issues (Environment Agency 1998b).

24.7 The way forward

Ensuring that the right questions are asked at the outset can fulfil much of the need for an improved interface between scientists and managers. It is important that managers clearly specify their requirements. For example, where this impinges upon field sampling, some forethought as to the actual accuracy and precision that is necessary (Bohlin *et al.* 1990) can enable best deployment of available resources. Scientists, for their part, must recognise that managers are obliged to make decisions, possibly against incomplete data sets, whilst they themselves might argue for further research. However, it is often feasible to reach a compromise whereby a best interim solution can harness a blend of fact and expert opinion. Invariably, it is better to have an approximate answer to the right question than an exact answer to the wrong question.

For the implementation of decisions, care is necessary to ensure that integration and acceptability of actions is achieved. Communication and education is paramount because best practice options can fail when operational staff and user groups are not aware, misunderstand or simply ignore managerial directives. Also, throughout the process of addressing fisheries objectives it is essential that an appreciation of the wider environmental, economic and social issues be maintained. Environmental practice is often structured around the notion of separable problems, whereas there should be recognition of interdependencies, including trade-offs amongst policies related to competing social ends (Davos 1977).

Where high level strategies are to be enacted, the best approach is to develop a framework of guiding principles, complete with interpretative criteria that allow local flexibility within pre-set boundaries. This encourages consistency of approach whilst permitting justified variations in detail between catchments. For example, international agreement (NASCO 1998) on management actions for Atlantic salmon, *Salmo salar* L. has, for England and Wales, culminated in the setting of river-based spawning targets (Milner *et al.*, Chapter 25).

In the future, fisheries science must continue to play a major and vital role in the management of freshwater fisheries. In some instances this may mean better recognition of the status of science in the overall process. Scientists and managers must strive to meet the modern-day demands placed upon them by the environment and society. Accordingly, it is recommended that those in authority should:

- improve cross boundary communication;
- ask the right questions to get the right answers;
- use a logical approach, e.g. state, pressures, options & response;
- consider all aspects, e.g. social, technical, economic & political;
- attempt to integrate attitudes and actions;
- establish guiding principles for consistency of approach;
- develop framework procedures with interpretative criteria for local flexibility;
- concentrate on outcomes rather than output.

Disclaimer

Note that the views expressed are those of the authors and not necessarily those of the Environment Agency.

References

Balon E.K. & Bruton M.N. (1986) Introduction of alien species or why scientific advice is not heeded. *Environmental Biology of Fishes* **16**, 225–230.

Balson W.E., Welsh J.L. & Wilson D.S. (1992) Using decision analysis and risk analysis to manage utility environmental risk. *Interfaces* **22**, 126–139.

Bohlin T., Heggberget T.G. & Strange C. (1990) Electric fishing for sampling and stock assessment. In I.G. Cowx & P. Lamarque (eds) *Fishing with Electricity*. Oxford: Fishing News Books, Blackwell Science, pp. 112–139.

Cowx I.G. (1994a) Strategic approach to fishery rehabilitation. In I.G Cowx (ed.) *Rehabilitation of Freshwater Fisheries* Oxford: Fishing News Books, Blackwell Science, pp. 3–10.

Cowx I.G. (1994b) Stocking strategies. *Fisheries Management and Ecology* **1**, 15–31.

Cowx I.G. (1999) An appraisal of stocking strategies in the light of developing country constraints. *Fisheries Management and Ecology* **6**, 21–35.

Davos C.A. (1977) Towards an integrated environmental assessment within a social context. *Journal of Environmental Management* **5**, 297–305.

Densen W.L.T. van, Steinmetz B. & Hughes R.H. (eds) (1990) *Management of Freshwater Fisheries*. Wageningen: Pudoc, 649 pp.

Eccles D.H. (1985) Lake flies and sardines – a cautionary note. *Biological Conservation* **33**, 309–333.

Environment Agency (1997) *An Environmental Strategy for the Millennium and Beyond*. Bristol: Environment Agency, 28 pp.

Environment Agency (1998a) *An Action Plan for Fisheries*. Bristol: Environment Agency, 24 pp.

Environment Agency (1998b) Species management in aquatic habitats: Compendium of project outputs – species action plans and management guidelines, *R&D Project Record W1/i640/3/M*. Bristol: Environment Agency, 247 pp.

Environment Agency (1999) A scoping study to establish the issues on coarse fish and fisheries in England and Wales, *R&D Technical Report W137*. Bristol: Environment Agency, 117 pp.

European Union (EU) (1997) Fifth framework programme for research and technological development (1998–2002), *Commission Working Paper on the Specific Programmes: Starting Points for Discussion*. COM(97) 553.

Food and Agriculture Organization of the United Nations (FAO) (1994) Report of the eighteenth session of the European Fisheries Advisory Commission, Rome, 17–25 May 1994. *FAO Fisheries Report*, No. 509, 78 pp.

FAO (1995). *Code of Conduct for Responsible Fisheries*. Rome: FAO, 41 pp.

Hickley P. (1996) Fish population survey methods: a synthesis. In I.G. Cowx (ed.) *Stock Assessment in Inland Fisheries*. Oxford: Fishing News Books, Blackwell Science, pp. 3–10.

Hickley P. & Tompkins H. (eds) (1998) *Recreational Fisheries: Social, Economic and Management Aspects*. Oxford: Fishing News Books, Blackwell Science, 310 pp.

Lane D.E. & Stephenson R.L. (1998) A framework for risk analysis in fisheries decision-making. *ICES Journal of Marine Science* **55**, 1–13.

Langford T. (1997) A worrying episode! *Fisheries Society of the British Isles Newsletter, September 1997*, p. 6.

Langton R.W., Steneck R.S., Gotceitas V., Juanes F. & Lawton P. (1996) The interface between fisheries research and habitat management. *North American Journal of Fisheries Management* **16**, 1–7.

Lévêque C. (1997) *Biodiversity Dynamics and Conservation: The Freshwater Fish of Tropical Africa*. Cambridge: Cambridge University Press, 426 pp.

Mastrorillo S., Dauba F. & Lek S. (1999) Predictive model of fish species richness in the Garonne river basin (SW France) using Artificial Neural Networks. *Proceedings of the Ninth International Congress of European Ichthyologists*, Trieste, Italy.

National Rivers Authority (NRA) (1994) The NRA National Fisheries Classification Scheme – A guide for users, *R&D Note 206*. Bristol: National Rivers Authority, 102 pp.

North Atlantic Salmon Conservation Organization (NASCO) (1998) *North Atlantic Salmon Conservation Organisation Agreement on a Precautionary Approach*, CNL(98)**46**, 4 pp.

Noss R. (1996) The naturalists are dying off. *Conservation Biology* **20**, 1–6.

Pauly D. (1994) *On the Sex of Fish and the Gender of Fisheries Scientists: A Collection of Essays in Fisheries Science.* London: Chapman & Hall, Fish and Fisheries Series 14, 250 pp.

Postle M. & Moore L. (1998) Economic valuation of recreational fisheries in the UK. In P. Hickley & H. Tompkins (eds) *Recreational Fisheries: Social, Economic and Management Aspects.* Oxford: Fishing News Books, Blackwell Science, pp. 184–199.

Raven P.J., Fox P., Everard M., Holmes N.T.H. & Dawson F.H. (1997) River Habitat Survey: a new system for classifying rivers according to their habitat quality. In P.J. Boon & D.L. Howell (eds) *Freshwater Quality: Defining the Indefinable?* Edinburgh: The Stationary Office, pp. 215–234.

Reynolds J.E. & Greboval D.F. (1988) Socio-economic effects of the evolution of Nile perch fisheries in Lake Victoria, *CIFA Technical Paper* No. 17. Rome: FAO, 160 pp.

Rose G.A. (1997) The trouble with fisheries science! *Reviews in Fish Biology and Fisheries* **7**, 365–370.

Turner J.L. (1982) Lake flies, water fleas and sardines. Studies on the pelagic ecosystem of Lake Malawi, *FAO FI:DP/MLW/019,Technical Report* **1**, 165–73.

Wager R. & Jackson P. (1993) The action plan for freshwater fishes. *Environment Australia Web Site*, 'www.anca.gov.au/plants/threaten/plans/action-plans/freshwater-fishes/13.htm'.

Welcomme R.L. (1998) Framework for the development and management of inland fisheries. *Fisheries Management and Ecology* **5**, 437–457.

Zar J.H. (1996) *Biostatistical Analysis.* Englewood Cliffs: NJ: Prentice-Hall, 662 pp.

Zuboy J.R. (1980) The Delphi Technique: A potential methodology for evaluating recreational fisheries. In J.H. Grover (ed.) Allocation of fisheries resources, *Proceedings of the Technical Consultation of Fishery Resources, Vichy, France, April 1980*, pp. 518–527.

Chapter 25
The use of spawning targets for salmon fishery management in England and Wales

N.J. MILNER, I.C. DAVIDSON and R.J. WYATT

National Salmon and Trout Fisheries Centre, Environment Agency, St Mellon's, Cardiff CL3 0LT, UK
(e-mail: nigel.milner@environment-agency.gov.uk)

M. APRAHAMIAN

Fisheries Department, Environment Agency, Warrington, UK

Abstract

Salmon fisheries management in England and Wales has, until recently, lacked clear objectives that could be translated into practical planning, monitoring and control. A national salmon strategy has established strategic objectives for the resource underpinned by biological targets, based on spawning requirements, to maintain recruitment in individual rivers. This approach has become feasible through improved understanding of stock dynamics and habitat relationships, supported by enhanced monitoring procedures.

This chapter describes the history and rationale behind spawning targets and outlines how they are being set for the principal rivers in England and Wales. Compliance against targets is currently based on rod catch statistics, supported by parameter estimation from intensively studied catchments. The approach is critically reviewed and further developments are discussed. Spawning targets represent only one component of the assessment required to fully evaluate stock and fishery status. For example, social and economic objectives also have to be considered, but their adoption is contributing to a more objective, structured approach to management, as well as better definition of monitoring and research priorities.

Keywords: Atlantic salmon, assessment, management, spawning targets.

25.1 Introduction

Salmon, *Salmo salar* L., fisheries management in England and Wales has undergone a change in recent years, with the introduction of targets linking management objectives to stock assessment (National Rivers Authority 1996). The principles of target-based management are not new, but the underlying science has developed extensively since the basis of stock and recruitment relationships was established in the 1950s (e.g. Ricker 1954; Beverton & Holt 1957). However, most progress has been applied to large, and mostly marine, fisheries (e.g. Smith *et al.* 1993). In this chapter the background to the use of spawning targets for Atlantic salmon is outlined and some of the issues and development needs behind the approach are discussed.

25.2   Development of target-based management

Prior to 1996 the broad aim of fisheries management in England and Wales was to improve, maintain and develop fisheries, but this raised many questions such as: at what point should improvement or development be started, at what level should a fishery be maintained and how should management be evaluated? Target setting and evaluation should be key stages in the management of any resource and informal targets are probably set intuitively by fishery managers when selecting options for salmon management. However, the target-setting approach was only formally recognised through the Salmon Strategy for England and Wales (National Rivers Authority 1996), which included amongst its objectives the intention that 'Individual salmon stocks and the environment in which they live should be managed to optimise recruitment to home water fisheries.' The practical implementation of this strategy is through the preparation of Salmon Action Plans (SAPs) for individual rivers which involve, amongst a range of assessment and management practices, the formulation of river-specific spawning targets. The quantitative framework of the Salmon Strategy was strengthened by the Environment Act in 1996, which introduced duties requiring greater support from fisheries science. These included provision for sustainable resource management, adopting the precautionary approach where applicable, the inclusion of costs and benefits into decisions and the protection of biodiversity.

An important factor was development in the management of high seas salmon fisheries, where spawning targets for North America had been used since 1977 to provide catch advice to the North Atlantic Salmon Conservation Organisation (NASCO) for the West Greenland fishery. Following a workshop to develop spawning targets for salmon stocks (ICES 1994), NASCO recommended that targets be set for all principal rivers in its North East Commission area, including England and Wales (NASCO 1996). Spawning targets have been set for some French rivers (Prevost & Porcher 1996) and are the basis for proposed catch controls in Irish salmon fisheries (Anon. 1996).

25.3   Options for spawning targets

25.3.1   *Spawning targets*

A central feature of the salmon life cycle is population regulation through a stock–recruitment (S–R) relationship (e.g. Fig. 25.1) driven by density-dependent processes, although, depending on the point at which recruitment is estimated, density-independent variables may mask the stock effect (Gibson 1993). It is convenient to think of smolts as recruits, being at the end of the freshwater phase during which density-dependent regulation occurs. Most mortality at sea is thought to be density-independent (Jonsson *et al.* 1998), so adult spawners and their potential egg deposition can be estimated from smolts, knowing parameters for marine

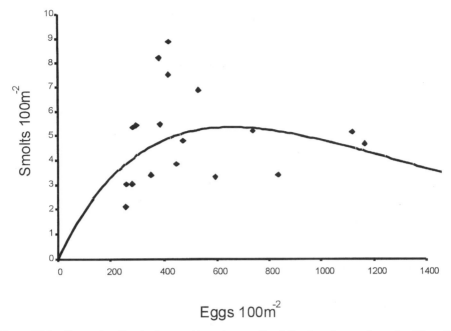

Figure 25.1 Example of a stock recruitment curve for Atlantic salmon, from the River Bush, Northern Ireland (adapted from Kennedy & Crozier 1993)

survival, age structure and fecundity. The direct relationship between smolts and the number of spawners (or eggs those spawners produce) is known as the replacement line. The combination of egg to smolt S–R curves and replacement line represents a complete life cycle model and gives three main reference points that can establish potential spawning targets (Fig. 25.2).

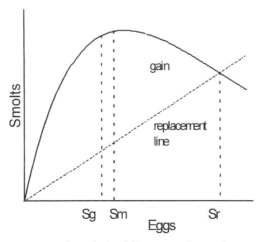

Figure 25.2 Spawning target options derived from a stock recruitment curve and replacement line. *Sg* = maximum gain, *Sm* = maximum smolts and *Sr* = maximum spawning stock

Replacement stock (Sr)

For all S–R curves there is a point where recruits on average replace the spawners that generated them. Under equilibrium conditions the population oscillates around the point where the replacement line intersects the S–R curve. In an unexploited stock this replacement point represents an upper limit to spawners. Such a target would imply no fishing, so on its own is of little practical value, but could be used to provide a base for other targets (e.g. expressing exploitation as some proportion of *Sr*).

Maximum recruitment (Sm)

With dome-shaped S–R curves definition of maximum recruitment is straightforward and mathematically unambiguous. Maximum smolt output offers an attractive definition because it has direct meaning for the productive capacity of a river. Even for asymptotic curves, a reproducible point could be set as some function of the asymptote or initial survival rate (the initial slope of the line where the influence of stock is negligible), but rational definition of such functions is contentious (ICES 1994).

Maximum gain (Sg)

At all escapement levels below *Sr* more recruits are produced than are required to replace their parents. This surplus is the difference between the replacement line and the recruits indicated by the S–R curve. The surplus recruits represent the potential for harvesting by fisheries and is termed variously the surplus yield, reproduction or gain. For all S–R curves the net gain – escapement curve is dome-shaped, so a point of maximum gain (*Sg*) can always be unambiguously defined. *Sg* defines an escapement level, and thus exploitation rate, that maximises potential catch under the life-cycle characteristics applying to a stock, and *Sg* is always less than *Sm*.

25.3.2   *Assessment constraints*

The selection of practical targets is constrained by the availability of practicable assessment methods. Smolt output reflects the production potential and environmental quality of each catchment so is of great potential value in managing the freshwater phase. But there are no long-term, direct assessment data on wild smolt outputs in England and Wales, and the costs of trapping programmes to derive these are very high. Constant effort (timed) electric fishing surveys of fry abundance have been used to provide indices of recruitment to the River Bush, Northern Ireland (Kennedy & Crozier 1993). Parr abundance stratified by habitat characteristics has been used to estimate smolt outputs in French rivers (Bagliniere *et al.* 1993). When combined with the measurement of biological variables, such as age and growth, and habitat measurements (Milner *et al.* 1998), juvenile population data provide essential insights into the factors controlling freshwater production and have an important role in salmon stock assessment. However, cost and precision issues render large-scale quantitative juvenile surveys difficult to sustain as long-term, absolute stock assessment methods.

Escapement is estimated by subtracting in-river rod catch (legal and illegal) and natural mortality from runs into each river. Run estimates are preferably made from direct counts using counters and traps, but these are available on comparatively few rivers. Rod catches are recorded through licence returns in all the principal rivers of England and Wales, so a target procedure based on these has immediate relevance to available data, providing the relationship with run can be adequately described. Escapement estimates do not distinguish between impacts arising at any preceding stage and are subject to the full range of stochastic variables acting in freshwater and marine environments; hence the need for complementary assessment tools.

25.4   Developing the spawning target approach

The approach involves four main processes: selecting the target, setting targets for individual rivers, assessing egg deposition, and testing compliance against targets. A final stage involves selecting and implementing options to achieve management aims. Guidelines for target setting and compliance assessment have been established (Environment Agency 1996), and are summarised below.

25.4.1   *Spawning target selection*

NASCO has adopted maximum gain (Sg) as the preferred reference point defining the minimum biological acceptable level (*MBAL*) of spawning stock. A stock that is on average at *MBAL* has a 50% chance of being below it and, depending on its variability, could intermittently be at spawning levels which compromise survival of the stock. In recognition of this, NASCO termed MBAL (Sg) a conservation limit, a threshold below which stocks should not fall, and recommended that to achieve this managers should aim to hold escapement at a rather higher (but unspecified) level, termed the management target.

Any changes in life cycle characteristics will alter the Sg value, so it would be lower under conditions of low carrying capacity (Fig. 25.3), for example. If targets are set based on the historical conditions that established the S–R curve, but these are no longer relevant, then the limit may be too high or low for present-day conditions (Hilborn & Walters 1992). Even if a freshwater habitat is severely degraded, it is still possible to set and manage around a maximum gain target, which simply exploits the depleted stock most efficiently. This illustrates a potential conflict between targets set for purely catch management purposes (which might be satisfied with Sg as the target) and those set for purely stock or environmental management (which might prefer Sm as a target). In practice, under present-day circumstances of low marine

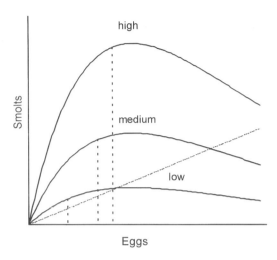

Figure 25.3 Effect of changing carrying capacity on maximum gain targets. In this hypothetical example high, medium and low carrying capacities correspond to Sg values of 528, 435 and 244 eggs 100 m^{-2} respectively

survival in many stocks (NASCO 1997) and degraded carrying capacity in some rivers, Sg conservation limits based on former conditions are likely to be conservative.

Maximum gain, whilst having some drawbacks, is conceptually linked to a key objective of the Environment Agency's Salmon Strategy and corresponds to common practice emerging through NASCO (NASCO 1998), so is the conservation threshold currently adopted by the Environment Agency.

25.4.2 *Estimating spawning targets for individual rivers*

Wyatt and Barnard (1997a) reviewed the technical problems in establishing S–R curves and associated targets. Stock–recruitment relationships are normally based on whole river assessments but, because they require long-time series, few good data sets are available. In the UK, only the River Bush in Northern Ireland has a data set that allows derivation of a stock–recruitment curve for a whole river, based on direct trap counts of smolts and returning adults.

Stock–recruitment curves are thought to be specific to different habitat types (Symons 1979; Elliott 1987; Gibson 1993; Kennedy & Crozier 1995), implying that individual catchments would also have unique S–R curves, as determined by the availability and distribution of habitat types. Applying an S–R curve from one river to another without some form of correction would therefore be questionable, unless their productivity was known to be similar. In Canada, a spawning target for maximum smolt production (*Sm*, see above) in fluvial habitat has been set at 240 eggs 100 m^{-2} (ICES 1994), and applied across all Canadian rivers, incorporating corrections for lacustrine habitat (O'Connell & Dempson 1995). However, this value is regarded as probably too low for many European rivers where, on the basis of observed smolt outputs, productivity appears to be generally higher (ICES 1994).

A whole river S–R curve can be thought of as a weighted mean of a large number of curves characteristic of the constituent habitat types of the catchment (Fig. 25.4). This premise provides a way to transport S–R relationships between rivers. The slope of the replacement line is a function of the marine survival, age of return, size and fecundity, so these parameters also have to be estimated for individual rivers. Wyatt and Barnard (1997a, b) have developed a method using a simple fish–habitat model to apply weighted corrections to the River Bush S–R curve (using a Ricker model), based on the habitat features of recipient rivers. Applying this procedure, corrected S–R relationships and targets are being estimated for the principal salmon rivers in England and Wales.

Provisional targets have been set for 66 rivers, 13 of which are based upon river-specific parameters and developed for SAPs (Anon. 1998). The average target value of all the rivers is 340 eggs 100 m^{-2} (standard deviation 79.1) ranging from 190 to 556 eggs 100 m^{-2}.

25.4.3 *Assessment of egg deposition*

Annual egg deposition is estimated by calculating escapement from best estimates of run into the river, and in practice rod catches are often the only data available. The calculations are performed separately for 1SW and MSW fish after an initial split

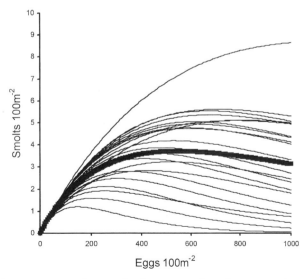

Figure 25.4 Target transporting method assumes that the composite target for a whole catchment (heavy line) is the weighted mean of constituent S–R curves for the variety of habitat types found in the catchment

and are finally added to give total eggs. The following definitions apply:

$$\text{'true' rod catch} = C_t = C_d \times 1/r$$

where C_d is the declared annual catch and r the proportion of true catch declared through licence returns. Rod catch is split into 1SW (C_{tg}) and MSW (C_{tm}) components using the proportion of 1SW (P_g) in the catch, derived from scale or weight data (Table 25.1), then for *each* sea-age component:

$$\text{river annual run} = R = C_t/U$$
$$\text{spawners} = S = (R - C_t) \times s$$
$$\text{spawners} = C_t((1/U) - 1) \times s$$
$$\text{egg deposition} = E = (S \times P_f) \times f$$

where U is the extant rod exploitation rate, expressed as proportion of the total annual run, s is the proportion of fish surviving the in-river phase, P_f is the proportion of females in escapement and f the mean number of eggs per female.

Overall total spawners (S_T) and egg deposition (E_T) are estimated by combining data for 1SW and MSW components:

$$S_T = S_g + S_m \text{ and } E_T = E_g + E_m$$

and for comparison against the original escapement reference point, knowing the accessible stream area (A):

$$\text{Egg deposition rate} = E_D = E_T/A$$

Default values for the parameter estimates have been defined (Table 25.1), which can be altered for each river depending on available data (Environment Agency 1996).

25.4.4 *Compliance testing*

Using the above procedures, runs of egg deposition values can be estimated over several years and evaluated against the conservation limit. Variability in the data has three main components: random error, measurement error and trends including the effects of auto-correlation. An analysis of several rivers in England and Wales (Environment Agency 1996) showed that salmon catch data are strongly auto-correlated (lag-one auto-correlation ranging from 0.3 to 0.6), although it is difficult to distinguish this effect from genuine trends, due to external factors acting on the stock. Statistical rules were developed to account for auto-correlation and random error effects (Environment Agency 1996), which set a 20 percentile standard for egg deposition (i.e. on average, egg deposition should be above the Sg value for 4 years in 5) and accepted Type 1 error occurrence of 5% (i.e. there should be a false alarm only once every 20 years). This procedure means that in conditions of a stock just achieving compliance the mean egg deposition is actually somewhat higher than *Sg*

Table 25.1 Summary of default values used in estimating egg deposition from rod catch (details in Environment Agency 1996)

Parameter	Value	Comments
R, rod catch reporting rate	0.91 for 1994 *et seq.*	Adapted from Small (1991); value has varied with licence and reminder system.
Pg, proportion of 1SW fish in catch	Approx. range: 0.4–0.8	Several alternative methods; analysis of monthly weight frequency preferred.
U, annual extant exploitation rate	Range of mean values for different rivers: 0.09–0.42	Variable between years and run groups; default method is based on relationship with fishing intensity.
S, in-river survival to spawning	0.91	Based on radio-tracking studies.
Pf, proportion of females	*Pf* (1SW) range: 0.3–0.6 *Pf* (MSW) = 0.69	*Pf* (1SW) estimated from river size relationship.
f, eggs per female	1SW = 3766 MSW = 7278	Preferred method applies equation of Pope *et al.* (1961) to spawner weight distribution in 1 lb weight categories.

(Fig. 25.5), and this level is equivalent to NASCO's management target (see Section 25.4.1).

25.5 Discussion

The term 'spawning targets' has become convenient shorthand for a suite of reference points that might be used to guide management, but which can be confusing because 'target' implies a level around which management aims to maintain spawning stock and this is not strictly true. The current Agency procedure uses two reference points: a conservation limit (*Sg*) and, linked to this by a statistical compliance scheme (see Section 25.4.4), a higher value management target. The aim is to keep stocks at or above the management target, below which compliance failure is triggered. This approach is a precautionary trade-off, accepting some reduction in catch opportunity in order to be risk averse in the face of uncertainties surrounding reference points and assessments.

The use of spawning targets brings scientists and managers up against fundamental issues facing the underlying science, and development needs can be identified for both target setting and compliance assessment. Salmon stocks and catches are naturally variable for many reasons. Recruitment estimates are especially variable, resulting from a combination of measurement, random and process errors

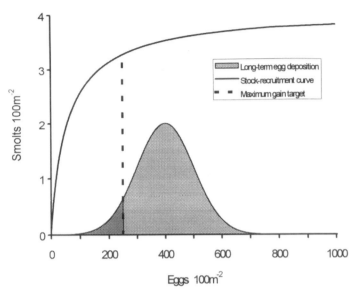

Figure 25.5 Environment Agency spawning target compliance scheme, showing the escapement frequency distribution for a stock just passing its management target, in which 20% of annual values fall below the maximum gain conservation threshold

(Hilborn & Walters 1992), introducing a risk of assessments and targets being wrong. This risk, the degree of error and the impact it has on management responses to target failures needs to be incorporated into the decision procedures. The next phase of development will see a move from the current deterministic analysis to a modelling approach where the focus will be on the error and variability of the targets and stock estimates.

The use of rod catch to assess run size presents problems because links between the two are complex. Exploitation rates vary between rivers and annually (Environment Agency 1996; Davidson *et al.*, Chapter 26), and where default values have to be used, greater caution should be applied to interpretation of compliance. However, the use of variables such as river flows and fishing effort to model the relationship between run and catch offers ways to improve escapement estimates. The current procedures assume that salmon egg deposition is independent of other species, but sea trout, *Salmo trutta* L., are a major source of eggs in some rivers and interactions with salmon might be expected, and need to be explored. The uneven and variable distribution of recruitment in a catchment resulting from annual variation in spawner penetration and the possibility of genetic structuring in large catchments are two further factors to consider. However, attention to the relative performance of juveniles and run composition changes around a catchment should inform managers about the impact of these effects. The habitat models currently employed in transporting targets are still basic and their performance is linked to the available methods for habitat assessment. There is trade-off between precision and cost in such work, but this field is currently receiving a lot of research attention to improve habitat survey and evaluation methods.

It is essential to recognise that compliance with just one target cannot provide a comprehensive picture meeting all management information needs. It seems hardly necessary to point this out, yet the short history of targets indicates that managers and users can become so focussed on the perspective of one target that other aspects of assessment become excluded or are given much lower prominence in decision making. This seems to be a consequence of setting formal targets, compounded by difficulties in communication and understanding of the process. Scientists have a duty to make the process intelligible and to be explicit about the unavoidable variability and risks inherent in assessment and management of wild biological resources. The final management decisions will need to take social and economic factors into account. Further issues to resolve include the choice between contrasting strategies of fixed harvest rate or fixed escapement management, as well as the time-scale of assessment and control (in-season, annual or longer periods). Resolution of these will require amendments to the legislative powers controlling management flexibility.

The procedures outlined here represent current best practice, which is continually improving as new information and understanding are acquired. Although some technical and presentational problems remain with targets, these are outweighed by the benefits which the approach offers. The quantitative framework allows a rational setting of management goals and enables managers to make the best use of available data. Targets embody the principles of sustainable development and, with appropriate compliance schemes and risk analysis, enable precautionary management to be applied. The data and understanding needed for robust procedures define the priorities for future research. Finally, through Salmon Action Plans, targets open up the management process to scrutiny and oblige users, managers and scientists alike to face up to choices about the quality of assessment they need and can pay for.

Disclaimer

The views expressed in this chapter represent those of the authors and not necessarily those of the Environment Agency.

References

Anon. (1996) Making a new beginning in salmon management. *Report by the Salmon Management Task Force*. Dublin: Department of the Marine, June 1996, 61 pp.

Anon. (1998) Annual assessment of salmon stocks and fisheries in England and Wales, 1997. Preliminary Assessment for ICES, *CEFAS/EA Joint Report*, April 1998, 43 pp.

Bagliniere J-L., Maisse G. & Nihouarn A. (1993) Comparison of two methods of estimating Atlantic salmon, *Salmo salar*, wild smolt production. In R.J. Gibson & R.E. Cutting (eds) Production of juvenile Atlantic salmon in natural waters. *Canadian Journal of Fisheries and Aquatic Sciences* **118**, 189–201.

Beverton RJ. & Holt S.J. (1957) On the dynamics of exploited fish populations. UK Ministry of Agriculture and Fisheries. *Fisheries Investigations (Series 2)* **19**, 533 pp.

Elson P.F. (1975) Atlantic salmon rivers smolt production and optimal spawning. An overview of natural production. *International Atlantic Salmon Foundation, Special Publication*, Series **6**, 96–119.

Elliott J.M. (1987) Population regulation in contrasting populations of trout *Salmo trutta* in two Lake District streams. *Journal of Animal Ecology* **57**, 49–60.

Environment Agency (1996) Salmon Action Plan Guidelines, Version 1, 11/96. *Fisheries Technical Manual Series*, No. 3, 125 pp.

Gibson R.J. (1993) The Atlantic salmon in freshwater: spawning, rearing and production. *Reviews in Fish Biology and Fisheries* **3**, 39–73.

Hilborn R. & Walters C.J. (1992) *Quantitative Fish Stock Assessment. Choice, Dynamics and Uncertainty*. New York: Chapman & Hall, 570 pp.

ICES (1994) E.C.E. Potter (ed.) *Report of the Workshop on Salmon Spawning Stock Targets in the North-East Atlantic, Bushmills, 7–9 December 1993*. C.M.1994/M:7, 74 pp.

Jonsson N., Jonsson B. & Hansen L.P. (1998) The relative role of density-dependent and density-independent survival in the life cycle of the Atlantic salmon *Salmo salar*. *Journal of Animal Ecology* **67**, 751–762.

Kennedy G.J.A & Crozier W.W. (1993) Juvenile Atlantic salmon (*Salmo salar*) – production and prediction. In R.J.Gibson & R.E.Cutting (eds) Production of juvenile Atlantic salmon, *Salmo salar*, in natural waters. *Canadian Journal of Fisheries and Aquatic Sciences* **118**, 179–187.

Kennedy G.J.A & Crozier W.W. (1995) Factors affecting recruitment success in salmonids. In D.M. Harper & A.J.D. Ferguson (eds) *The Ecological Basis for River Management*. Chichester: Wiley, pp. 349–362.

Milner N.J., Wyatt R.J. & Broad K. (1998) HABSCORE – applications and future development of related habitat models. *Aquatic Conservation: Marine and Freshwater Ecosystems* **8**, 633–644.

National Rivers Authority (1996) *A Strategy for the Management of Salmon in England and Wales*. Bristol: National Rivers Authority, 36 pp.

North Atlantic Salmon Conservation Organization (NASCO) (1996) *Report of the ICES Advisory Committee on Fishery Management*. NASCO CNL(96) **15**, 30 pp.

NASCO (1997) *Report of the ICES Working Group on North Atlantic Salmon*. CNL(97) **12**, 242 pp.

NASCO (1998) *Report of the ICES Working Group on North Atlantic salmon*. NASCO CNL(98) **11**, 293 pp.

O'Connell M.F. & Dempson J.B. (1995) Target spawning requirements for Atlantic salmon, *Salmo salar* L., in Newfoundland rivers. *Fisheries Management and Ecology* **2**,161–170.

Prévost E. & Porcher J.P. (1996) Méthodologie pour l'élaboration de totaux autorisés de captures (TAC) pour le saumon atlantique (*Salmo salar* L.) dans le Massif Armorican. Propositions et recommandations scientifiques. *GRISAM, Evaluation et Gestion des Stocks de Poissons Migrateurs, Document Scientifique et Technique No.* **1**, 18 pp.

Ricker W.E. (1954) Stock and Recruitment. *Journal of the Fisheries Research Board of Canada* **11**, 559–623.

Smith S.J., Hunt J.J. & Rivard D. (1993) Risk evaluation and biological reference points for fisheries management. *Special Publication of Canadian Journal of Fisheries and Aquatic Sciences* **120**, 442 pp.

Symons P.E.K. (1979) Estimated escapement of Atlantic salmon (*Salmo salar*) for maximum smolt production in rivers of different productivity. *Journal of the Fisheries Research Board of Canada* **36**, 132–140.

Wyatt R.J. & Barnard S. (1997a) Spawning |escapement targets for Atlantic salmon, *R&D Technical Report W64*. Bristol: Environment Agency, 124 pp.

Wyatt R.J. & Barnard S. (1997b) The transportation of the maximum gain salmon spawning target from the River Bush (N.I) to England and Wales, *R&D Technical Report W65*. Bristol: Environment Agency, 80 pp.

Chapter 26
The effectiveness of rod and net fishery bye-laws in reducing exploitation of spring salmon on the Welsh Dee

I.C. DAVIDSON and R.J. COVE
National Salmon and Trout Fisheries Centre, Environment Agency, Chester Road, Buckley CH7 3AJ, UK

N.J. MILNER
National Salmon and Trout Fisheries Centre, Environment Agency, St. Mellons, Cardiff CL3 0LT, UK

Abstract

In 1995, rod and net fishery bye-laws were introduced on the Welsh Dee to protect declining stocks of 'spring' (pre-1 June) running Atlantic salmon, *Salmo salar* L. These delayed the start of season on both the rod and net fisheries, and on the rod fishery, restricted fishing methods to 'fly only' in the spring months. This chapter examines changes in exploitation rates on both fisheries in the 3 years pre- and post-bye-law and estimates the numbers of salmon saved from capture as a consequence. Analysis concentrates on the rod fishery where mark–recapture estimates of exploitation have been obtained for monthly entrant groups of salmon since 1992. Average rates for January–May entrants have fallen from 26% to 15% between the pre- and post-bye-law periods, saving an estimated 36 fish per year (11% of the spring run). However, factors unrelated to the bye-law may have influenced exploitation in the post-bye-law period (e.g. flow, run size) and these are incorporated into a multiple regression model, which predicts smaller savings of 10 fish per year. Examination of the patterns of exploitation in the pre- and post-bye-law period indicate that both elements of the bye-law – season closure and the 'fly only' method restriction – have been effective in reducing exploitation.

Keywords: Atlantic salmon, bye-law assessment, spring salmon, net and rod exploitation.

26.1 Introduction

In 1995 the Welsh Region of the National Rivers Authority (NRA) introduced rod and net fishery bye-laws on the rivers Dee, Usk and Wye to protect early or 'spring' (pre-1 June) running Atlantic salmon, *Salmo salar* L.

The Dee, Wye and Usk were singled out for attention because of their historic reputation as spring salmon rivers (indicated, for example, by the proportion of the rod and net catch taken pre-June) and because of concern about the marked decline in this run component in recent years. Such declines have been widely reported in rivers throughout the salmon's range (e.g. Gough *et al.* 1992; Youngson 1994).

This chapter explores the effectiveness of the spring fish bye-laws on the Dee by examining changes in net and rod exploitation rates in the 3 years before and after their introduction, and quantifies the benefits in terms of the numbers of salmon saved from capture.

26.1.1   *Historic abundance of spring salmon on the Dee*

Regular monthly catch records for the Dee are available back to the 1930s and indicate a fishery dominated by pre-June catches for the best part of this century. For example, salmon catches on the rod fishery (Fig. 26.1) were at their peak in the 1960s (10-year mean catch of 744 fish), when around 60% of the catch was taken pre-June.

Prior to the 1960s the spring rod catch formed an even greater proportion of the total, but throughout the 1970s and 1980s this declined steadily to only 15% in the current decade, when the catch has rarely exceeded 100 fish in a season. A similar pattern of decline was also apparent on the net fishery.

Scale samples from net caught salmon from the late 1960s to the early 1980s indicated a pre-June catch almost entirely composed of 2 sea winter (SW) and 3SW salmon. Throughout this period, catches were dominated by 2SW fish but increasingly so in recent times, as the proportion of 3SW fish fell from almost 40% of the catch in the 1960s and 1970s to less than 10% in the 1980s.

26.1.2   *Spring salmon byelaws*

As a consequence, the following changes in the fishery bye-laws of the River Dee were introduced in 1995 to protect the spring fish:

 (i) a later start to the net fishing season – delayed from 1 March to 1 May (season ends 31 August);
(ii) a later start to the rod fishing season – delayed from 26 January to 3 March (season ends 17 October);
(iii) angling methods restricted to fly fishing only from 3 March to 1 June, when previously fly and spinner were permitted up to 15 April, and any method thereafter.

The bye-laws introduced on the Wye and the Usk incorporated similar measures to protect spring running fish (Winstone & Gough 1996).

26.2   Materials and methods

In the summer of 1991, the National Rivers Authority (NRA) began the Dee Stock Assessment Programme (or DSAP) – a comprehensive monitoring programme for salmon and sea trout, *Salmo trutta* L., on the River Dee. In 1996, the duties of the

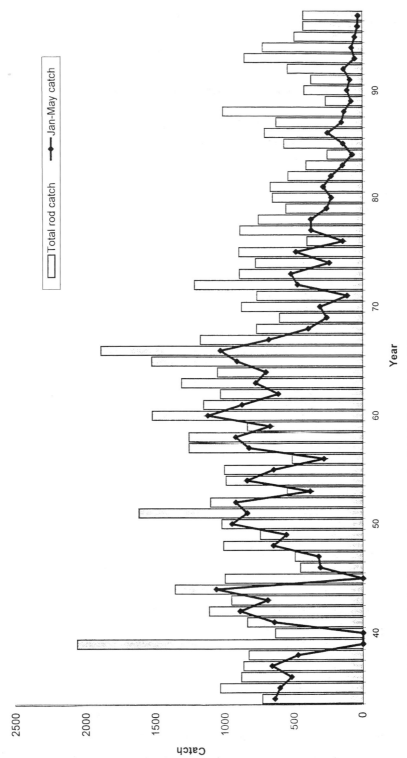

Figure 26.1 Spring (Jan–May) salmon rod catch on the Welsh Dee, 1934–1997

NRA – including the running of DSAP – were passed on to the Environment Agency.

A key component of the DSAP is the operation of a partial, head-of-tide trap at Chester Weir (SJ 407 658), where adult salmon are sampled and tagged for subsequent recapture by the rod fishery upstream (Fig. 26.2). The trapping programme provides detailed information on the timing and composition (age, size and sex) of the run at Chester, while fishery recaptures of tagged salmon allow monthly mark–recapture estimates of run size and angling exploitation.

A fuller description of the Dee programme, including details of the fisheries and derivation of run and angling exploitation estimates, is given in Davidson *et al.* (1996). For the net fishery, nominal exploitation rates were calculated from the declared catch (assumed to be less than the actual catch) and estimated run at Chester Weir (which lies immediately upstream of the net fishery) (Fig. 26.2).

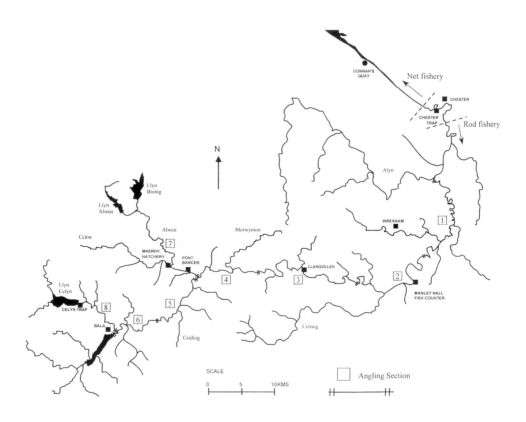

Figure 26.2 River Dee catchment

26.3   Results

26.3.1   *Current status of the spring salmon run*

Estimates of run size and composition obtained from DSAP (1992–1997) indicate a spring salmon run which continues to be dominated by 2SW fish (88%) but with 3SW fish averaging only 9% of the total (the remainder of the spring run is made up of previous spawners). These estimates correspond to a period when the mean run of spring salmon (397 fish) formed only 7% of the total run (5978 fish) and when around 75% of fish returned as late summer/autumn 1SW salmon (Environment Agency, unpublished data).

26.3.2   *Impact of the net fishery bye-law*

Between 1992 and 1997 inclusive, up to four trammel and 30 draft net licences were available on the Dee, although only around half the draft net licences were taken up in any one year. The trammel nets operated in the estuary downstream of Connah's Quay to the river mouth while the draft nets fished between Connah's Quay and Chester (Fig. 26.2). In recent years, most of the fishing effort and salmon catch for both gears occurred in the last 3 months of the season. For example, in the pre-bye-law period (1992–1994), 94% of the net fishing effort and 97% of the catch were recorded from June to August.

Following introduction of the bye-law in 1995, closure of the net fishery in March and April eliminated exploitation of salmon entering the river in these months. In the 3 years prior to 1995, nominal exploitation rates on fish entering in March and April averaged 2.2% (range 1.1–5.4%). These rates were calculated from the estimated run of salmon at Chester Weir in March and April (i.e. upstream of the net fishery) and the declared net catch in these months. They assume that there were no other losses of fish between the net fishery and the trap and that salmon exposed to the net fishery in any one month arrived at Chester Weir in the same month. The latter assumption is supported by radio-tracking studies of fish migrating through the estuary upstream of Connah's Quay, which indicated that most salmon arrived at Chester Weir within 24 h of release – a distance of up to 16 km (Purvis *et al.* 1994).

Declared March–April catches of salmon in the pre-bye-law period (1992–1994) averaged four fish (range 3–5), with runs of salmon at Chester Weir in these months ranging from 86–375 fish (average 184). A similar March–April average salmon run of 175 fish (range 149–192) was estimated for the post-bye-law period (1995–1997). Assuming that, in the absence of the bye-law, the Dee nets would have continued to exploit fish at the pre-bye-law average rate of 2.2%, then around four salmon per year would have been saved from capture.

26.3.3 *Impact of the rod fishery bye-law*

Assessing the effect of the rod fishery bye-law on exploitation is less straightforward than for the net fishery. For example, although the rod bye-law (like the net bye-law) prolongs the close season (3 March start compared to 26 January previously) it has the additional complication of the 'fly only' method restriction to the 1 June. In addition, salmon initially protected by the bye-law on the rod fishery remain available for capture (although perhaps less vulnerable to capture) for the rest of the season.

The question of vulnerability to capture is important when considering method restrictions. The justification for introducing a 'fly only' rule was based on this appearing to be one of the least effective methods of catching salmon in the spring. This was judged from the proportion of the spring catch taken by fly as opposed to bait and spinner, although in practice, it may not reflect the outcome following introduction of the bye-law. For example, distributing the catch among different methods may simply reflect the popularity of those methods, rather than their effectiveness – such that the fish that would have been caught on bait or spinner will now be caught on fly. Indeed, the effectiveness of fly fishing itself may increase as anglers are forced to improvise with the methods that remain open to them.

Exploitation estimates

Angling exploitation estimates for monthly entrant groups of spring (January–May) salmon generally fell markedly in the post-bye-law period (Fig. 26.3). The overall

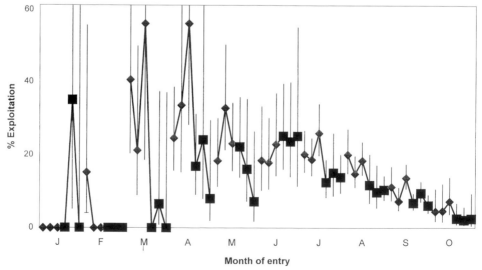

Figure 26.3 Angling exploitation rates on monthly entrant groups of salmon for the 3 years pre-bye-law (◆) and post-bye-law (■) (95% confidence limits shown around each estimate)

rate for fish entering at this time declined from 26.0% (95% confidence limits 21.7%–31.3%) in the 3 years pre-bye-law to 15.1% since (95% confidence limits 11.2%–20.4%). Wide confidence limits around the estimates for the earliest entrant groups (Fig. 26.3) arise from the small numbers of fish available for tagging and recapture at this time of year (Table 26.1).

The patterns of capture for spring entrants also appear to demonstrate the effectiveness of the bye-laws. For example, average cumulative exploitation rates for early run salmon in the pre- and post-bye-law period indicated both reduced exploitation of January to April fish in the months when the bye-law was operating and showed little evidence of a marked rise in exploitation outside this period (Figs 26.4a–d). This suggests that fish protected by the bye-law from early season capture do not become more heavily exploited later on and hence initial benefits are not negated. Furthermore, the post bye-law reduction in exploitation is at least as marked in March and April entrants as it is in February fish, indicating that the 'fly only' rule as well as season closure is providing some degree of protection.

The reduction in exploitation rate in May fish (Fig. 26.4e) is less marked in the post-bye-law period but greatest in May itself – when the bye-law still offers protection. In June, the cumulative rate rises sharply to just below the pre-bye-law level (although this falls away subsequently), and indicates that May fish are still highly vulnerable to capture 4–8 weeks after entering the fishery. Beyond this time the flattening of the cumulative rate graph suggests that fish become much less 'catchable' – probably as they enter the quiescent phase described in many behavioural studies (see Milner 1990). A similar decline in 'catchability' is also evident among the other entrant groups.

In contrast, for June fish, post-bye-law exploitation rates exceed pre-bye-law rates (Fig. 26.4f). However, this is not true of other groups of later running fish (i.e. July–October entrants), where post-bye-law rates are generally below the pre-bye-law rates (Fig. 26.3).

Benefits of the bye-law

Run estimates at Chester Weir in the post-bye-law period averaged 334 fish in the spring (range 274–373). Assuming that, in the absence of the bye-law, a pre-bye-law exploitation rate of 26% applied to these fish rather than the observed rate of 15% (see above), then around 36 fish per year were protected from capture by the bye-law.

This simplistic estimate of the bye-law's effectiveness assumes that the observed reductions in average exploitation rate entirely resulted from the bye-law change rather than any other factor. Figure 26.5 examines relationships between exploitation rates for monthly entrant groups (pre- and post-bye-law) and various fishery and environmental factors, including:

- *flow* – mean daily flow at Manley Hall in the month of entry;
- *cumulative run size* – the total number of fish in the river up to the month of entry;

Table 26.1 Pre- and post-bye-law angling exploitation estimates

	Jan	Feb	Mar	Apr	May	Jan–May	Jun	July	Aug	Sept	Oct	Jun–Oct	Unknown	All
Pre-bye-law angling exploitation rates – 1992–1994 combined														
No. tagged, adjusted for tag loss	1	18	48	109	258	**434**	211	735	895	1009	319	**3168**	0	3602
No. recaptured, adjusted for tag loss	0	2	16	33	62	**113**	42	157	158	114	18	**489**	13	615
Exploitation estimate (%)	0.0	11.3	33.0	30.3	24.1	**26.1**	19.9	21.4	17.7	11.3	5.7	**15.4**		17.1
Lower 95% confidence interval	4.3	3.0	20.3	21.6	18.8	**21.7**	14.7	18.3	15.1	9.4	3.6	**14.1**		15.8
Upper 95% confidence interval	413.9	41.3	53.7	42.6	30.9	**31.3**	26.9	25.0	20.6	13.6	8.9	**16.9**		18.5
Post-bye-law angling exploitation rates – 1995–1997 combined														
No. tagged, adjusted for tag loss	6	6	30	96	141	**278**	158	477	621	1509	412	**3177**	0	3455
No. recaptured, adjusted for tag loss	1	0	1	15	25	**42**	39	65	66	115	10	**295**	21	358
Exploitation estimate (%)	17.5	0.0	3.3	15.6	17.8	**15.1**	24.7	13.6	10.6	7.6	2.4	**9.3**		10.4
Lower 95% confidence interval	2.6	–0.7	0.5	9.5	12.0	**11.2**	18.1	10.7	8.3	6.4	1.3	**8.3**		9.3
Upper 95% confidence interval	99.9	68.1	18.7	25.8	26.3	**20.4**	33.8	17.4	13.5	9.1	4.5	**10.4**		11.5

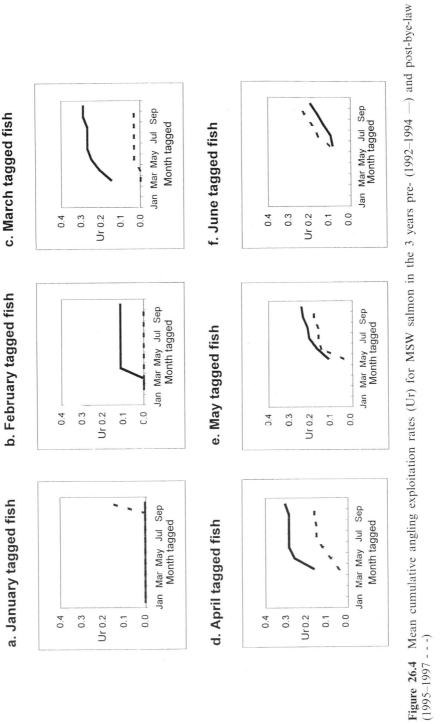

Figure 26.4 Mean cumulative angling exploitation rates (Ur) for MSW salmon in the 3 years pre- (1992–1994 —) and post-bye-law (1995–1997 - - -)

a. Exploitation and flow

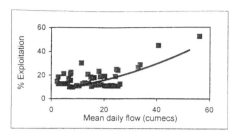

b. Exploitation and cumulative run

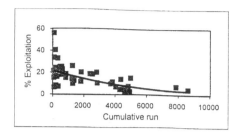

c. Exploitation and angling effort

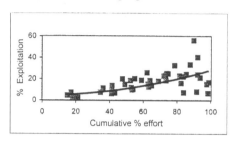

d. Exploitation and month of entry

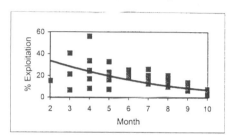

e. Exploitation and time available to the fishery

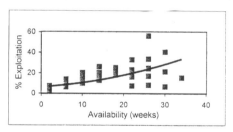

Figure 26.5 Relationships between monthly salmon angling exploitation rate and environmental, stock and fishery variables on the Welsh Dee, 1992–1997

- *cumulative percentage fishing effort* – the % of total angling effort fish are exposed to in the season. Information on angling effort was taken from a logbook scheme which targets an unknown proportion of Dee anglers – hence the absence of absolute measures of fishing effort;
- *month* – the month tagged (as an index of any seasonal effect);
- *availability* – the number of weeks fish are available for exploitation.

A regression was fitted to each relationships (Fig. 26.5) using a linearised logistic equation:

$$\log_e((K - N)/N) = a - rt$$

This was then converted to the usual logistic form (see Krebs 1978):

$$N = 100/(1 + e^{a-rt})$$

where the assymptote K is assumed to be 100%, N is the observed exploitation rate, t is the independent variable and a and r are constants. Using this logistic model, significant correlations ($r^2 = 0.18$–0.48; $P < 0.01$) were found for each of the relationships (Fig. 26.5).

A multiple regression model ($r^2 = 0.82$; $P < 0.001$) was derived from pre-bye-law data only using the variables mean daily flow (ADF) and cumulative run (cmN) to predict expected exploitation rates in the post-bye-law period – and thus give a better indication of the impact of the bye-law alone on exploitation:

$$\log_e((K - N)/N) = 1.48710ADF + 0.00024cmN$$

Observed and predicted exploitation rates from this model are shown in Figure 26.6. For spring salmon, observed exploitation rates showed the most extreme deviation from the predicted range, with observed rates for March fish below the lower 95% prediction limit in all 3 years and for April and May fish close to or below the limit in 2 of the 3 years. Observed rates for other monthly entrant groups also tended to fall in the lower half of the prediction interval, but, with the exception of October fish, they do not fall outside this interval.

If the most extreme case that only deviation below the lower 95% prediction limit represents any real saving of fish which can be ascribed to the bye-law is assumed, then around 10 salmon per year would have fallen into this category – approximately 30% of the previous estimate.

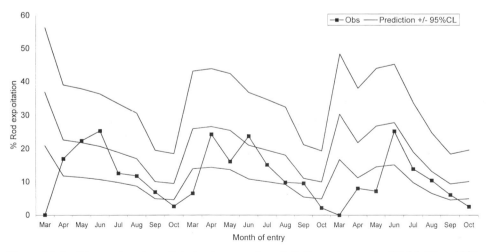

Figure 26.6 Observed and predicted salmon rod exploitation rates on the Welsh Dee, 1995–1997

Impact of under-reporting of tags

One possible cause of the post-bye-law reduction in rod exploitation is a fall-off in tag reporting by anglers disillusioned with the scheme because they disagreed with the bye-law. This could explain a decline in exploitation of late season fish, which, at face value, is almost equivalent to that observed in early run salmon (Table 26.1). The exploitation estimates given here assume full reporting based on values derived from radio-tracking studies at the beginning of the DSAP. An alternative approach to exploring trends in reporting rate (illustrated in Fig. 26.7) compares cumulative exploitation estimates produced from all tag returns with estimates based on the total rod catch and run estimates. Run estimates are not reliant on all tags being reported, but assume an accurate measure of the ratio between tagged and untagged fish obtained from the returns of logbook anglers only.

The greatest differences between these two methods of estimating exploitation tended to occur in the first 3–4 months of the season when rod catch estimates were usually greater than tag-recapture estimates (Fig. 26.7). However, this is likely to be due to the misreporting of rod-caught kelts as fresh salmon – a theory which is supported by the large proportion of fish which are released in the spring. For example, 49% of the pre-June rod catch on the Dee was released in 1997 (Environment Agency, unpublished data). By the middle of the season, rod catch estimates often fell slightly below the tag-recapture estimates, but by the end of the season the two estimate were usually very similar. This would indicate that, over the season, there is no marked under-reporting of tagged salmon. Similarly, there is no strong evidence to suggest a systematic under-reporting of tagged salmon in the post-bye-law period.

26.4   Discussion

Evaluation of the benefits arising from changes to the management of a fishery is essential if we are to have confidence that such changes are effective. On the Dee, the existence of an established monitoring programme (DSAP) at the time the spring salmon bye-laws were introduced provided an ideal opportunity to do this. However, as the results presented here suggest, arriving at a definitive measure of success, e.g. in terms of the numbers of salmon saved, is not a straightforward process.

For example, strong correlations between angling exploitation rate and fishery and environmental variables (Section 26.3) indicated that factors other than the bye-law could have influenced the observed 'post-bye-law' decline in spring fish exploitation. Hence, estimated savings of 36 salmon per year based on average exploitation rates are probably too simplistic.

Using a multiple regression model to account for the effects of two of these variables – flow and run size – on post-bye-law exploitation may give a more realistic evaluation of the benefits attributable to the bye-law alone, although an estimated saving of 10 salmon per year (based on observations below the 95% prediction interval) is probably too conservative.

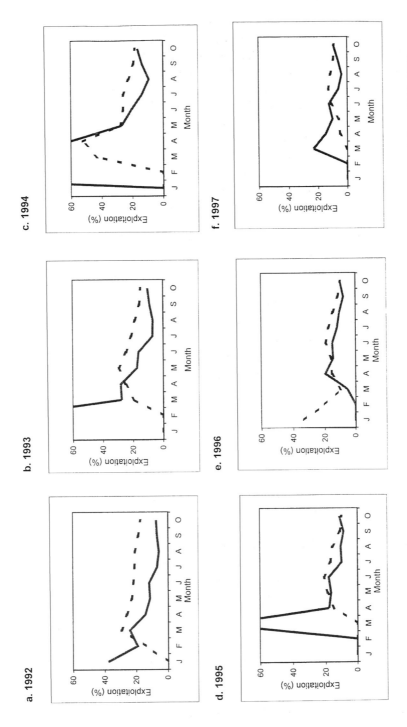

Figure 26.7 Cumulative rod exploitation estimates derived from declared rod catch (—) and tag recapture (- - -) data, 1992–1997

The variables selected for the multiple regression model were considered among the most important of those examined. For example, a number of studies reported an inverse relationship between exploitation rate and stock size (Petermen & Steer 1981; Mills *et al.* 1986; Beaumont *et al.* 1991). Similarly, rod catches, within varying limits, are often positively correlated with flow (Milner 1990), although an increased catch may or may not reflect an increase in exploitation rate.

Angling effort was also positively correlated with exploitation rate but was excluded from the multiple regression model because it may itself have been influenced by the bye-law change, i.e. the bye-law may have deterred some anglers from fishing in the early season, although logbook measures of relative angling effort provided no evidence of this.

Aside from estimating the numbers of fish saved by the bye-law, patterns of exploitation pre- and post-bye-law (Fig. 26.4) provided insight into the effectiveness of its different components, i.e. season closure and the fly only method restriction. Such detailed information on salmon exploitation is rare (see Solomon & Potter (1992)) and indicates that both components helped reduce exploitation – especially in fish entering before May. For May salmon, the bye-law offers less protection, as these fish (like other monthly entrant groups) still appear highly vulnerable to capture within 4–8 weeks of entering fresh water. In hindsight, it may have been more prudent to extend the fly only restriction to the middle of June to enhance the protection of these fish.

From the outset, the Dee bye-laws were never expected to result in immediate savings of large numbers of spring salmon – especially while pre-June runs averaged fewer than 400 fish. Indeed their main benefit was seen as helping to conserve what little stock remained and to protect it during recovery. This recognised that the current low levels of spring salmon stocks may be part of a long-term cyclical trend partly driven by marine factors (see for example Youngson (1994)).

At the same time, environmental degradation in fresh water may also have contributed to the decline in spring salmon. Acidification, impoundment, flow regulation, land use practices and waste disposal have all been identified as possible factors adversely affecting salmon production on the Dee. The Dee Salmon Action Plan (Environment Agency 1997), like other river specific Salmon Action Plans produced by the Agency, recognised that environmental improvements – along with exploitation control – will be necessary if spring salmon stocks and salmon stocks in general are to recover to former levels. Whatever approaches are adopted, the need to evaluate their effectiveness remains if salmon management is to progress.

References

Beaumont W.R.C., Welton J.S. & Ladle M. (1991) Comparison of rod catch data with known numbers of Atlantic salmon (*Salmo salar* L.) recorded by a resistivity counter in a southern chalk stream. In I.G. Cowx (ed.) *Catch Effort Sampling Strategies: Their Application in Freshwater Fisheries Management.* Oxford: Fishing News Books, Blackwell Science, pp. 49–60.

Davidson I.C., Cove R.J., Milner N.J. & Purvis W.K. (1996) Estimation of Atlantic salmon (*Salmo salar* L.) and sea trout (*Salmo trutta* L.) run size and angling exploitation on the Welsh Dee using mark-recapture and trap indices. In I.G. Cowx (ed.) *Stock Assessment in Inland Fisheries.* Oxford: Fishing News Books, Blackwell Science, pp. 293–307.

Environment Agency (1997) *River Dee Salmon Action Plan Consultation Document.* Cardiff: Environment Agency, Welsh Region, 47 pp.

Gough P.G., Winstone A.J. & Hilder P.G. (1992) Spring salmon – A review of factors affecting the abundance and catch of spring salmon from the River Wye and elsewhere, and proposals for stock maintenance and enhancement. *Technical Fisheries Report* No. 2. Cardiff: National Rivers Authority, Welsh Region, 57 pp.

Krebs C.J. (1978) *Ecology – The Experimental Analysis of Distribution and Abundance.* New York: Harper International Edition, 678 pp.

Mills C.P.R., Mahon G.A.T. & Piggins D.J. (1986) Influence of stock levels, fishing effort and environmental factors on anglers' catches of Atlantic salmon (*Salmo salar* L.) and sea trout (*Salmo trutta*, L). *Aquaculture and Fisheries Management* **17**, 289–297.

Milner N.J. (1990) *Fish Movement in Relation to Freshwater Flow and Quality.* Moulin, Pitlochry: Atlantic Salmon Trust, 51 pp.

Petermen R.M. & Steer G.J. (1981) Relation between sport fishing catchability coefficients and salmon abundance. *Transactions of the American Fisheries Society* **110**, 585–595.

Purvis W.K., Crundwell C.R., Harvey D. & Wilson B.R. (1994) Estuarial migration of Atlantic salmon in the River Dee (N. Wales), *Energy Technology Support Unit (ETSU), ESTU T/04/ 00154/REP.* London: Dept. of Trade and Industry, 53 pp.

Solomon, D.J. & Potter E.C.E. (1992) *The Measurement and Evaluation of Exploitation of Atlantic Salmon.* Moulin, Pitlochry: Atlantic Salmon Trust, 37 pp.

Winstone A.J. & Gough P.G. (1996) A review of factors affecting the abundance and catch of spring salmon from the River Wye and proposals for stock maintenance and enhancement. In D.H. Mills (ed.) *Enhancement of Spring Salmon.* Moulin, Pitlochry: Atlantic Salmon Trust, pp. 26–44.

Youngson A. (1994) *Spring Salmon.* Moulin, Pitlochry: Atlantic Salmon Trust, 53 pp.

Chapter 27
From sector to system: towards a multidimensional management in the Lower Amazon floodplain

F. CASTRO

ACT Student Building 331, Indiana University, Bloomington, Indiana, USA (e-mail: fdecastr@indiana.edu)

D. McGRATH

Nucleo de Altos Estudos da Amazonia, Federal University of Para, Belem, PA Brazil

27.1 Introduction

To a large extent, fisheries management models have been based on maritime fisheries, in which the fish is treated as a separate resource which can be managed without regard to other resource use activities (Rothschild 1973; Larkin 1978). However, this sector-based approach to fishery management is less appropriate for systems where there is a high degree of ecological interdependence between aquatic and terrestrial environments and where resource users employ diversified economic strategies involving more than one resource.

Floodplain fisheries are a case in point. In recent years, ecological analysis of the floodplain has shown the importance of habitat protection for the conservation of floodplain fisheries. Such an analysis improves the system approach to floodplain fisheries. Yet, it assumes that human impact on floodplain habitats has been a major cause of the degradation of Amazon fisheries (Goulding *et al.* 1996), while a broader picture of the social dimensions of the floodplain resource use is still missing.

Floodplain fisheries are strongly influenced by the social, economic and political environment in which they are carried out. This social context must be integrated into the analysis of Amazonian floodplain fisheries for at least three reasons. First, the floodplain is occupied by human groups with different capacities for exploiting natural resources. For example, both local smallholders and itinerant commercial fishers exploit floodplain fish. In addition, habitat degradation on the floodplain is related not only to the fishing activity of the smallholder, but also to those of large-scale ranchers (Goulding *et al.* 1996). Therefore, generalisations concerning human impact tend to obscure the different processes through which floodplain resources are exploited. Second, human populations have occupied the floodplain for more than 10 000 years (Roosevelt 1989) and have developed strategies of floodplain resource use which may be well adapted to the ecological characteristics of the floodplain (Meggers 1971; Moran 1993). Finally, ecological interdependence between floodplain subsystems is reflected in the socio-economic dimension. Small-scale fishing carried out by thousands of floodplain families is only one part of their management strategies. Smallholders have traditionally exploited a series of floodplain resources over the course of the year which integrate farming on the higher levees, cattle ranching on the seasonally inundated grasslands, and fishing in the lakes and river channels into household management systems. Therefore, fishing patterns are closely related to the range of economic options available to local users.

In summary, in addition to the need to account for the ecological heterogeneity and interdependence between subsystems, there is a need to integrate the social heterogeneity (different actors involved) and the interdependence between natural resource use systems into management policy. The purpose of this chapter is to contribute to the systematic analysis of fish resource use in the Lower Amazonian floodplain through the integration of the social dimension of fishing activity. A brief analysis of the ecological, historical, economic, and socio-political dimensions of the Lower Amazon fishery are presented and the role of local management in a system approach to fisheries management discussed.

27.2   The ecological dimension of the fish resource

Floodplains are areas on either side of a river channel which are periodically inundated by the lateral overflow of rivers or lakes and/or by direct precipitation or groundwater (Junk *et al.* 1989). This definition then considers the river channel and its surrounding floodplain as an indivisible unit that shares physical and biological characteristics (Junk 1997). Through the annual flood regime, this ecosystem alternates between aquatic and terrestrial phases, and the dynamic and patchy structure of the landscape strongly influences the characteristics of the biological community.

Large extensions of the Amazon floodplain are fed by nutrient-rich sediments derived from the recent geological formations of the Andean and pre-Andean zones (Junk 1997). This particular environment – locally called *varzea* – covers approximately 160 000 km^2 (Junk 1997), and was traditionally the most densely populated portion of the Amazon Basin. The Lower Amazon *varzea*, which covers a total area of approximately 18 000 km^2, is characterised by a seasonal, monomodal flooding pattern, the amplitude of which varies between 5 and 7 m. The annual flooding regime creates a dynamic pattern: during the flood season, the lakes may expand to cover the entire floodplain, whereas during the dry season, only the deeper lakes remain and four main subsystems emerge: river channels, natural levees bordering river channels, seasonally inundated grasslands and permanent lakes.

Fishing productivity in the lakes tends to be higher during the dry season when non-migratory fish species become concentrated in deep lakes. During the flood season, fishing productivity is extremely patchy, tending to be higher in habitats dominated by aquatic macrophyte communities and in the flooded forest. Fish adaptation to the flood pulse has led to an intimate relationship between fish behaviour (such as feeding habits, reproduction and protection strategies) and spatial and temporal patchiness of the habitat (Goulding 1980; Junk 1984; Junk *et al.*

1997). Therefore, the sustainability of fish resources is closely related to the degree of ecological integrity of the floodplain.

27.3   The human dimension of fish resource use

27.3.1   *The floodplain occupants and their use of the* varzea *system*

Two main groups occupy the Amazon *varzea*: (i) smallholders; and (ii) cattle ranchers. Over many generations, smallholders have developed strategies of resource use that are adapted to the floodplain environment. Their diversified management systems take advantage of the flood pulse (Moran 1993); annual crops such as beans, corn, squash and fast-growing varieties of manioc are cultivated on the natural levees during the dry season. Animal protein is obtained primarily from fish, but also from small animals such as chickens and, to a lesser degree, from game. The *varzea* forest provides firewood, medicinal plants, bait and some edible fruits and nuts. Smallholder settlements are distributed along the river levees and organised into communities that typically range from 30 to a few hundred households. The political structure of these communities is usually based on a leadership system formerly organised by the Catholic Church. Despite such an organisation, communities are not homogeneous. There is often considerable variation in wealth and political power within communities, and this diversity affects the degree of decision-making power of individual households.

   The second major stakeholder group on the floodplain is large-scale ranchers, who occupy the major part of the floodplain area. Câmara and McGrath (1995) found that in a study area of 20 000 ha, 70% of the land was owned by ranchers, with the remainder split between a large number of smallholders living in four different communities. Ranchers have great influence over the decisions of local smallholders. Local smallholders are not eligible for government credit programmes because they lack land title. As a result, economic support for cash crops is often only available through ranchers. The dependence which has often developed between these two groups gives ranchers the ability to maintain their economic and political dominance over smallholders.

27.3.2   *The history*

The intensification of the commercial fisheries began in the early 1960s as a result of several interrelated factors (McGrath *et al.* 1993). The introduction of new fishing technology increased the efficiency of fishing (synthetic monofilament gillnets), storage (ice factories and styrofoam boxes), and transportation (diesel motor boats) of fish products (Mello 1989). In addition, with the decline of jute cultivation, the main cash crop of the floodplain through the 1970s, *varzea* smallholders turned to fishing as their main income source (Gentil 1988). Finally, the demand for fish products increased due to urban growth and to the construction of export-oriented fish processing plants (Bayley & Petrere 1989).

Financing for commercial fisheries has supported mainly large-scale, export-oriented fishing operations (Shoenenberg 1994), while local smallholders were ignored due to their lack of institutional support. As a result, two main groups of commercial fishers have emerged in the region: (i) more sedentary groups of fishers who tend to concentrate effort in lakes close to their communities and use smaller scale fishing technologies based on canoes, small gillnets and other traditional fishing gear; and (ii) itinerant fishers, mainly based in urban areas, who use larger-scale fishing technology, including motorised boats, larger gillnets and, in some areas, small seines.

The small-scale sedentary fishers are more sensitive to a decrease in yield of fishing effort, since they are more geographically and technologically limited compared to itinerant fishers, who can overcome the decline in fishing productivity by adding more gillnets and/or moving to more distant, but more productive lakes. Since the ability of local fishers to adapt to a decline in fishing productivity is limited, they have responded to the intensification of the fisheries by creating informal rules at the community level – fishing accords – to regulate fishing effort in lakes close to their communities, so as to maintain the productivity of local fisheries (McGrath *et al.* 1994).

Fishing accords focus on the lake system as the unit of management. Lakes are considered to be the collective property of the community and part of a broader system, which includes surrounding grassland and forested levees. In some cases, fishing accords have succeeded in increasing the productivity of lake fisheries (McGrath *et al.* 1994). In addition, this local institution may explicitly recognise the interdependence of these different habitats and include rules to regulate the use of other floodplain resources, such as seasonally inundated grasslands and *varzea* forests (Castro 1999). Therefore, fishing accords provide a basis for the establishment of a management strategy that recognises the interdependence between the different resources and habitats exploited by the floodplain population.

27.3.3 *Socio-economy*

The recent increase in the importance of commercial fisheries in the Amazon is related to economic factors that range from global to local markets. As noted earlier, the decline in value of natural fibres on the world market was one of the main factors leading to the intensification of commercial fishing in the local economy. Commercial fishing was the best economic alternative to replace jute cultivation for floodplain smallholders, because fishing is a traditional activity with a growing local market.

Fishing has been economically more attractive than farming because it is more productive in terms of labour than agriculture in the short-term. Fishing demands a smaller labour force than agriculture because it usually involves trips lasting between one and several days, whereas farming involves investments of several months before any return can be realised. If a fishing trip fails, only one workday is lost; however, when a harvest fails, which can occur quite often given the risky nature of floodplain farming, the whole agricultural season is lost (Chibnik 1994; Futemma 1995). The smaller investment in terms of time and labour, and the daily or weekly income provided by fishing, combined with local markets to absorb the fish product are important factors in the increased importance of fishing in the household economy.

In addition to fishing, cattle ranching is another important economic activity that has been intensified on the floodplain. Smallholders and large-scale ranchers have developed different strategies for raising cattle and water buffalo (Castro 1999). The former usually have a few animals (up to 30), and cattle raising is regarded as a means of saving money to invest in productive activities or for emergencies. Ranchers have much larger herds (upwards of 200–300), and raise both cattle and water buffalo as their main source of income. They are rarely involved in other economic activities on the *varzea*.

Although fishing and ranching are less risky and possibly more lucrative than farming, the intensification of these activities has led to an unsustainable strategy of exploiting floodplain resources. Ranching, as currently practiced, leads to the degradation of floodplain habitats and so conflicts with fishing. Likewise, as the productivity of lake fisheries declines, fishing becomes less economically attractive. Under these conditions, a return to the integrated use of different resources on the floodplain is necessary, not only to maintain the fish resources but also to conserve the system as a whole, including the human population.

27.3.4   *Regional socio-political structure*

Several formal organisations influence the use of natural resources on the *varzea*. The Catholic Church continues to play a major role in the political organisation of *varzea* communities, and more generally in the development of grassroots organisations in the region.

Grassroots organisations such as the Fisher's Union (FU) and the Rural Worker's Union (RWU) play an important role in addressing local grievances and needs, and both have been instrumental in organising peasant resistance to government control of their organisations during the 1970s and 1980s (Leroy 1991). Since the 1980s, the RWU has become the representative of upland peasants, while the FU became the representative of floodplain peasants. This division has created an artificial dichotomy (upland/agriculture and floodplain/fishing), that has weakened the political organisation of the floodplain farmers.

In addition to grassroots organisations, government agencies are also important actors on the floodplain. IBAMA, the government agency responsible for the management of renewable natural resources, has a limited capacity to address local resource use issues. Its highly centralised structure of decisions limits its ability to mediate conflicts and monitor the exploitation of local resources. A second governmental institution relevant to the floodplain system is the Department for the Patrimony of the Union (DPU). This agency is responsible for administering federal property, and in this case is the agency that grants use rights to floodplain properties. Although there are legal mechanisms for obtaining rights to floodplain properties, the process is complicated and highly bureaucratic, with decisions made in the state capital. As a result, obtaining formal recognition for floodplain property is difficult for *varzea* smallholders and tends to favour ranchers. Since IBAMA regulates the use of natural resources and the DPU regulates formal access to floodplain land, an efficient strategy for management of the *varzea* system will depend on the degree of collaboration between these two agencies.

27.4   Towards a multidimensional management approach

The sector-based model of fisheries management that is developing in the Amazon has resulted in legislation based on fishing technology restriction and fishing closure seasons that does not take into account social and ecological heterogeneity, and the interdependence of the natural resources of the system. While the ecological studies have revealed the importance of habitat protection (Goulding *et al.* 1996; Junk 1997), human ecological studies have emphasised the importance in accounting for local institutions, such as fishing accords (McGrath *et al.* 1999), for floodplain con-servation. Nevertheless, the relationship between fishing accords and habitat protection is not straightforward. Castro (1999) showed that fishing accords have the potential for conserving floodplain fisheries; but they also have limitations which must be taken into account if this approach is to be ecologically and socially effective.

To make lake reserves work, two main points should be taken into account. First, floodplain occupants who organise lake reserves usually have a common interest in expelling itinerant fishers, but do not always share similar interests with regard to the strategies of resource use among themselves. Floodplain communities are usually socially heterogeneous, and thus the potential for collective decisions regarding resource use should not be taken for granted. Rather, strategies to reconcile conflicting interests must be worked out by community members if some mutually acceptable system of management is to be adopted. Second, while the lake reserve model addresses the interests of floodplain communities, it potentially neglects the interests of other stakeholders, such as urban fishers and, in some cases, ranchers. Thus, like ecological systems, the heterogeneity and interdependency of different stakeholding groups with regard to use of natural resources must be taken into account in developing management systems.

27.4.1   *Integration of local management*

One of the most important contributions of social scientists to the issue of natural resource management has been their effort to integrate local management into a larger policy framework (McCay & Acheson 1987; Pinkerton 1989; Ostrom 1990; Bromley 1992). In the Lower Amazon, a similar approach has been taken to demonstrate the potential of local management for conserving fish resources (Hartmann 1989; McGrath *et al.* 1993, 1999).

Although community-based management systems have the potential to achieve ecologically and socially sustainable use of natural resources, they are not a panacea for the conservation of natural resources. One problem related to local management systems is their instability due to social and economic pressures and lack of institutional support (Ostrom 1990). In many cases, fishing accords are a result of the collective efforts of local smallholders and ranchers to restrict outsider access to local lake systems. However, although fishing accords may be effective for excluding itinerant fishers and cattle rustlers, they may not be as efficient for regulating local resource use. Therefore, a co-management model (Fig. 27.1) that is based on the integration of local management into the formal management structure requires careful analysis of the social context in which the development and implementation of fishing accords takes place.

A second limitation of local management systems is related to the congruence of the local rules with the ecological system (Ostrom 1990). Since fishing accords focus on the lake system, locally based management is an effective way to protect sedentary fish species but is likely to be less effective for migratory species. In addition, similar rules may result in different degrees of ecological effectiveness, depending on the extent of degradation of the ecosystem. Thus, the performance of fishing accord is not only a matter of local interest to conserve the resource, but also a function of the ecological resilience of the system being managed.

Therefore, despite the interest of local communities in implementing lake reserves, the participation of governmental agencies is crucial to monitor the performance of local management systems. Nevertheless, IBAMA has limited ability to monitor the dozens, perhaps hundreds, of fishing accords that Amazon floodplain communities create each year. Given these limitations, a co-management system should start with a small number of local management systems that have the potential to succeed. International donors have followed such a strategy by funding research/extension projects on the Amazon floodplain which can serve as 'demonstration projects', such as the Mamiraua Ecological Reserve in the Upper Solimões and the Ituqui Island in the Lower Amazon. Nevertheless, the criteria for screening local collaboration must be carefully defined because should these initiatives fail, the co-management strategy could lose credibility and support. To avoid this scenario, it is essential that the communities selected demonstrate interest in participating, and the ability to address successfully the diverse and complex array of challenges involved in the collective management of local fisheries.

Castro (1999) proposed a minimum data set as a guideline to help in decisions regarding collaboration, which includes the ecological features related to ecosystem resilience, economic alternatives available to the local population, demographic patterns, local organisation, level of conflicts, land tenure, and institutional support, among others. In particular, it is important to evaluate the biophysical characteristics that smallholder management affects, and the pace of recovery of ecosystem integrity. In addition, the social analysis of the local population must be undertaken to evaluate the diverse interest related to the establishment of the local

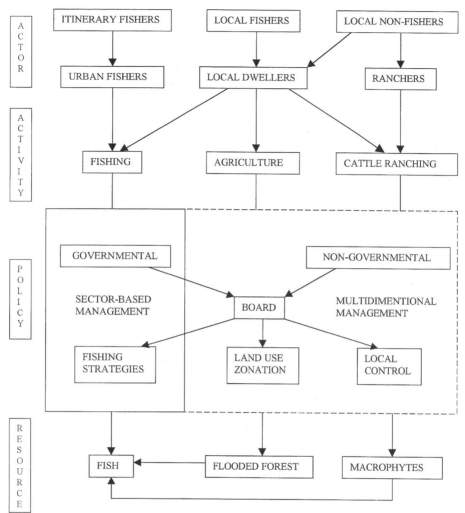

Figure 27.1 Scheme of the shifting from a sector-based management system to a multi-dimensional management of fisheries in the Amazonian floodplain

management, as well as the ability of the population to coordinate the local management appropriately. To conclude, an assessment of social and ecological features of the system should be generated to evaluate the potential of a community to carry out successfully the tasks related to the co-management system.

27.4.2 *Relationship between different resources and actors in the system*

The nature of the commitment of different stakeholders is a key factor in the success of a co-management system (Ostrom 1990). In general, local communities demand the right to manage the system, but rarely offer any commitment to the State in terms of ensuring the sustainability of the way they manage the resource. A co-management system should require a matrix of rights and duties of collaborators, and provide mechanisms for enforcing any such agreement (Pinkerton 1989). Therefore, the right to manage the local lake system should be given in return for the commitment to follow a set of rules which are not based solely on fishing activity, but on floodplain resource use in general. Such a commitment would seek to ensure the conservation of the environment by regulating the use of the subsystems (such as grazing rights, forest timber extraction, burning of grasslands) and thereby ensure the ecological integrity of the floodplain ecosystem. The increasing prevalence of rules focusing on the system as a whole in fishing accords is evidence of local interest in a system approach to floodplain resource management (Castro 1999).

In this regard, policies should reflect the different needs and capabilities of the three main stakeholder groups – smallholders, urban fishers and large-scale ranchers – exploiting floodplain resources in the Lower Amazon. For smallholders who use the floodplain for many purposes, regulation of fishing, farming and cattle raising should take into account their economic opportunities and constraints. In particular, technical and economic assistance for floodplain farming is a key factor that could contribute to the efficacy of fishing accords (McGrath *et al.* 1994). However, intensification of agricultural systems could lead to the degradation of *varzea* forests and pollution of lake waters, which in turn would affect the integrity of the floodplain system (Goulding *et al.* 1996). Therefore, the manner in which individual economic activities are carried out must take into account its impact on the floodplain ecosystem. In this regard, well-organised communities are better able to carry out collective decisions related to the creation, coordination and monitoring of local rules. Once a lake reserve is established, periodic evaluation of the social and ecological performance of the local management system should provide the basis for decisions on continuity or termination of the collaborative exercise.

If lake reserves are to be recognised by the government, mechanisms should also be developed to integrate urban fishers who are not part of lake reserves and do not have access to land for farming. However, access to lake reserves should involve mechanisms to ensure that urban fishers are held accountable to the same rules of use as other lake reserve members. One solution is to create a zoning system within the lake reserve that specifies lakes to which urban fishers would have access. This kind of approach would require a major effort on the part of regional organisations, such as Fishers' Unions and governmental agencies such as IBAMA.

Finally, policies should be developed to regulate land use by ranchers, since the expansion and intensification of cattle ranching has been one of the major causes of floodplain degradation and of conflicts between ranchers and smallholders. This essentially involves regulating access to, and use of, the natural grasslands. However, since ranchers mostly control natural grasslands there is no incentive for this group to avoid any effect of overgrazing on the floodplain system. In general, ranchers have supported fishing accords not to conserve resources but to ensure that outsiders who might be cattle rustlers do not have access to the system. In recent years, however, there has been a trend towards the development of grazing accords to regulate access to, and use of, natural grassland habitat. The involvement of ranchers in the lake reserves is essential not only to reduce habitat degradation, but also because they are politically and economically powerful and control a large proportion of floodplain area, including lakes critical to local smallholders. In summary, for a co-management policy to succeed, resource management policies must take into account the different needs and interests of these different stakeholder groups.

27.5 Conclusion

The floodplain system has long been exploited through diversified strategies involving a variety of natural resources. Through this strategy a higher population density was sustained on the *varzea* in the past than exists today, demonstrating that population density *per se* is not as problematic as the way in which resources are used. Management strategies which focus on individual resources separately have not succeeded, mainly because a system of multiple resource use demands a multi-dimensional approach to management. Cattle ranching *per se* is not a harmful activity, nor are farming or commercial fishing. However, when these activities are intensified without adequate regulation, they affect other aspects of the environment and enter into conflict with each other. Given the importance of the resources of the Amazon floodplain to the *varzea* population, and the interdependence of the different components of the social and ecological systems, a multidimensional management strategy must be developed to ensure the sustainability of the natural resources and in turn, of the economic base which supports the different stakeholders who depend on these resources (Fig. 27.1).

Acknowledgements

This study is based upon research carried out between 1991 and 1997, sponsored by WWF, Heinz Foundation (1991–1993), ODA/WWW (1994–1997) and Fundação Nacional do Meio Ambiente (FNMA) (1996–1997). F. Castro was funded by CNPq for his doctoral research. We would like to thank members of local communities and the grassroots institutions in Santarém for their support during the fieldwork period.

References

Bayley P.B. & Petrere M. (1989) Amazon fisheries: assessment methods, current status and management options. In D. Dodge (ed.) *Proceedings of the International Large River Symposium. Canadian Special Publications, Fisheries and Aquatic Science* **106**, 385–98.
Bromley D.W. (ed.) (1992) *Making the Commons Work: Theory, Practice, and Policy*. San Francisco, CA: ICS Press, 339 pp.

Câmara E. & McGrath D. (1995) A viabilidade da reserva de lago como unidade de manejo sustentável dos recursos da varzea amazônica. *Boletim do Museu Parense Emílio Goeldi, Ser. Antro* **11**, 87–132.

Chibnik M. (1994) *Risky Rivers: The Economic and Politics of Floodplain Farming in Amazonia.* Tucson, AZ: The University of Arizona Press, 267 pp.

Castro F. (1999) Fishing accords: the political ecology of fishing intensification in the Amazon. PhD dissertation, Indiana University, USA, 356 pp.

Futemma C. (1995) Agriculture and Caboclo household organization in the Lower Amazon Basin case studies. MS dissertation, Tulane University, 104 pp.

Gentil J.M.L. (1988) A juta na agriculturale de várzea na area de Santarem – Medio Amazonas. *Boletim do Museu Paraense Emilio Goeldi, Serie Antropologia* **4**, 119–99.

Goulding M. (1980) *The Fishes and the Forest: Explorations in Amazonian Natural History.* Berkeley: University of California Press, 280 pp.

Goulding M., Smith, N. & Mahar D.J. (1996) *Floods of Fortune: Ecology and Economy along the Amazon.* New York: Columbia University, 193 pp.

Hartmann W. (1989) Conflitos de pesca em águas interiores da Amazônia e tentativas para sua solução. In A.C. Diegues (ed.) III Encontro de Ciências Sociais e o Mar no Brasil. *Pesca Artesanal: Tradição e Moderinidade.* São Paulo, Brasil, pp. 103–118.

Junk W.J. (1984) Ecology, fisheries and fish culture in Amazonia. In H. Sioli (ed.) *The Amazon: Limnology and Landscape Ecology of a Mighty Tropical River and its Basin.* Dordrecht: Monographiae Biologicae. Junk, pp. 215–43.

Junk W.J. (ed.) (1997) *The Central Amazon Floodplain Ecology of a Pulsing System.* Ecological Studies 126. Berlin: Springcr, 525 pp.

Junk W.J., Soares M.G.M. & Saint-Paul U. (1997) The fish. In W.J. Junk (ed.) *The Central Amazon Floodplain Ecology of a Pulsing System, Ecological Studies* **126**. Berlin: Springer, pp. 385–405.

Larkin P.A. (1978) Fisheries management: an essay for ecologists. *Annual Reviews in Ecology and Systematics* **9**, 57–73.

Leroy J.P. (1991) *Uma Chama na Amazonia.* Petropolis: FASE, Ed. Vozes, 213 pp.

McCay B.J. & Acheson J.M. (eds) (1987) *The Question of the Commons: The Culture and Ecology of Communal Resources.* Tucson, AZ: The University of Arizona Press, 439 pp.

McGrath D., Castro F., & Futemma C., Amaral B.D. & Calabria J. (1993) Fisheries and the evolution of resource management on the Lower Amazon floodplain. *Human Ecology* **21**, 167–195.

McGrath D., Castro F. & Futemma C. (1994) Reservas de lago e o manejo comunitário da pesca no Baixo Amazonas: uma avaliação preliminar. In M.A. D'Incao & I.M. Silveira (eds) *Amazônia e a Crise da Modernização.* Belém, Pará: Museu Paraense Emílio Goeldi, pp. 389–402.

McGrath D., Castro F., Camara E. & Futemma C. (1999) Community management of floodplain lakes and the sustainable development of Amazonian fisheries. In C. Padoch, J.M. Ayres, M. Pinedo-Vasquez & A. Henderson (eds) Várzea: Diversity, development, and conservation of Amazonia's whitewater floodplains. *Advances in Economic Botany, Vol. 13.* New York: The New York Botanical Garden Press, pp. 59–82.

Meggers B. (1971) *Amazonia: Man and Culture in a Couterfeit Paradise.* Atherton: Aldine, 182 pp.

Mello A.F. (1989) Contribuições para uma teoria dos conflitos pesqueiros no Brasil: partindo do caso Amazônico. In A.C. Diegues (ed.) III Encontro de Ciências Sociais e o Mar no Brasil. *Pesca Artesanal: Tradição e Moderinidade.* São Paulo, Brasil, pp. 63–76.

Moran E.F. (1993) *Through Amazonian Eyes: The Human Ecology of Amazonian Populations.* Iowa City: University of Iowa Press, 230 pp.

Ostrom E. (1990) *Governing the Commons: The Evolution of Institutions for Collective Action.* New York: Cambridge University Press, 280 pp.

Pinkerton E. (ed.) (1989) *Cooperative Management of Local Fisheries: New Directions for Improved Management and Community Development*. Vancouver: University of British Columbia Press, 289 pp.

Rothschild B.J. (1973) Questions of strategy in fishery management and development. *Journal of the Fisheries Research Board of Canada* **30,** 2017–2030.

Roosevelt A. (1989) Resource management in Amazonia before the Conquest: beyond ethnographic projection. *Advances in Economic Botany* **7**, 30–62.

Shoenenberg R. (1994) *As Formas Institucionais e Organizacionais de Articular Interesses na Area da Pesca no Baixo Amazonas em Particular, e na Amazônia em Geral*. Santarém, Pará: Relatório Preliminar para Projeto IARA/IBAMA, 32 pp.

Chapter 28
The role of Nkhotakota Wildlife Reserve, Malawi, Africa, in the conservation of the mpasa, *Opsaridium microlepis* (Günther)

D. TWEDDLE

J.L.B. Smith Institute of Ichthyology, P/Bag 1015, Grahamstown 6140, South Africa (e-mail: unecia@infocom.co.ug) (Present address: Lake Victoria Fisheries Research Project, PO Box 2145, Jinja, Uganda)

Abstract

The mpasa, *Opsaridium microlepis* (Günther), is endemic to Lake Malawi. It is threatened by overfishing and habitat degradation as a result of its dependence on affluent rivers for spawning. Annual research visits to the Bua River, Nkhotakota Wildlife Reserve, from 1977 to 1995 provided information on growth rates and breeding behaviour, and showed that abundance of spawners in the river depended on river levels. A decline in catches was noted in the last decade, with apparent total failures of the spawning run in years when low river levels resulted in blockage of the river by nets and weirs. Villagers adjacent to the river below the reserve were generally aware of the problems, but had conflicting opinions on how to manage the fishery. Recommendations for conservation of the fishery include involving the communities in control to allow adequate escapement of spawners, particularly in low rainfall years, and continued protection of the spawning habitats within the Nkhotakota Wildlife Reserve, i.e. no further erosion of the reserve's boundaries into the catchment area of the Bua River.

28.1   Introduction

Species of the genus *Opsaridium* are widespread in the rivers of western, central and southern Africa. They are generally small species not exceeding 20 cm in length. They have vertical barring on the body, pink or orange tinted fins and resemble salmon parr or small trout. In Malawi, there are two such species, *O. zambezensis* (Peters) in the Lower Shire River, a tributary of the Lower Zambezi, and *O. tweddleorum* Skelton in the affluent rivers of Lake Malawi and east bank tributaries of the Lower Shire.

In addition, the formation of Lake Malawi has resulted in the evolution of two large endemic species adapted to the large lacustrine environment, the mpasa, *O. microlepis* (Günther), and the sanjika *O. microcephalus* (Günther). *Opsaridium microlepis,* in particular, has specialised in the role of openwater predator on small pelagic zooplanktivorous fishes and has remarkable similarities to salmon, *Salmo salar* L., in both appearance and behaviour (Tweddle & Lewis 1983). It reaches over

70 cm TL and 4 kg in weight. The mpasa ascends rivers from the lake during and just after the rainy season to spawn in gravel and cobble stretches of rivers, and thus is vulnerable to fishing and habitat degradation. It is considered to be a threatened species. The other endemic species, *O. microcephalus*, reaches a weight of 0.5 kg.

Initial concerns that the mpasa fishery was in decline in a major spawning river, the Bua River in central Malawi (Fig. 28.1), led to brief surveys during the breeding season in the mid-1970s, which continued annually between 1977 and 1995, except for the 1979 season.

Nkhotakota Wildlife Reserve (NWR) covers an area of 1800 km^2 in the Central Region of Malawi (Fig. 28.1). It protects the steep slopes of the Great Africa Rift escarpment and thus conserves water catchment areas. Protection is given to the Bua River, which passes through the reserve for a distance of about 35 km. The Bua rises

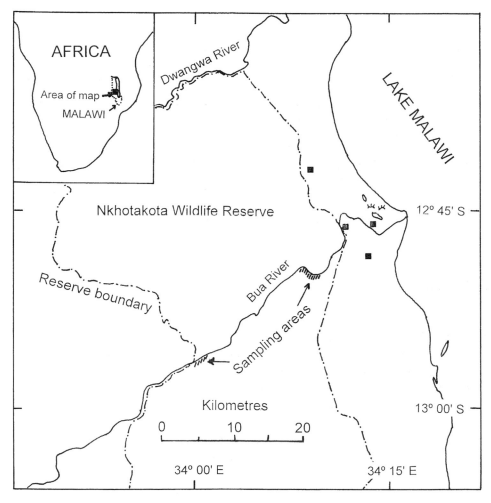

Figure 28.1 The Bua River, showing the sampling areas and villages selected for the questionnaire

on the Malawi–Zambia border and flows for approximately 200 km before entering the NWR.

Biodiversity is influenced by the character of the rivers and the range of habitats available within the systems. As the NWR protects rift valley slopes, the rivers are generally fast flowing and rocky within the reserves, although there is a considerable variety of habitat. The fast-flowing rocky stretches are interspersed with deep pools. There are also extensive areas of reedy shoreline with other patches of weed providing good fish cover. Tributaries arising within the reserve provide fast flowing, mountain stream habitat. On the Bua River, the base of the rift escarpment coincides with the boundary of the reserve, and from there to Lake Malawi, the river is unprotected over a distance of approximately 15 km on the flat lakeshore plain.

With the emphasis on the importance of the NWR to protect the mpasa's spawning grounds (Tweddle 1985), conservation became an important component of management policy of the Malawi Government Department of National Parks and Wildlife. Subsequently, management of the fishery was included in the objectives of a project to develop and manage the NWR, funded by the Japanese International Cooperation Agency. With this support it was possible to involve the communities bordering the river below the reserve to consider the best ways to manage the resource.

This chapter summarises knowledge of the fish fauna of the Bua River, the biology of mpasa, and the attitudes of villagers and fishermen towards the mpasa fishery and the NWR. Options for managing the fishery aimed at sustainable exploitation and conservation of this unique, but threatened, species are presented.

28.2   Methods

28.2.1   *Fish species composition in the reserve*

The fish populations in the Bua River in the NWR were sampled at several sites by electric fishing in 1976, 1992 and 1995. Sites were selected to investigate as many habitats as possible. Full details of the sampling gear, sites and fish caught were given by Tweddle (1996).

28.2.2   *Mpasa biology*

The mpasa fishery of the Bua River was monitored from 1977 to 1995 during an annual visit in mid-May to the stretch around Bua Camp in the NWR. Three or four sampling days were usually spent on each visit. Extra visits were made in some years. Sampling was carried out by angling with spinning tackle. The size of the sampling team varied, but was usually three or four people.

Sampling started daily at 8 am and continued to dusk at 17.30 pm. The same stretches of river were fished each year using the same techniques. All mpasa caught were measured (TL and FL, nearest cm), weighed to the nearest 50 g, and released alive. From 1979 to 1985, approximately 1000 mpasa were tagged using Floy spaghetti tags inserted in the dorsal musculature between the neural spines. A financial reward was offered for return of the tagged fish to a Fisheries Department office or field assistant for measurement and examination.

Mpasa breeding behaviour was examined by direct observation from vantage points above spawning grounds (Tweddle 1983, 1995), while measurements of fecundity were made by Tweddle (1983) and Msiska (1990).

Differences in abundance in the river from year to year were compared using catch per unit effort (CPUE) data, expressed as mean number caught per day per angler over the sampling period. Because of variations in anglers' ability, only data collected by two consistent anglers (D. Tweddle and D.S.C. Lewis) were used in the comparison (these catches were always within 5% of each other in joint sampling programmes and mean CPUE was recorded for such visits).

Data on Bua River levels were obtained from the Water Department to compare with mpasa abundance in the river.

28.2.3   *Village questionnaire*

A questionnaire was designed to cover aspects of the river fisheries in the area between the reserve and the lake. These included fish consumption and value, degree of dependence on fish in the diet and fishing in the household economy, knowledge of the target fish species' habits and biology, fishing methods used, changes in the fishery over time, and opinions on possible methods of regulation of the fisheries.

In July and August 1995, 80 household heads were questioned by teachers in four villages (20 per village); two alongside the Bua River and two further away from the river, one at 5 km distance and one at 10 km distance. The interviewers were thoroughly briefed in the aims of the survey and the approach needed to gain the confidence and cooperation of the target group. The questionnaire was reviewed to resolve any difficulties or misunderstandings that might arise in the course of the interviews.

Little difference was found between responses from villages on the river and further away, and therefore all data were pooled for analysis.

28.3   Results

28.3.1   *Fish species diversity of the Bua River within Nkhotakota Wildlife Reserve*

In the reserve, the river has frequent rapids, small waterfalls, deep rocky pools, reed-lined stretches (although these are currently much reduced after being destroyed by a flood in 1989), and sandy and gravely glides. The only fish habitat not represented by this large variety is swamps and weedy lagoons. The fauna in the reserve was fairly uniform and dominated by the larger species *Barbus johnstonii* Boulenger, *Labeo cylindricus* Peters, *O. microcephalus, Clarias gariepinus* (Burchell) and *Oreochromis shiranus* (Boulenger), all of which are of some commercial importance. The small *Barbus trimaculatus* Peters and *Astatotilapia calliptera* (Günther) are also abundant in the river. At the lowest sampled site in the reserve, small lakeshore stream species occurred, including *Hemigrammopetersius barnardi* (Herre), *Barbus arcislongae* Keilhack, *B.* cf. *lineomaculatus* (A), *B.* cf. *lineomaculatus* (B), *B. macrotaenia* Worthington, *B. atkinsoni* Bailey and an unidentified Lake Malawi endemic cichlid species.

The drought of the early 1990s severely affected the flat lakeshore area, with little to no flow, a sandy bed, and limited, almost stagnant shallow pools. The numerous small lakeshore species were absent in 1995 samples, which contained only *A. calliptera* in any number, while *C. gariepinus* and *O. shiranus* also survived in what was left of the river.

A full annotated checklist and key to the species found in the reserve was given by Tweddle (1996).

28.3.2 *Mpasa biology*

Breeding behaviour

The breeding behaviour of *O. microlepis* was described by Tweddle (1983), based on observations made in the Bua River in June 1982. Eggs are buried at the gravel/sand interface, having filtered down through the interstices in the clean gravel near the surface. Spawning appears to be primarily a nocturnal and early morning activity with no fish on the exposed shallows during the main daylight hours.

Mpasa is a fractional spawner. They breed in the larger rivers, where they remain for several months laying many batches of a few thousand eggs at a time over the period. Their breeding season extends well into the dry season, when the eggs and fry are not at risk of destruction in flash floods. The fecundity of mpasa from the Bua River was estimated by Msiska (1990) at 34 000 to 66 000 eggs for fish 54–58 cm in length.

More evidence of a flexible breeding strategy is gained from the number of different year classes which take part in spawning. In the North Rukuru River, for example, up to five year classes are represented (Tweddle 1987), while in the Bua, three year classes are usually present, with few smaller fish. Moreover, the length frequency structure of the spawning population suggests that not all fish leave the main river to run up to spawn every year.

Probably because of the flexibility in breeding strategy, mpasa recruitment runs did not suffer the catastrophic collapse seen in the Lake Malawi fishery for nchila, *Labeo mesops* Günther, in the early 1960s (Tweddle *et al.* 1994a). The latter is a total spawner which uses small temporary streams to breed during the rains. Even a complete spawning failure in one year may not have a devastating effect on mpasa stocks. However, a series of bad years could be devastating, and this occurred in the 1990s, giving cause for concern for the future of mpasa in the Bua River.

Growth

Evidence from juveniles taken by electric fishing in the Bua and Linthipe Rivers, and by fine-meshed beach seining in Lake Malawi adjacent to the North Rukuru River mouth (Tweddle 1987) suggested that the species reached about 11 cm total length in its first year. North Rukuru River breeding season data showed strong modes at 20–25 cm, with the next peak at over 30 cm. The evidence suggests these were 2- and 3-year-old fish, respectively. Four-year-old fish were over 40 cm and 5-year-old fish were over 50-cm.

Good data on the younger age groups are not available for the Bua River, where most of the catch was of fish over 50 cm (Fig. 28.2). A bimodal distribution was apparent in several years in catches from the Bua, e.g. in 1986 and 1989, with modes in the mid-50 cm range and 60 cm or just over, while strongly skewed distributions were apparent in other years, e.g. 1985 and 1990, again suggesting the presence of more than one year class. In some years, particularly more recently, e.g. 1980, 1986, 1990 and 1993, there was a strong presence of fish in the upper 40 cm range. Comparing these data with the North Rukuru data suggests that the Bua fish population consists largely of 5- and 6-year-old fish, with 4-year-olds also present in many years.

28.3.3   *Bua River Mpasa Research And Monitoring Programme, 1977–1995*

Catches in the 1970s were erratic (Fig. 28.3) but 1978 was a good year for mpasa spawning, with a healthy stock in the river and large numbers of fry to be seen. There was no sampling in 1979 but it was known to be a good year. While there were fewer fish in 1980 and 1981, 1982 and 1983 had excellent runs of fish upriver. The relatively low figure for 1982 (Fig. 28.3) was due to sampling being a month later than the optimum of mid-May. Catches in 1985 were exceptional but then declined steadily. There are several possible reasons for the sequence of good years followed by the decline.

The Fisheries Department was first established in Nkhotakota in 1978. Because of the concerns expressed about the mpasa, emphasis was given to proper management and in 1978 and 1979 the law requiring a gap to be left in fishing weirs to allow a percentage of fish to get upriver was enforced. Large numbers of fish ascended the river to spawn, resulting in very good recruitment and excellent catches in the mid-1980s.

Successful spawning in the 1980s continued to give good recruitment, but a failure to prevent weirs blocking the river led to a steady decline and eventually a collapse in

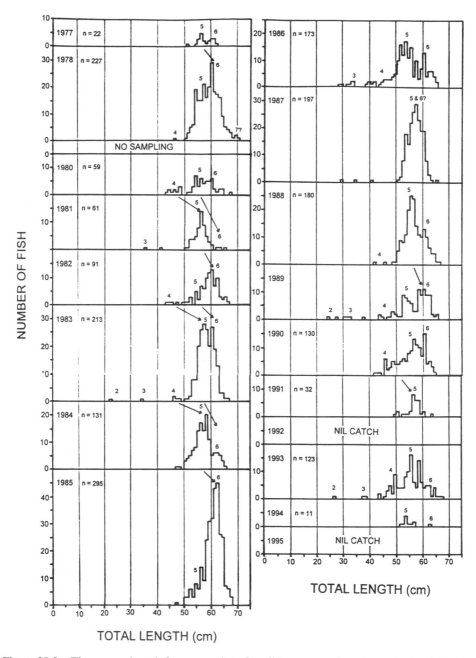

Figure 28.2 The mpasa length frequency data for all Bua research and monitoring trips since 1977, showing year classes and fluctuations in abundance over the years

1991, when generally low river levels allowed the fishermen to effectively block the river. In 1992, the most severe drought year, no mpasa, either adult or fry, were detected in the river.

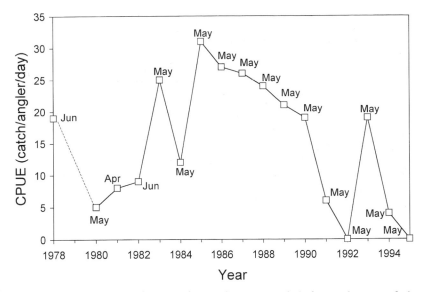

Figure 28.3 CPUE (expressed as catch per day per angler) for each year of the mpasa research and monitoring programme, showing erratic data in the early years, high catches in the mid-1980s and a collapse in the 1990s. The month of sampling is also shown. In early years, the time of sampling was variable, with standardisation on mid-May from 1983. Catches in April may be lower because of high river levels, turbid water due to the rains, and many fish still in the lake. In June, catches are lower than May because of heavy predation and return of spent fish downstream

Spawning failure in two successive years led to a decision that the Departments of Fisheries and of National Parks and Wildlife would take joint action to keep the river open in 1993. Reasonable numbers of fish were thus able to spawn. Shortly after the monitoring visit, poachers poisoned the river above Bua Camp and wiped out all the fish from there to the lake.

In 1994, mpasa moved upriver after early rains but the very early end to the rainy season led to the extensive use of weirs again, completely blocking the river. The few fish caught in the river in May were in very poor condition and had been in the river for several months. There were no fresh-run fish and very few fry to be seen. In 1995, there was again a total spawning failure, with no fish caught whatsoever. The electric fishing survey conducted in July confirmed this, with a total absence of fry. Electric fishing in the same areas in 1976 yielded large numbers of fry.

In essence, therefore, there were five spawning failures in a row. Given that most fish spawn at 5 years old, the situation was viewed with great concern. In addition to the probable impact of fishing in the river on the stocks, other factors may be implicated in the poor catches.

The initial approach to the Fisheries Department by the Department of National Parks and Wildlife followed concern about low numbers of fish in the Bua River in the early 1970s. From the start of the research programme, catches improved to very high levels in 1978 and in the 1980s before declining. Tweddle (1983) found relationships between catch rates and river levels in the year of fishing, and also river levels some years previously, when the fish were hatched. The low catches in the early 1970s followed generally low river levels in the 1960s, which would have allowed more intensive fishing effort. The 1970s was a decade of high rainfall with high river levels preventing weir construction until well after the rainy season. High flow levels also mean there was more water in the river until late in the year and therefore better habitat for the fry. High flows also dilute fish poison and lessen its impact on the fry. These factors combined to generate the large runs of mpasa in the 1980s. Lower rainfall and river levels in the decade to 1995 contributed to the decline in stocks by allowing more intensive fishing and limiting suitable fry habitat.

CPUE was compared with river level using the 14 May water level each year as this date falls in the majority of annual sampling trips. The relationship shows that low river levels result in poor catches ($P < 0.01$) (Fig. 28.4).

Village questionnaire

Of the 80 villagers questioned, 95% originated locally in Nkhotakota District. Farming was the primary occupation of 91% of respondents, while 8% listed fishing or fish trading. Fishing was a primary or secondary occupation in 91% of the households. Fishing grounds mentioned were the Bua River (89% of fishermen), Lake Malawi (38%), swamps (15%), dams (1%) and other rivers (2%). Traps were used by 66% of the fishermen, hooks by 59%, gillnets by 44%, weirs by 36%, seine nets by 8% and scoopnets by 4%. Twenty-six per cent of the fishermen admitted to

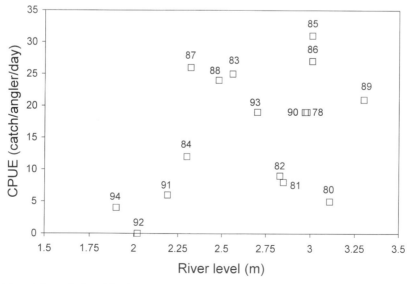

Figure 28.4 The relationship between CPUE and the Bua River level on 14 May at the lakeshore road bridge

fishing in the NWR, although one claimed he only did so prior to the establishment of the reserve. The rains and early dry season were cited as the peak fishing period.

Fish is an important component of the local diet, with 75% of those who responded reporting that fish constitutes 50% or more of the protein in the diet. With mpasa costing approximately £1 per kilo, the value to the local rural economy is high.

The opinions of the fishermen about the quality of fishing in the Bua River in 1995 and in previous decades are shown in Table 28.1. Over 80% of villagers stated that the fishery in the river had declined during the 1990s. Reasons suggested for the decline and opinions about the quality of the river itself are shown in Table 28.2. Answers show wide recognition that the drought years of the 1990s had an impact on the number of spawning fish in the river. There was also widespread, although not universal, recognition that the quality of the river had deteriorated. The river is now wider, shallower and sandier, and rises and falls more quickly than in the past. The loss of trees along the river banks was noted as a problem. It was recognised that destructive fishing methods were being used and that the Bua River Irrigation Scheme has had an impact on fish migration.

Fish poisoning was recognised as a major problem. The respondent quoting fuel smell as a reason for the fishery decline may have been referring to a major poisoning episode using agricultural chemicals in 1993. All respondents were aware of the use of poison and 96% condemned it in various degrees, although 5% stated that poisoning could be allowed using traditional herbs. 26% admitted to using poison themselves. One interviewee noted that poison was used in anger at enforcement of government regulations.

Knowledge of the biology of mpasa was very weak. Over half thought that breeding occurred in slow-flowing pools. Only one respondent stated that the fish breed in fast-flowing water. The majority thought that the fish breed when less than a year old.

Table 28.1   The opinions of villagers near the Bua River about the quality of fishing in 1995 and in the last five decades. Figures are expressed as percentage of respondents ($n = 80$). The scale used is from 1 = excellent to 5 = very poor

	Scale					
	5	4	3	2	1	No reply
Quality of fishing in 1950s					85	15
Quality of fishing in 1960s				3	91	6
Quality of fishing in 1970s			10	46	41	3
Quality of fishing in 1980s		24	54	11	10	1
Quality of fishing in 1990s	65	26	8			5
Quality of fishing in 1995	51	29	11			8

Table 28.2   Opinions of villagers on the reasons for the decline in mpasa catches and on the quality of the river[a]

Reasons suggested for decline in mpasa fishery	Number
Drought	71
Overfishing and/or overpopulation	28
Change of river course by Bua River Irrigation Scheme	20
Use of bad fishing methods (including small-meshed seine nets, drift nets and nets completely blocking the river)	15
Tree removal along banks	11
Poison	8
Lamp fishing in Lake Malawi at night	4
Fishing at the river mouth	3
Fertiliser use	3
Use of nylon nets	1
Use of weirs	1
River too wide	1
Fuel smell	1
Bua River Irrigation Scheme dam blocking river	1

Opinions on changes in quality of river	
Dirtier	61
Not dirtier	16
Sandier	88
Rises and falls more quickly than in past	65
Wider	68
Narrower (refers to river itself in drought, not channel width)	11
Shallower	5
Increased algae	5

[a]The number given is the percentage of villagers presenting that opinion ($n = 80$).

The opinions of the villagers on the mpasa fishery and the role of NWR in its conservation and management are shown in Table 28.3. The questions were in three sections. General questions about the importance of the reserve for the fishery were followed by proposals for possible regulations, and finally villagers were asked for their own suggestions. The answers showed wide recognition that the NWR protects the fish and fishery, but there were conflicting opinions on how best to manage the fishery. Most were in favour of giving the fish some protection when they are running upriver, with 54% in favour of a closed season and 80% in favour of banning weirs until the fish have gone upriver. Construction after June will intercept fish running down river after spawning.

Opinions varied on fishing in the reserve itself. While the majority was in favour, a significant minority, 16%, was against the idea. The number in favour of fishing in the reserve increased if the proviso was added that it be linked to community-enforced bans on destructive fishing methods.

A wide range of ideas is presented in the third part of Table 28.3. In general, they showed a desire to stop fishing methods which prevent fish getting upstream to spawn, stop the diversion of river flow by the Bua River Irrigation Scheme, and protect river banks from vegetation clearance. A desire for better enforcement of laws was evident in some answers.

28.4   Discussion

Extensive clearance of the Bua watershed above the NWR for agriculture, and particularly for tobacco-growing, has occurred in the last three decades, with a marked impact on river quality. The village survey included questions on changes in the quality of the river and there was almost universal recognition of the changes that have occurred, i.e. the river is more turbid, wider and sandier, and it rises and falls much more quickly when it rains. A decline in riverbank vegetation and excessive tree clearance were also noted by several interviewees.

These observations by fishermen who have lived in the area all their lives support observations made on the river in annual visits since 1975. In 1989, a major flood destroyed much of the riparian vegetation and the river banks themselves in many places. At Namadinga Pool, for example, approximately 15 m of well-consolidated, heavily-overgrown bank was removed by the flood, together with high gravel ridges, which were covered in *Phragmites* reeds in the middle of the pool. The deep, fast-flowing pool between the banks was replaced by a very wide, shallow, gently flowing silt-bottomed river bed. The banks were gradually re-consolidated with new vegetation over 5 years, but the banks were 15 m further back than previously and the reedbeds were not replaced. Mpasa spawn in clean gravel beds and possibly also in stony rapids (Tweddle 1983, 1995). The extent of these was reduced by the altered character of the river following the 1989 flood, which may have been exacerbated by the increased land clearance for agriculture above the reserve. The impact of tobacco herbicides and pesticides on fish breeding is unknown.

Many fishermen claimed that catches from the Bua River had been declining since the 1960s, and a survey of five river mouth fishermen (N.G. Willoughby, unpublished data) agreed with this. However, the results of the angling survey inside the NWR suggest that the fishery was thriving in the early 1980s and was almost certainly better then than at any time in the 1970s. In the present survey, a number of fishermen agreed that the 1980s fishing was still good.

The fish populations of the Bua River suffered in the series of drought years. Without the protection of the NWR, the situation would be much worse. The reserve protects not only the main river spawning reaches for a number of important commercial species, but also a number of tributaries which are utilised by fish for spawning and by fishermen for food. A series of normal rainfall years and improved management of the fisheries for the potamodromous species, including the mpasa, is necessary for recovery.

Table 28.3   The opinion of villagers near the river about management options for the mpasa fishery in and near the Nkhotakota Wildlife Reserve. 'Other' answers include 'don't know', 'no answer' and alternative opinions. Answers are given in percentages ($n = 80$)

Question	Yes	No	Other
Is the existence of NWR important for protection of mpasa spawning?	89	8	3
Should the protection given by the reserve be maintained?	84	10	6
Are weirs blocking the river harmful to mpasa migration?	78	21	1
Proposals for regulations			
Closed season in the river until May	54	45	1
Ban on weirs until June	80	19	1
Allow fishing in the reserve	76	16	8
Complete ban on poisoning	96	3	1
Ban nets blocking the river	91	8	1
Stop all fishing in the reserve	16	83	1
Allow reserve fishing in exchange for community-enforced bans on destructive fishing methods	91	5	4
Other suggestions made by villagers			
Control fishing in lake (includes removing recent settlers, stopping lamp fishing, stopping river mouth blocking with nets and traps, stopping river mouth fishing in general)	15		
Encourage use of hooks, not nets and weirs	13		
Encourage large meshes in nets	8		
Encourage non-damaging methods/stop bad methods (unspecified)	10		
Stop drift net fishing	1		
Encourage use of scoop nets.	1		
Build dams near river for fishing during proposed river closed season	6		
Restore original course of river	11		
Deepen the river mouth to allow fish easier access	1		
Stop Bua River Irrigation Scheme damming river	1		
Stricter law enforcement and more patrols	9		
More education of fishermen	1		
Scrap the reserve completely	1		
Stop tree cutting/plant trees along river banks	4		
Impose short (3–4 week) closed season	1		
Assist people with acquisition of gear to fish in lake instead	1		
Allow even small meshed nets in reserve	1		
Go back to 1920s-style traditional control (NB suggestion of village headman)	1		
Stop corruption by enforcement officers who accept bribes to allow fish poaching	1		
No other ideas presented	9		

Fishermen claimed that the decline in the fishery in the 1990s was due to drought (Table 28.3). The spawning grounds, however, still had good river flows and could support a large spawning population. The decline was probably due to the greater vulnerability of the fish to capture in low water conditions and thus overfishing of the migrating fish. Many fishermen were aware of the problem and blamed overfishing, the increasing human population, and bad fishing methods, in addition to the drought (Table 28.2).

Management of the Bua fishery must aim to modify the fishermen's behaviour to ensure that enough fish get up the river to spawn even in low rainfall years, i.e. a spawning escapement as in salmon (see Milner *et al.*, Chapter 25). The last few questions in the village survey sought opinions on how best to control the fishery. Many suggestions were put forward in addition to the options expressed in the questionnaire (Table 28.3).

For sustainable catches and long-term conservation of the mpasa fishery, closure of the river to commercial fishing while allowing a lake fishery would be the ideal management option. It is a system increasingly used in salmonid fisheries in the northern hemisphere, where river net fisheries are bought out by angling syndicates to allow more fish upstream to provide sport and to breed (Milner *et al.*, Chapter 25). Commercial fisheries are increasingly restricted to the open sea away from river mouths. Mpasa roam widely in the lake and are rarely caught except when gathering near the river mouths prior to running upriver to spawn. The main fishery is in the lake near the river mouths. Accurate data on the size of the lake fishery for mpasa are not available. The statistical data routinely collected by the Malawi government Fisheries Department do not include specific data on these two species, which are lumped with several other species in the 'other species' category. Mpasa is an important component of this category (Tweddle *et al.* 1994b). Catches of 'other species' ranged from 130 to 740 t annually, from 1976 to 1989, with a mean of 290 t. At approximately £1 per kg in 1995, this is a very valuable fishery.

If the fish which succeed in running the gauntlet of nets and entering the river are subsequently left undisturbed to breed, and provided river water quality is not allowed to deteriorate and siltation of the gravel spawning grounds does not become excessive, the future of both the fish and the fishery would be assured. Extending the NWR boundary to the river mouth would help in achieving this aim. The reaction of the great majority of the Bua fishermen to the suggestion that the river is closed and fishing allowed only in the lake was negative and it is unlikely that such a policy could be implemented.

If complete closure is not feasible, one option is to prevent the use of the most damaging gears during the critical period when the fish are moving upstream. Two suggestions were included in the village survey; a ban on fishing in the river until May, and a ban on weir construction until June. Fifty-five per cent were in favour of a closed season, 44% opposed. A much higher number – 80% – were in favour of banning weirs until June. Several interviewees were against closed seasons because the period in question is the best fishing time.

As 80% were in favour of a ban on weirs until June, and 66% believed that nets blocking the river should be banned, there is a possibility that regulations to achieve such measures would have some backing, provided the communities are fully involved in the decision-making process.

There was a widespread agreement that poison should be banned. One chief in the area has already banned the use of poison in his area. However, many of the people who use poison stated that it is a bad method, which kills juvenile fish, and yet they continue to use it. There was also a prominent minority of villagers who stated that the method was not harmful and should be allowed. Katupe (*Tephrosia* sp.) is not only a fish poison but is also an important traditional medicine. It is used in solution to wash dogs to remove parasites such as ticks and fleas, and it is also used to cure disease in chickens. A ban on growing the plant is therefore not feasible, although it may be possible to distinguish between plantations large enough to poison extensive areas of river and the small number of plants necessary for medicinal use.

Most villagers were in favour of allowing fishing in the reserve, with community involvement in management and cooperation with the Departments of Fisheries and of National Parks and Wildlife. The programme would link concessions to the villagers with an agreement to ban destructive gears such as weirs, blocking nets and fish poison. Several villagers commented enthusiastically on this proposal, while most other questions simply elicited 'yes' or 'no' responses.

Local knowledge of the biology of mpasa was assessed. The destructive fishing methods used by many fishermen suggested that they had limited knowledge of the damage they were causing to the fish stocks. The results demonstrated their limited awareness. The majority thought that mpasa grow so fast that they reach maturity and breed at less than 1 year old. In reality, most of the breeding population consists of 5- and 6-year-old fish. The fish are therefore vulnerable to overfishing as they take so long to grow and mature. There was also a surprising lack of knowledge of the breeding behaviour of the species, given that most fishermen have lived in the area and exploited the fishery all their lives. Mpasa can often be seen spawning over gravel-bottomed, swift-flowing stretches of river. Almost all fishermen, however, said slow flow or pools were necessary, yet the eggs, buried in the gravel, need a strong flow to carry oxygen to them through the gravel and to keep the gravel scoured free of smothering silt.

An education programme is therefore necessary to persuade fishermen to use non-destructive fishing methods. This programme should educate the fishermen about the fish, the fishery, the problems of destructive fishing methods, the possibilities of community-based management, and the problems of environmental degradation, including the importance of retaining vegetation buffer zones along the river banks.

Options for this programme include the involvement of either the two government departments, Fisheries and National Parks and Wildlife, both of which are developing community-based management programmes, or the Wildlife Society of Malawi Dwangwa Branch.

Enforcement of fisheries laws has worked in the past for this fishery. The success of the early 1980s was a result of enforcement of the law requiring a gap to be left in weirs in the late 1970s. In 1995 also, gaps in weirs were enforced and fish were able to spawn. In theory, enforcement should be relatively straightforward. In practice, however, there are major difficulties in this approach. One is the ill-feeling generated in the local communities by strong-arm enforcement tactics, particularly if inconsistently applied. This can lead to retaliatory action by the villagers, including large-scale poisoning. Another is the financial cost of regular joint Fisheries/Parks patrols, which would be necessary to be effective.

Community participation is of great value in the conservation of natural resources. To be effective, enforcement must have the cooperation of the communities. This can only be achieved if a series of management measures acceptable to the majority (not necessarily all) fishermen is negotiated between the communities (or fishermen's clubs) and both government departments. The village survey demonstrated that most fishermen are in favour of curbing poisoning and river-blocking gears. It should therefore be possible for the two government departments and a fishermen's representative organisation to negotiate regulations acceptable to all, with joint community and government control of the fishery. This may be extended to include the question of fishing in the NWR itself.

There was a very strong opinion in the communities that fishing in the reserve should be allowed, although a significant minority (13 interviewees) were opposed, stating that it would be difficult to control.

It is understandable that the fishermen would be keen to gain access to the excellent fish-holding pools of the NWR. If allowed, it would have to be strictly controlled, with limited access. In favour of allowing fishing is that it should greatly improve relations between the Department of National Parks and Wildlife and the villages adjacent to the reserve, provided that the fishing is properly managed, with no favouritism or bribery. Also in its favour is that cooperation should then be possible with the villages to impose restrictions on damaging methods such as weirs, blocking nets and poison.

There are, however, strong arguments against allowing fishing in the reserve. It has been stressed here that the reserve is the main reason for the success of the Bua River mpasa fishery compared to other fisheries for this species in unprotected rivers. In periods of drought, the fish are vulnerable to excessive fishing pressure when running upstream. Opening the river in the reserve to fishing would increase pressure on the stocks. Rather than giving the fishermen short-term good catches now with the very real risk of further decline in future, the emphasis should be on long-term sustainable catches, even if this means not acceding to the fishermen's current wishes in their entirety. Any fishing allowed in the reserve would have to be linked to total cooperation in eliminating destructive methods downstream to allow more fish to reach the spawning grounds.

If fishing is allowed in the reserve, its impact on potential tourism developments would have to be taken into consideration. With good rains and good management of the fishery, the mpasa should recover. Angling of 1980s quality has potential for tourism if facilities in the reserve are improved. Ecotourism and local fishing in the same area are incompatible. The rules for angling in the reserve are that all fish must be returned alive and barbless hooks have to be used to minimise damage to the fish on capture. If fishing by villagers for food were allowed, it should only be in clearly demarcated areas away from the main tourism areas, which would be designated as a permit angling only area. The areas to be exploited by the villagers could be negotiated with the fishermen's representatives.

In conclusion, the management measures available are: (1) education on the importance of allowing enough fish to grow and to gain access to breeding grounds in the NWR; (2) improved enforcement, with the risk of unsustainable and inconsistent actions and lack of cooperation by the communities; (3) joint agreement between government and communities on rules and regulations to be introduced or retained, and cooperation between the parties on management measures to be instituted; (4) the introduction of permits to allow a limited number of villagers to fish in the reserve in exchange for full cooperation in banning damaging methods of fishing, such as poisoning and fish weirs.

A vital recommendation is that no further erosion of the reserve's boundaries into the catchment area of the Bua should ever be considered. The value of the mpasa fishery in Nkhotakota District is substantial, dependent entirely on the Bua River spawning grounds in the reserve. In addition, the reserve protects extensive areas of catchment for other streams which supply the downstream communities with water.

Mpasa is endemic to Lake Malawi. Its uniqueness, large size, and high economic value makes it a very high profile species in both fisheries management and in the conservation of biodiversity of Lake Malawi. Spawning populations in other rivers running into the lake are seriously threatened by man's activities. The protection of Nkhotakota Wildlife Reserve is vital to the fishes' long-term survival.

Acknowledgements

Anglers too numerous to mention individually assisted in the monitoring programme. Drs D.S.C. Lewis, A.G. Seymour and J.G.M. Wilson were major contributors. The Directors of National Parks and Wildlife (Mr M. Matemba) and Fisheries (Mr B.J. Mkoko) and their staff, both past and present, are thanked for their long-term backing of the mpasa research. The Dwangwa branch of the Wildlife Society of Malawi carried out the village questionnaire. The JICA study team, in particular Ms Y. Kato, is thanked for assistance in conducting the 1995 survey. Professor P.H. Skelton provided office and laboratory facilities at the J.L.B. Smith Institute of Ichthyology.

References

Msiska O.V. (1990) Reproductive strategies of two cyprinid fishes in Lake Malawi and their relevance for aquaculture development. *Aquaculture and Fisheries Management* **21**, 67–75.
Tweddle D. (1983) Breeding behaviour of the mpasa, *Opsaridium microlepis* (Gunther)(Pisces: Cyprinidae), in Lake Malawi. *Journal of the Limnological Society of Southern Africa* **9**, 23–28.
Tweddle D. (1985) The importance of the national parks, game reserves and forest reserves of Malawi to fish conservation and fisheries management. *Nyala* **11**, 5–11.

Tweddle D. (1987) An assessment of the growth rate of mpasa, *Opsaridium microlepis* (Günther, 1864) (Pisces: Cyprinidae) by length frequency analysis. *Journal of the Limnological Society of Southern Africa* **13**, 52–57.

Tweddle D. (1995) Observations on territorial behaviour and spawning of mpasa in the middle of the day. *Nyala* **18**, 44–45.

Tweddle D. (1996) Fish survey of Nkhotakota Wildlife Reserve, *Investigational Report No. 53.* J.L.B. Smith Institute of Ichthyology 79 pp.

Tweddle D. & Lewis D.S.C. (1983) Convergent evolution between the Lake Malawi mpasa (Cyprinidae) and the Atlantic salmon (Salmonidae). *Luso: Journal of Science and Technology, Malawi* **4**, 11–20.

Tweddle D., Alimoso S.B. & Sodzapanja G. (1994a) Analysis of catch and effort data for the fisheries of the South East Arm of Lake Malawi, 1976–1989, with a discussion of earlier data and the inter-relationships with the commercial fisheries. Traditional Fisheries Assessment Project Working Paper, TFAP/2(1991). *Malawi Fisheries Bulletin* **13**, 34 pp.

Tweddle D., Alimoso S.B. & Sodzapanja G. (1994b) Analysis of catch and effort data for the fisheries of Nkhotakota area: Lake Malawi, 1977–1989. Traditional Fisheries Assessment Project Working Paper, TFAP/8(1991). *Malawi Fisheries Bulletin* **16**, 25 pp.

Chapter 29
Conservation status, threats and future prospects for the survival of freshwater fishes of the Western Cape Province, South Africa

N.D. IMPSON and K.C.D. HAMMAN

Cape Nature Conservation, Private Bag X5014, Stellenbosch, 7599, South Africa
(e-mail: impsond@cncjnk.wcape.gov.za)

Abstract

The freshwater fishes of the Western Cape Province of South Africa are highly threatened, with nine of the 13 species endemic to river systems of the province now endangered or critically endangered. The major threats are the impact of predatory exotic fishes, especially smallmouth bass, *Micropterus dolomieu*, (Lacépède) and habitat degradation and destruction, principally as a result of unsustainable agricultural development. Urgent intervention by the state, landowners, research organisations, the private sector and the public is needed to halt their slide towards extinction.

Keywords: Threatened fishes, South Africa, alien fishes, habitat degradation, conservation.

29.1 Introduction

The conservation status of Western Cape freshwater fishes was revised in 1996 using the latest IUCN Red List categories (Mace & Lande 1994) as part of a national evaluation. The Western Cape Province of South Africa (Fig. 29.1) is recognised as a centre for a distinct Cape component of southern Africa's freshwater ichthyofauna, as its old mountain chains have contributed to the evolution of fish assemblages with high levels of endemicity, relatively inflexible life-history strategies and a low resilience to disturbance (Skelton 1987).

Anecdotal information (Harrison 1952a–c) and more detailed studies (Van Rensburg 1966; Gaigher *et al.* 1980) indicated that most Western Cape freshwater fish species were abundant and more widespread until introduced species such as carp, *Cyprinus carpio* L., bluegill sunfish, *Lepomis macrochirus* (Rafinesque), *Micropterus dolomieu* (Lacépède) and rainbow trout, *Oncorhynchus mykiss* (Walbaum), began to dominate the local ichthyofauna. The freshwater biota of southern Africa has been heavily impacted by introductions of aquatic organisms (Bruton & van As 1986; De Moor & Bruton 1996). At least 93 introduced aquatic species are now established in this region, of which 44 are exotic species and 49 are translocated indigenous species (Bruton & Merron 1985). Hey (1995) explained why various fish species were imported, cultured and stocked in the Western Cape. A further major

Figure 29.1 Map of South Africa showing the major river systems of the Western Cape Province. The insert shows the location of South Africa on the African continent

impact has been the rapid expansion and intensification of agricultural production this century.

In response to these impacts, population numbers and distribution ranges of several species exhibited rapid declines. Increasing concern at national level resulted in the publication of two Red Data Books for South African fishes (Skelton 1977, 1987). These and subsequent evaluations (e.g. Skelton *et al.* 1995) revealed that the Western Cape had become a prominent hotspot for threatened endemic fish species in South Africa.

The aim of this chapter is to present the revised conservation status of Western Cape fishes and to compare this with their 1977 and 1987 status. This comparison shows that the region is now home to one of the highest concentrations, percentage-wise, of threatened endemic fishes worldwide, and hence should become the subject of international attention. This chapter also highlights threats to these fishes, identifies some research needs and looks at positive trends that could improve their changes of survival.

29.2   Methods

Cape Nature Conservation (CNC), the statutory provincial conservation agency for the Western Cape Province, has undertaken regular freshwater fish surveys since 1962. The Berg, Breede, Olifants and Gourits river systems (Fig. 29.1) have received the most attention as they contain the majority of the freshwater fish species found within the province. The aim of the surveys was to obtain updated records of distribution, population status and habitat information of fish species, which were then entered into a dedicated database.

For this study, updated distribution and population status records were compared to historical records to determine an appropriate conservation status for each species, based on the new IUCN Red List categories. The main criteria used for listing species in the different categories was reduction in population size, extent of occurrence, whether populations were fragmented, estimated total population size, and analysis of the probability of extinction in the wild over a given period. Species listings for 1977, 1987 and 1996 were then compared.

29.3   Results

The conservation status of Western Cape freshwater fishes in 1977, 1987 and 1996 is shown in Table 29.1. The results show three trends over this 20-year period. First, overall conservation status has rapidly deteriorated. For example, eight endemic species were listed in 1977 compared to 12 in 1987 and 13 in 1996. A further trend is that the species listed as endangered have grown from one in 1977 to three species in 1987 to nine species (endangered and critically endangered) currently.

Second, the conservation status of most species has deteriorated, dramatically in several cases. For example, the Clanwilliam sandfish, *Labeo seeberi* (Gilchrist & Thompson) was unlisted in 1977, listed as rare in 1987 and is now categorised as critically endangered. Its declining distribution range over this period is illustrated in Figure 29.2. Finally, some Western Cape populations of more widespread species (e.g. chubbyhead barb *Barbus anoplus* (Weber)) are regarded as highly threatened. These populations are isolated from other populations and may be genetically distinct, and, therefore, of high conservation value.

29.4   Discussion

29.4.1   *Principal threats*

The principal threats are the impact of introduced exotic fishes and habitat alteration, degradation and destruction, principally as a result of unsustainable agricultural development (Gaigher *et al.* 1980; Skelton 1987).

Table 29.1 Conservation status of Western Cape fishes in 1977 (Skelton 1977), 1987 (Skelton 1987) and 1996 (Baillie & Groombridge 1996). The conservation status of many species has rapidly deteriorated (*denotes endemic to Western Cape rivers)

Species	Distribution	1977 status	1987 status	Current status
*Austroglanis barnardi**	Olifants River catchment	Not described	Endangered	Critically endangered
*Austroglanis gilli**	Olifants River catchment	Rare	Rare	Vulnerable
*Barbus andrewi**	Berg and Breede catchments	Not listed	Vulnerable	Vulnerable (CNC regards the Berg population as critically endangered)
Barbus anoplus	Widespread in S. Africa (in Olifants & Gourits catchments of W. Cape)	Not listed	Not listed	Not listed (some W. Cape populations regarded as endangered by CNC)
*Barbus calidus**	Olifants River catchment	Rare	Rare	Endangered
*Barbus capensis**	Olifants River catchment	Rare	Rare	Vulnerable
*Barbus erubescens**	Olifants River catchment	Vulnerable	Vulnerable	Critically endangered
*Barbus serra**	Olifants River catchment	Not listed	Vulnerable	Endangered
*Galaxias zebratus**	Widespread in W. Cape	Not listed	Not listed	Lower risk (nt)
*Labeo seeberi**	Olifants River catchment	Not listed	Rare	Critically endangered
Labeo umbratus	Widespread in S. Africa (in Gourits catchment of W. Cape)	Not listed	Not listed	Not listed (Gourits population regarded as threatened by CNC)
Pseudobarbus afer	Coastal rivers of W. Cape and E. Cape	Not listed	Not listed	Lower risk (nt)
Pseudobarbus asper	Gamtoos and Gourits river catchments	Not listed	Not listed	Vulnerable
*Pseudobarbus burchelli**	Breede and adjacent river catchments	Rare	Rare	Endangered
*Pseudobarbus burgi**	Berg River catchment	Rare	Endangered	Critically endangered
*Pseudobarbus phlegeton**	Olifants River catchment	Endangered	Endangered	Endangered
*Pseudobarbus tenuis**	Gourits and Keurbooms river catchments	Rare	Rare	Endangered
Sandelia capensis	Widespread in W. Cape	Not listed	Not listed	Not listed

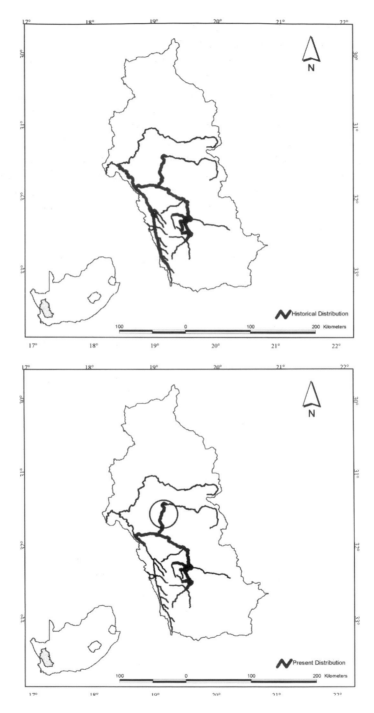

Figure 29.2 Historical (top, from Skelton (1987)) and present distribution (bottom) of the critically endangered cyprinid *Labeo seeberi*, endemic to the Olifants River system. The only large population remaining, the Oorlogskloof population (encircled) is situated within a nature reserve

Alien species such as *O. mykiss* and brown trout, *Salmo trutta* L., dominate the headwater and foothill zones of many Western Cape streams to the probable detriment of two redfin species (Berg River redfin, *Pseudobarbus burgi* (Boulenger), and Burchell's redfin, *Pseudobarbus burchelli* (Smith)). The transitional and lower zones of these catchments are dominated by other alien species such as *C. carpio*, *Lepomis macrochirus*, *Micropterus dolomieu* and *Micropterus salmoides* (Lacépède), together with translocated species such as sharptooth catfish, *Clarias gariepinus* (Burchell), Mozambique tilapia, *Oreochromis mossambicus* (Peters), and banded tilapia, *Tilapia sparrmani* (Smith). *M. dolomieu* is implicated as a cause for greatly reduced whitefish *Barbus andrewi* (Barnard), Clanwilliam yellowfish, *Barbus capensis* (Smith), and sawfin, *Barbus serra* (Peters), populations in mainstream areas (Harrison 1952a; Van Rensburg 1966; Scott 1982; I.G. Gaigher, unpublished data), as well as for the elimination of mainstream redfin and Cape kurper, *Sandelia capensis* (Cuvier), populations (Harrison 1952b, 1952c).

There are thus few zones in major Western Cape rivers free of the impact of introduced fishes. Perhaps the greatest impacts are experienced in the transitional and lower zones of rivers, where indigenous fish encounter cumulative impacts due to the presence of several predators (e.g. *C. gariepinus, L. macrochirus, M. dolomieu* and *M. salmoides*) and competitors for food and other resources (e.g. *C. carpio, O. mossambicus* and *T. sparrmani*). Other deleterious effects of introduced fishes on indigenous aquatic communities include introduction of parasites and diseases, habitat alterations, trophic alterations, hybridisation and water quality reductions (Bruton & Merron 1985). These associated impacts have, however, not been quantified. To date, no study on the impact of *M. dolomieu* has been done in South Africa. *O. mykiss* was implicated in the localised extinction of *P. burgi* in the Eerste River (Gaigher *et al.* 1980) and the syntestid damselfly *Ecclorotestes peringueyi* in certain Kwazulu/Natal streams (Samways 1994).

The impact of burgeoning, and to an extent uncontrolled, agricultural development is likely to have been severe on fish species inhabiting Western Cape rivers. The waters of this province are usually clear, acidic, cool, poorly mineralised and oligotrophic in character (King *et al.* 1979). In addition, they drain a winter rainfall region and, hence, often flood in winter and have low or no flow in summer (Allanson *et al.* 1990). These characteristics make them especially susceptible to impacts from irrigation-dependant agricultural development.

The impacts of modern agriculture in the Western Cape include the intensive use of fertilisers and pesticides, high levels of water abstraction often for offstream dams, the construction of instream dams without fishways, large-scale clearing of land for crops and orchards, and river bed modification by bulldozing to protect crops planted within floodline areas. For a review of threats to Cape and South African threatened fishes see Gaigher *et al.* (1980) and Skelton (1987) respectively.

Introduced species are likely to have been the main beneficiaries of modified flow regimes, instream barriers and reduced water quality in Western Cape rivers. The impact of introduced fishes compounds the effects of habitat degradation and pollution (Bruton 1995), forming a lethal combination that has driven mainstream species such as B. *andrewi*, B. *capensis*, B. *serra* and L. *seeberi* towards extinction.

Importantly, Baltz and Moyle (1993) found that indigenous fishes that inhabit relatively undisturbed streams in California are able to resist invasion by introduced fishes. In the Western Cape, the only habits where indigenous fish are still abundant are where there are no or few introduced fish species and where agricultural impact is low or absent. These habitats are primarily within mountain catchment areas in the transitional river zone.

29.4.2 *Research needs*

The freshwater fish species of the Western Cape are highly threatened and inhabit rivers that are accessible to research institutions, yet their biology and ecology and our understanding of impacting forces are still poorly understood. Their taxonomy and distribution ranges are known (Skelton 1993), but only the biology and ecology of two endemic species, the Twee River redfin, *Barbus erubescens* (Skelton), (Marriot 1997) and *P. burchelli* (Cambray & Stuart 1985) have been adequately addressed. Research is hence needed to determine priority requirements for their effective conservation (Skelton 1987). This research should address:

- biology and ecology of threatened endemics;
- effects of habitat degradation and fragmentation, especially from agricultural development;
- quantifying the impacts of introduced invasive fish species on indigenous fishes, amphibians and aquatic invertebrate communities;
- restoration ecology, especially to evaluate whether populations of introduced fishes can be permanently removed from otherwise pristine areas.

29.4.3 *Other issues of concern*

Cape Nature Conservation has seen its scientific and technical staff involved in scientific work on aquatic ecosystem conservation reduced from eight to two since 1991. Capacity has been reduced due to repeated budget reductions in real terms. Further consequences of budget shortfalls in the 1990s have been the closure of operations at the Department's two fish hatcheries, the Jonkershoek and Clanwilliam Yellowfish hatcheries. These developments have reduced the Department's ability to manage and rehabilitate freshwater and estaurine aquatic systems at a time when these systems and their biota are under increasing threat.

29.4.4 *Positive developments*

Several recent developments should improve management of habitats and indigenous freshwater fishes in the Western Cape, as well as enhance public awareness. These include the establishment of a river ecosystem display featuring indigenous Western Cape fishes at the Two Oceans Aquarium in Cape Town, and associated educational programmes. In addition, research projects have been recently completed or are being undertaken on *Austroglanis* spp. (biology and ecology), *B. andrewi* (genetics), *B. capensis* (genetics, flow releases and induced spawning), *B. erubescens* (biology and ecology) and *P. burgi* (genetics).

A national initiative has been the National Water Conservation Campaign of South Africa's Department of Water Affairs & Forestry which uses Reconstruction and Development Programme funds to clear invasive alien plants from catchment areas. Approximately 33 000 ha were cleared in 1995/1996 throughout South Africa (DWAF 1996). Benefits for rivers should be increased stream flow, less bank erosion and reduced shading by alien plants of normally open canopy streams.

The recently proclaimed Matjies River Nature Reserve in the Cederberg mountains includes a substantial component of the transitional zone of the Matjies River catchment, where three of the larger endemic fish species of the Olifants River System occur (*B. capensis, B. serra* and *L. seeberi*). The Western Cape, however, does not have a representative series of reserves for conserving the entire spectrum of river zones. Only headwater zones and, to a lesser extent, foothill zones are well conserved at present. Most are within state land in proclaimed Mountain Catchment Areas.

Legislation has recently been introduced and is being developed to promote sustainable utilisation of natural resources. Impact Assessment Regulations were promulgated in 1997 under Section 21 of South Africa's Environmental Conservation Act (Act No. 73 of 1989). Impact assessments are required for the construction or upgrading of dams or weirs that affect river flow and canals and channels, including diversions of water flow in a river bed and interbasin water transfers. Furthermore, South Africa's outdated Water Act of 1956 is being rewritten, with extensive input from the environmental lobby. A White Paper on a National Water Policy for South Africa was recently published (DWAF 1997), allocating the only right to water use to the aquatic ecosystem and basic human needs (identified as the 'Environmental Reserve'). This is a paradigm shift in legislative focus compared with the original legislation and should ultimately provide South Africa with advanced water resource legislation.

A nationwide survey for WWF-SA on 'Conservation priorities in Southern Africa' accorded 'Research on aquatic systems' as the highest priority rating under the category 'Research and Monitoring' (Macdonald *et al.* 1993).

29.5   Conclusion

The freshwater fishes of the Western Cape, South Africa, need urgent conservation attention as nine of the 13 species endemic to its river systems are now internationally listed as endangered or critically endangered (Baillie & Groombridge 1996). Most of these species are endemic to single catchments. The primary causes for their threatened status are thought to be the impact of predatory alien fish and habitat alteration, fragmentation and destruction, principally by agricultural development.

The future outlook for these fishes is both positive and negative. A dedicated conservation effort is required, embracing all major role players, to halt the slide of several species towards extinction.

Acknowledgements

The Acting Director of Cape Nature Conservation is thanked for permission to publish this chapter. Mr S.C. Thorne of CNC is thanked for field assistance and reviewed the conservation status of several species. The authors would also like to thank Professors M.N. Bruton and P.H. Skelton and Dr J.A. Cambray for comments on a draft of this chapter.

References

Allanson B.R., Hart R.C., O'Keefe J.H. & Robarts R.D. (1990) *Inland Waters of Southern Africa: An Ecological Perspective.* Delft: Kluwer, 458 pp.

Baillie J. & Groombridge B. (1996) 1996 *IUCN Red List of Threatened Animals.* Gland, Switzerland: IUCN, 368 pp.

Baltz D.M. & Moyle P.B. (1993) Invasion resistance to introduced species by a native assemblage of Californian stream fishes. *Ecological Applications* **3**, 246–255.

Bruton M.N. (1995) Have fishes had their chips? The dilemma of threatened fishes. *Environmental Biology of Fishes* **43**, 1–27.

Bruton M.N. & Merron S.V. (1985) Alien and translocated aquatic animals in southern Africa: a general introduction, checklist and bibliography. *South African National Scientific Programmes Report No. 113*, 71 pp.

Bruton M.N. & van As J.G. (1986) Faunal invasions of aquatic ecosystems in southern Africa, with suggestions for their management. In I.A.W. Macdonald, F.J. Kruger & A.A. Ferrar (eds) *The Ecology and Management of Biological Invasions in Southern Africa.* Cape Town, SA: Oxford University Press, pp. 47–61.

Cambray J.A. & Stuart C.T. (1985) Aspects of the biology of a rare redfin minnow, *Barbus burchelli* (Pisces: Cyprinidae), from South Africa. *South African Journal of Zoology* **20**, 155–165.

De Moor I.J. & Bruton M.N. (1996) Alien and translocated aquatic animals in southern Africa (excluding Zimbabwe and Mozambique) – revised checklist and analysis of distribution on a catchment basis. *Annual of Cape Provincial Museums (Natural History)* **19**, 1–344.

Department of Water Affairs and Forestry (DWAF) (1996) *The Working For Water Programme.* Cape Town, SA: Department of Water Affairs and Forestry, 12 pp.

DWAF (1997) *White Paper on a National Water Policy for South Africa.* Pretoria, SA: Department of Water Affairs and Forestry, 37 pp.

Gaigher I.G., Hamman K.C.D. & Thorne S.C. (1980) The distribution, conservation status, and factors affecting the survival of indigenous freshwater fishes in the Cape Province. *Koedoe* **23**, 57–88.

Harrison A.C.H. (1952a) The Cape Witvis (*Barbus andrewi*). *Piscator* **21**, 25.

Harrison A.C.H. (1952b) The Cape Kurpers. *Piscator* **23**, 82–90.

Harrison A.C.H. (1952c) Cape Minnows ('Rooivlerks' and 'Gillieminkies'). *Piscator* **24**, 117–128.

Hey D. (1995) *A Nature Conservationist Looks Back*. Stellenbosch, SA: Cape Nature Conservation, 280 pp.

King J.M., Day J.A. & Van Der Zel D.W. (1979) Hydrology and hydrobiology. In J.A. Day, W.R. Siegfried, G.N. Louw & M.L. Jarman (eds) Fynbos ecology: a preliminary synthesis. *South African National Scientific Programmes Report 40*, Pretoria, SA: CSIR, pp. 27–42.

Macdonald I.A.W., Van Wyk K. & Boyd L. (1993) *Conservation Priorities in Southern Africa*. Stellenbosch, SA: WWF-SA, 37 pp.

Mace G.M. & Lande, R. (1994) *IUCN Red List Categories*. Gland, Switzerland: The World Conservation Union, 21 pp.

Marriot M.S. (1997) Conservation biology and management of the Twee River redfin *Barbus erubescens* (Pisces: Cyprinidae). Unpublished MSc thesis, Rhodes University, Grahamstown, 104 pp.

Samways M.J. (1994) Damsels in distress. The threatened synlestid damselflies of southern Africa. *Africa Environmental Wildlife* **2**, 86–87.

Scott H.A. (1982) The Olifants River system – unique habitat for rare Cape fishes. *Cape Conservation Series 2*. Cape Town, SA: Cape Nature Conservation, 15 pp.

Skelton P.H. (1977) South African Red Data Book – Fishes. *South African National Scientific Programmes Report No. 14*. Pretoria, SA: CSIR, 39 pp.

Skelton P.H. (1987) South African Red Data Book – Fishes. *South African National Scientific Programmes Report No. 137*. Pretoria, SA: CSIR, 199 pp.

Skelton P.H. (1993) *A Complete Guide to the Freshwater Fishes of Southern Africa*. Halfway House, SA: Southern, 388 pp.

Skelton P.H., Cambray J.A., Lombard A. & Benn G.A. (1995) Patterns of distribution and conservation status of freshwater fishes in South Africa. *South African Journal of Zoology* **30**: 711–781.

Van Rensburg K.J. (1966) *Die Vis van die Olifantsrivier (Weskus) met Spesiale Verwysing na die Geelvis* (Barbus capensis) *en Saagvin* (Barbus serra). Research Report No. 10. Cape Town, SA: Cape Provincial Administration, Department of Nature Conservation.

Chapter 30
Conservation of endangered fish species in the face of water resource development schemes in the Guadiana River, Portugal: harmony of the incompatible

I.G. COWX

University of Hull, International Fisheries Institute, Hull HU6 7RX, UK

M.J. COLLARES-PEREIRA

Centro de Biologie Ambiental, Dept de Zoologica e Antropologia, Faculdade de Ciencias, Campo Grande C2, 1749-016 Lisboa, Portugal

Abstract

The fish fauna of the Guadiana River in southern Portugal is characterised by 31 species of fish, of which 23 are native and the remaining eight were introduced. However, 13 of the indigenous species are considered endangered or critically endangered. The main reasons for the decline in the population status of these species are continuous alteration and destruction of the habitat, mainly through water resource development schemes, but also pollution, sand and gravel extraction and dispersion of exotic species. Despite this bleak scenario there is increasing pressure to further exploit the water resources of the catchment for domestic supply, agriculture and recreation. This chapter assesses the status of the fish populations and proposes options for addressing conflict between conserving and enhancing fish biodiversity and exploitation of the water resources.

Keywords: Conservation, endangered species, *Anaecypris hispanica*, Portugal, fisheries management.

30.1   Introduction

Demand for water for domestic supply, agriculture and power production has resulted in large-scale impoundment of many rivers throughout the world (Petts 1984). The situation in semi-arid countries is made worse because of extremes in climate, especially during the prolonged dry summers, when flow in rivers is often inadequate to meet needs. Consequently, numerous reservoirs have been built or are planned, to overcome the ever-increasing demand.

One such river facing this scenario is the Guadiana River in the south-west Iberian Peninsula. This is one of the major rivers of the region (Fig. 30.1), and is already heavily impounded in Spain and on several large tributaries in Portugal. Several reservoirs are also under construction and numerous others are planned (Fig. 30.1).

Unfortunately, during planning of these reservoirs impact studies (e.g. DRENA/EGF 1986; SEIA 1995) on the aquatic fauna and flora, including the fisheries, are generally

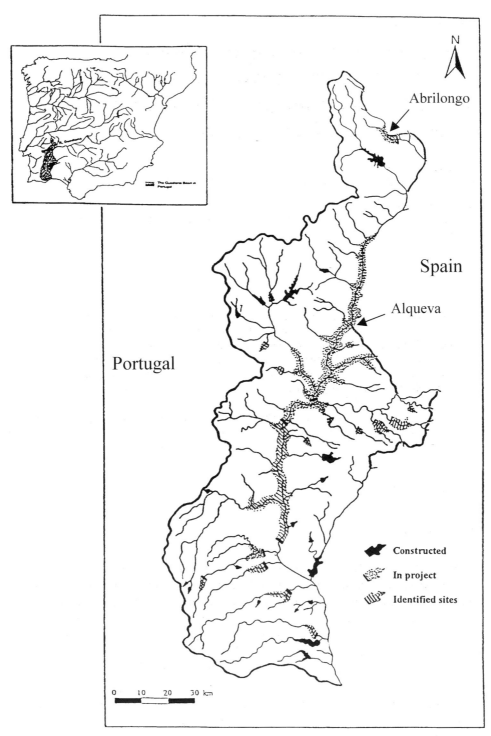

Figure 30.1 Map of the Guadiana catchment in Portugal showing location of resource development schemes

weak and involve no more than a superficial review of the basic biology, distribution and abundance. This is considered undesirable because the fish fauna has a high level of endemism and supports a number of endangered or critically endangered species (SNPRCN 1991; Baille & Groombridge 1996). This chapter examines the potential threats by water resource schemes to the fish and fisheries of the Guadiana River in Portugal, and suggests mechanisms for ameliorating or minimising possible impacts.

30.2   The Guadiana River

The Guadiana River basin is a southern Iberian catchment with a typical Mediterranean hydrological regime, i.e. very irregular with severe droughts and floods, and >80% of rainfall between October and March (mean annual rainfall in Portugal – 598 mm). The river flows over 810 km from its headwaters in Ruidera Lagoons (Spain) at an altitude of 1700 m (550 km in Spanish territory, 150 km in Portugal and 110 km of river forming the border between the two countries) to its mouth in the Atlantic Ocean in the Southern Portugal (Vila Real de St António) (Fig. 30.1). The middle and lower reaches of the Guadiana, located in Portugal, represent only around 17% of the overall catchment area (66 960 km^2). Its most important tributaries in Portugal are the rivers Ardila and Chança on the left bank, and the Caia, Degebe, Cobres, Vascão, Foupana and Odeleite on the right.

Human settlement in the Portuguese sector of the Guadiana is concentrated in the upper part of the catchment between the border near Elvas and the Moura/Beja region. There is a scattered pattern of urbanisation, with a total of around 230 000 inhabitants, ranging from 1000 to 15 000 per village. The catchment, especially in the central and lower zones, is considered poorly populated, having <20 inhabitants km^{-2} (INAG/COBA 1995). In the last 30 years the population has steadily declined, with migration of people out from the region away from traditional agricultural practices.

The main activities are concentrated in the northern region and include extensive agriculture, olive and tanning industries, cattle and some tourism. Water quality in most of the northern streams is lower than recommended (INAG 1996, 1997), especially in late spring and summer when the smaller tributaries start drying up, because of inadequate treatment of domestic and agricultural effluents. Intense eutrophication is also found in several reservoirs (Monte Novo, Caia, Lucefecit, Vigia and Beliche; SEIA 1995) and pools due to nutrient and pesticide run-off. This situation is more critical in the middle reaches of the Guadiana in Spain, due to the diffuse pollution (Doadrio *et al.* 1991), and regulation of flow in Spanish waters which has a severe negative impact on water quality downstream in Portugal.

30.3   Water resource development schemes in the Guadiana River

Demand for water resources in this region has increased dramatically in recent years, especially in late spring and summer, and this situation has been exacerbated due to a prolonged period of intense drought. The current estimates of water needs from the Guadiana in Portugal are 80 and 480 Mm^3 yr^{-1} for consumption and agriculture, respectively. It is predicted that these values will increase to 110 and 1570 Mm^3 respectively in the near future, i.e. a three-fold increase in demand (INAG/COBA 1995). This is a consequence of a strategy to increase the economy of the region, with particular focus on developing the tourism potential of the region and further south, and also changing the traditional agriculture practices.

This development is linked to distinct water resource schemes, including impoundment of the rivers and inter-basin transfers. In the Portuguese Guadiana, 13 dams have already been built (76% earth dams), of which 60% are for irrigation purposes. A further 25 dams and weirs have been proposed, of which two, Abrilongo and Alqueva, are under construction (Fig. 30.1). The latter will have a useable volume of 3150 Mm^3 over an area of 245 km^2 (215 km^2 in Portuguese territory). Its main objectives are to supply water to all the south of Portugal (including transfers into the Sado basin – 385–700 Mm^3 – and to the south – 315 Mm^3), hydropower generation, agriculture diversification, tourism development and the establishment of a 'business climate' to create 22 000 jobs.

In addition to this large-scale utilisation of the water resources, there are numerous small abstraction points throughout the river. These are mainly for agriculture and often exploit the water remaining in the permanent pools during the dry periods. To access this source of water the pools are often deepened and bankside vegetation is removed, leaving any fish which have sought refuge vulnerable to degraded habitat, elevated temperatures and in the worst case scenario, drying up.

The situation in Spain is also critical. In the Spanish Guadiana there are 38 dams plus 25 proposals for new constructions (INAG/COBA 1995), an existing transfer from the Segura basin to the upper Guadiana reaches, and a proposal to transfer water from the Chança River to the Huelva region. However, the demand for water in the near future is not as high as in Portugal (it will increase from 2370 to 2970 Mm^3 yr^{-1} for consumption and irrigation).

30.4   Status of the fish and fisheries

The Guadiana catchment is considered to have the most diverse fish fauna in Portugal. From the 31 species listed, five are diadromous fish with a Palearctic distribution (sea lamprey *Petromyzon marinus,* sturgeon *Acipenser sturio,* allis shad, *Alosa alosa,* thwaite shad, *A. fallax* and eel *Anguilla anguilla* (L.); seven belong to marine families (Atherinidae, Gasterosteidae, Mugilidae and Moronidae); and of the remaining, 19 are freshwater fish species.

The fish community in the middle and lower Guadiana River is dominated by two primary species: the *Leuciscus alburnoides* Steindachner complex and *Barbus steindachneri* Almaça. Nine other indigenous species: *Anaecypris hispanica, Chondrostoma willkommii* Steindachner, *Chondrostoma lemmingii* (Steindachner),

Leuciscus pyrenaicus Günther, *Barbus microcephalus* Almaça, *B. comisa* Steindachner, *B. sclateri* Günther, *Cobitis paludica* (De Buen) and *Salaria fluviatilis* (Asso),
and eight exotic fish species – pike, *Esox lucius* L., black bass, *Micropterus salmoides*
(Lacépède), *Lepomis gibbosus* (L.), *Gambusia holbrookii* Girard, *Cichlasoma facetum*
(Jenyns), *Fundulus heteroclitus* L., *Cyprinus carpio* L. and *Carassius auratus* (L.) are
also found, but their distribution and abundance varies considerably with
topographical and hydrological conditions (Collares-Pereira *et al.* 1997; Collares-
Pereira *et al.* 1998; Godinho & Ferreira 1998).

The high variation in river discharge levels resulting from the Mediterranean
hydrological conditions, causes significant seasonal shifts in fish community
structure (see Collares-Pereira *et al.* 1997, 1998; Godinho *et al.* 1997). Under the
severe drought conditions experienced during the summer the majority of the
tributaries are reduced to isolated pools or they dry up. These pools constitute
important refuges for fish populations until the beginning of the rainy season
(October or November). The dynamics of the fish assemblages also depend strongly
on recolonisation from the main river, illustrating the importance of maintaining
longitudinal connectivity (see Cowx & Welcomme 1998).

A major proportion of the Guadiana freshwater fish species (over 70% of native
non-migratory species, i.e. *Anaecypris hispanica*, *Chondrostoma willkommii*, *C.
lemmingii*, *Barbus steindachneri*, *B. microcephalus*, *B. comiza*, *B. sclateri* and *Salaria
fluviatilis*) are listed in the Portuguese Red Data Book (SNPRCN 1991) and other
international conventions, e.g. Habitats Directive (92/43/CEE). This is primarily due
to a marked decline of their distribution and abundance throughout Portugal (e.g.
Chondrostoma lemmingii, Barbus microcephalus) or within the Guadiana (e.g. *Salaria
fluviatilis, Anaecypris hispanica*). The latter species is endemic to the Guadiana basin
and is designated critically endangered (Collares-Pereira 1990; Collares-Pereira *et al.*
1999; B. Elvira, personal communication). Examples of the distribution of several of
the endangered species which were once common throughout the Portuguese waters
of the Guadiana are shown in Figure 30.2. The threatened status of these species,
and probably others, is considered to be the effect of habitat alteration and
destruction through flow regulation (especially by damming), pollution and
eutrophication, water and sand/silt extraction and dispersion of exotic fish species.
Overfishing is also responsible for the decline in the diadromous fish stocks and the
larger cyprinid taxa, and indirectly for those species exploited for live-bait (Collares-
Pereira *et al.*, 2000).

The commercial inland fisheries in the catchment concentrate on four diadromous
species which have high economic value, i.e. sea lamprey, allis shad, thwaite shad
and eel (mainly glass-eel), and the blackbass. There are also local markets for the
biggest cyprinid taxa (common carp, the endemic *Barbus* species and *Chondrostoma
willkommii*). Some exotic species, such as the pike and some small cyprinid species,
have a local value. Mullets are fished around Mértola (40 km from the mouth of
Guadiana River), where the tidal influence still persists. Fishing activity, which
frequently uses illegal gears and mesh sizes, mainly operates in the mainstem of
Guadiana, from the estuary to Pulo do Lobo (a natural barrier for upstream
migrations except under flood conditions) but also occurs in some of the bigger

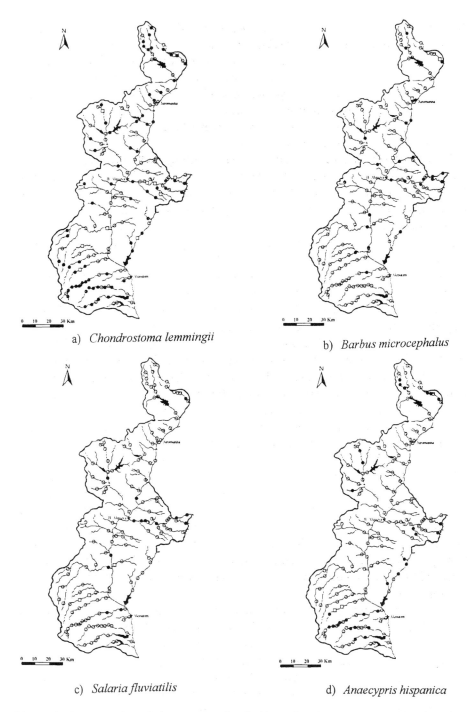

a) *Chondrostoma lemmingii*

b) *Barbus microcephalus*

c) *Salaria fluviatilis*

d) *Anaecypris hispanica*

Figure 30.2   Examples of the current distribution of endangered species in the Guadiana catchment in Portugal. (a) *Chondrostoma lemmingii*; (b) *Barbus microcephalus*; (c) *Salaria fluviatilis*; (d) *Anaecypris hispanica*. Closed circles show confirmed locations during 1997 surveys; open circles show sites where the species was absent during 1997 surveys

tributaries, e.g. the Ardila. Therefore, this activity seems to be more important in the lower reaches.

Recreational fishing is generally concentrated in lentic zones and in the big pools when the rivers start to dry up. The most important fish is the black bass, which has high commercial value in local restaurants (Godinho & Castro 1996), and, to a lesser extent, pike. Sport fishermen use live baits, which creates a high risk to the fish communities because they catch large numbers of stoneloach (*Cobitis paludica*) in small streams with small meshed nets during cyprinid recruitment periods.

30.5   Issues affecting the fish populations

The majority of the endemic species, although rheophilic, are well adapted to the severe conditions experienced in the intermittent rivers and streams of the Guadiana, especially flow reduction in dry periods (Pires *et al.* 1999). However, the distribution and abundance of many species are contracting rapidly, and are the cause of great concern (e.g. Collares-Pereira *et al.* 1999). There are a number of issues affecting the fish populations that must be addressed if the rich biodiversity of the catchment is to be conserved for future generations.

Most species appear to be highly vulnerable to alterations in the flow regime and water abstraction, especially in the refuge pools they occupy during the drought periods. Improper regulation of the rivers resulting in extremely low flows during drought periods and indiscriminate water abstraction from the permanent refuge pools is putting the fish communities under serious threat.

Existing and proposed impoundment of many tributaries and the main river has led or will lead to the loss of spawning and nursery grounds for most rheophilic species. These are being replaced by lacustrine species or those that are able to adapt to the changing environment, a situation that particularly favours the exotic species.

The populations have been fragmented by the numerous impoundments (dams and weirs) on almost all the tributaries. The implications of these barriers to free movement of fish are unknown, but they could result in reduction in the gene pools and instability in population structure (Carvalho & Cross 1998). This is particularly important in populations that are severely depleted through natural processes during drought periods.

The introduction of exotic species appears to be another contributor to the demise of many species (Collares-Pereira *et al.* 1998, 1999). The majority of exotics are piscivores, and the distribution of native fish in the Guadiana can be differentiated according to their presence or absence, as they are unable to tolerate their presence in large numbers (Collares-Pereira *et al.* 1998; Godinho & Ferreira 1998). Of particular relevance here is the widespread distribution of black bass, which is a voracious predator and is linked to the absence of certain species (e.g. *Anaecypris hispanica*) where it has colonised. The origin of the black bass appears to be through natural dispersal from reservoirs and other static water bodies where they have been stocked. The prospect of further impoundments throughout the catchment will probably exacerbate this problem unless concerted action is taken to minimise stocking of black bass.

Pollution from domestic, agricultural and industrial production sources has led to a deterioration in water quality in several rivers, e.g. the Degebe. This has increased the prospects of eutrophication, especially in the permanent pools during the drought periods, with inevitable adverse consequences for the fish communities occupying these pools.

30.6   Options for conserving endangered species

The current information indicates that there have been considerable changes in the distribution and abundance of many fish species in Portuguese waters of the Guadiana since the 1970s. Many species have declined in abundance and the fish populations have become fragmented over the past 30 years. Unfortunately, conservation of these endangered species in the Guadiana River is likely to be difficult since economic development and arresting of the emigration of people from the region are primary objectives of the government. Implementation of these objectives is intrinsically linked to increased availability of water resources to support agriculture and tourism in this semi-arid region. Consequently, there is likely to be conflict between social and economic development and conservation of the flora, fauna and landscape of the region. Furthermore, classical methods of conservation, e.g. setting up of nature reserves and specific legislation, do not appear to have worked in the past. This is partly due to the remoteness of the region making it difficult to police, but also because the fish and fisheries have been largely neglected in any development programmes, presumably because of the low perceived value of the resources. Thus, to conserve the fish and fisheries alone, a multi-faceted strategy, which will allow water resource development to take place in harmony with conservation of the natural environment, is required.

First there is a need to expand the nature reserves of the region to protect pristine habitat and regions where the species are relatively abundant. Rivers such as the Vascão and Ardila in the south and the Caia and Xévora in the north should be designated conservation areas and no further water resource development schemes should be allowed on these rivers. Which rivers are designated will have to be negotiated but environmental considerations must be heavily favoured in any deliberations. Appropriate legislation should also be put into place to prevent socio-political intervention in the areas chosen for conservation, i.e. political decisions to develop the region should not be allowed to override the conservation status of the designated area. Indeed, environmental legislation as a whole must be tightened up and a code of practice should be formulated to protect the fish and wildlife of the region.

Second, the deterioration of the fish stocks is such that active enhancement measures are required immediately. It is recommended that this focuses on rehabilitation of the rivers to remove the bottlenecks to recruitment and growth of the various fish populations where possible. Particular attention should focus on reconnecting fragmented populations by installation of fish passage facilities or other appropriate mechanisms. This approach is considered the most desirable because it implies long-term sustainability of the stocks (Cowx & Welcomme 1998) as opposed to short-term interventions, such as stocking and transfers of fish (Cowx 1994). Indeed transfer of fish from one river to another is not recommended in the Guadiana because stocks are too depleted in potential donor rivers and this has genetic implications (dilution of the gene pool of host populations; Carvalho & Cross 1998) which are not fully understood. It may be possible to establish artificial breeding programmes to restock the rivers, but the broodstock must be taken directly from the river to be stocked to prevent inbreeding and loss of genetic integrity typically found amongst farm-reared stocks (Cowx 1994). The broodstock should also be returned to the river after spawning to prevent any loss to the natural breeding population. Thus it is essential to adopt conservation strategies that aim to improve environmental conditions for all life stages of the threatened fish species, remove the bottlenecks to natural recruitment, survival and growth, and promote the stabilisation of the fish communities.

Unfortunately, these two approaches in themselves are unlikely to restrict politically and economically-driven water resource development schemes. The harmonisation of development implies the adoption of appropriate aquatic resource planning and management techniques, the establishment of a river consultation group which represents all users, and the development of strategies to resolve conflicts, based on the value of the fisheries and other aquatic resources (Cowx 1998). Consequently, there is a need to value the fisheries, both as recreational and commercial resources, in addition to the value of the endangered species in terms of their loss to society as a whole, both now and in the future. Some attempt to show the importance of the fishing to the region was attempted by Collares-Pereira *et al.* (2000), this work needs to be expanded to value the fish and fisheries *per se* using contingent valuation methods (Postle & Moore 1998) or similar economic tools. It is only when this information is available that the environmentalists can lobby for the preservation of the fish stocks and species diversity in the face of water resource development schemes.

One recent step forward in this direction is the formulation of the Guadiana River Basin Management Plan (GRBMP), which is soon to be implemented. This must guarantee an active public involvement and have the community recognition and support for the establishment of conservation measures. It is expected that the GRBMP will provide reference conditions against which change in the ecological status of these rich, although presently highly threatened, inland waters can be measured.

Essential to this approach is public involvement, which must be enhanced through awareness schemes and extension activities. Unless the local populace, particularly farmers, is aware of the problems they are creating, they cannot be held accountable for their activities, which have probably been ongoing for many decades.

Finally, correct fiscal measures, such as accurate and complete EIA studies, must be undertaken by independent organisations, before any major water resource scheme is promoted. In this process mitigation costs of any development activities must be imposed on the development agencies and the costs of rehabilitation should be internalise against end users (e.g. tourism).

In conclusion, aquatic resource planning should become an integral part of the conservation development process, and consultation with all user groups is essential to promote optimal, sustainable resource use, whilst affording protection to threatened fish species.

Acknowledgements

The authors thank the Empresa de Desenvolvimento e Infra-Estruturas de Alqueva (EDIA), the LIFE Nature Programme (contract B4–3200/97/280) and the Treaty of Windsor Joint Action B 63/97 for funding the research. Thanks are also due to J.A. Rodrigues, L. Rogado, L. Moreira da Costa, F. Ribeiro, P. Nunes, A.F. Filipe and T. Marques for their help in the above referred projects.

References

Baille, J. & Groombridge, B. (1996) *IUCN Red List of Threatened Animals.* Switzerland: IUCN, 257 pp + annexes.

Carvalho, G.R. & Cross, T.F. (1998) Enhancing fish production through introductions and stocking: genetic perspectives. In I.G. Cowx (ed.) *Stocking and Introduction of Fish.* Oxford: Fishing News Books, Blackwell Science, pp. 329–338.

Collares-Pereira, M.J. (1990) *Anaecypris hispanica* (Steindachner), a cyprinid fish in danger of extinction. *Journal of Fish Biology* 37 (Suppl. A), 227–229.

Collares-Pereira, M.J., Coelho, M.M., Rodrigues, J.A., Moreira da Costa, L., Rogado, L. & Cowx, I.G. (1997) *Anaecypris hispanica,* um endemismo piscicola em extinção. I. Caracterização da situação actual, *Report EDIA/ICN/FCUL, EDIA.* Lisbon, 83 pp + annexes.

Collares-Pereira, M.J., Pires, A.M., Coelho, M.M. & Cowx, I.G. (1998) Towards a conservation strategy for *Anaecypris hispanica,* the most endangered non-migratory fish in Portuguese streams. In I.G. Cowx (ed.) *Stocking and Introduction of Fish.* Oxford: Fishing News Books, Blackwell Science, pp. 437–449.

Collares-Pereira, M.J., Cowx, I.G., Rodrigues, J.A., Rogado, L. & Moreira da Costa, L. (1999) The status of *Anaecypris hispanica* in Portugal: problems of conserving a highly endangered Iberian fish. *Biological Conservation* 88, 207–212.

Collares-Pereira, M.J., Cowx, I.G., Riberio, F., Rodrigues, J.A. & Rogado, L. (2000) Threats imposed by water development schemes on the conservation of endangered fish species in the Guadiana River, Portugal. *Fisheries Management and Ecology* 6 (in press).

Cowx I.G (1994) Stocking strategies. *Fisheries Management and Ecology* 1, 15–31.

Cowx I.G. (1998) Aquatic resource management planning for resolution of fisheries management issues. In P. Hickley & H. Tompkins (eds) *Social, Economic and Management Aspects of Recreational Fisheries.* Oxford: Fishing News Books, Blackwell Science, pp. 97–105.

Cowx, I.G. & Welcomme, R.L. (eds) (1998) *Rehabilitation of Rivers for Fish.* Oxford: Fishing News Books, Blackwell Science, 260 pp.

Doadrio, I., Elvira, B. & Bernat, Y. (1991) *Peces Continentales Españoles. Inventario y Clasificacion de Zonas Fluviales.* Madrid: ICONA/CSIC, 221 pp.

DRENA/EGF (1986) *Estudos de Impacte Ambiental do Empreendimento de Alqueva.* Lisboa: Relatório Final, 252 pp. + annexes.

Godinho, F.N. & Castro, M.I.P (1996) Utilização piscícola de pequenas albufeiras do Sul. Estrutura da ictiocenose da Albufeira da Tapada Pequena (Mértola, Mina de São Domingos): diagnose e gestão potencial. *Silva Lusitana (no. especial)*, 93–115.

Godinho, F.N. & Ferreira, M.T. (1998) The relative influences of exotic species and environmental factors on an Iberian native fish community. *Environmental Biology of Fishes* **51**, 41–51.

Godinho, F.N., Ferreira, M.T. & Cortes, R.V. (1997) Composition and spatial organization of fish assemblages in the lower Guadiana basin, southern Iberia. *Ecology of Freshwater Fish* **6**, 134–143.

INAG (1996) *Programa de Despoluição da Bacia do Rio Guadiana (versão de trabalho), Vol. 1.* Lisboa, 107 pp + annexes.

INAG (1997) *Melhoria e Controle da Qualidade da Agua na Bacia do Rio Guadiana (Versão preliminar).* Lisboa: Direcção de Serviços de Recursos Hídricos, 89 pp.

INAG/COBA (1995). Recursos hídricos do rio Guadiana e sua utilização. Lisboa, 19 pp.

Martins, M.J., Collares-Pereira, M.J., Cowx, I.G. & Coelho, M.M. (1998) Diploids *v.* triploids of *Rutilus alburnoides*: spatial segregation and morphological differences. *Journal of Fish Biology* **52**, 817–828.

Petts, G.E. (1984) *Impounded Rivers.* Chichester: Wiley, 326 pp.

Pires, A., Cowx, I.G. & Coelho, M.M. (1999) Seasonal changes in fish community structure of intermittent streams in the middle reaches of the Guadiana basin, Portugal. *Journal of Fish Biology* **54**, 235–249.

Postle, M. & Moore, L. (1998) Economic valuation of recreational fisheries in the UK. In P. Hickley & H. Tompkins (eds) *Social, Economic and Management Aspects of Recreational Fisheries.* Oxford: Fishing News Books, Blackwell Science, pp. 184–199.

SEIA (1995) *Estudo Integrado de Impacte Ambiental de Empreendimento de Alqueva, Vol. V.* Qualidade da água, caudal ecológico, comunidades dulçaquicolas, bilharziose. Lisboa, 162 pp + annexes.

SNPRCN (ed.) (1991). *Livro Vermelho dos Vertebrados de Portugal, Vol. II – Peixes Dulciaquícolas e Migradores.* Lisboa: Secretaria de Estado do Ambiente, 55 pp.

Subject index

Abstraction, 177, 180, 183, 190, 192, 197, 218, 219, 221, 224, 225, 228, 229
Acidification, 268, 349, 386
Adaptive management, 279, 315
Afforestation, 268, 352
Agriculture, 252, 295, 334, 420, 423, 424, 426, 428
Amazon, 333, 388, 389, 390, 391, 393, 394, 396
Angling, 90, 386, 402
Aquatic resource planning, 436, 437, 350
Artisanal fishing, 232, 252, 338, 339
Aude River, 170–172
Augmentation, 177, 180, 183, 221

Bag limits, 339
Biodiversity, 201, 202, 216, 225, 262, 285, 326, 349, 350, 353, 357, 362, 402, 428, 434
Biological control, 331
Biomanipulation, 268, 355
Boulder clusters, 307
Bua River, 400, 401, 402, 403, 405, 407, 408, 411, 415, 416
Bye-law, 373, 374, 377, 384

Carrying capacity, 101, 126, 127, 130, 134, 137–140, 158, 159, 166, 169–171, 268, 306, 366
Cattle ranching, 392, 396, 397
Channelisation, 101, 144, 152, 269, 307
Climate, 292, 293, 303, 349
Closed seasons, 53, 343, 349, 378, 410, 413
Co-management, 331, 332, 394, 395, 396, 397
Commercial fisheries, 63, 98, 232, 233, 262, 264, 270, 274, 334, 337, 338, 390, 391, 413, 432
Community-based management, 215, 394, 414
Compensation flows, 219, 220, 227–229
Competition, 139, 226, 312
Conservation, 56, 58, 62, 67, 98, 152, 155, 180, 195, 228, 252, 253, 264, 274, 279, 286, 291, 292, 326, 334, 335, 336, 343, 346, 353, 355, 366, 388, 393, 394, 396, 400, 402, 410, 413, 415, 418, 419, 420, 424, 425, 426, 428, 435, 436, 437
Contingent valuation, 436

Cost–benefit analysis, 273
Counters, 196, 354, 365
Cover, 138, 146, 150

Deforestation, 269
Deschutes River, 115, 116, 118, 122
Disease, 224, 336
Dissimilarity index, 4–7, 10
Dissolved oxygen, 66, 102, 103, 106, 195, 224, 225
Draft net licences, 377
Dredging, 66, 144, 340
Drought, 218, 219, 220, 221, 224, 225, 226, 227, 228, 232, 240, 346, 409, 411, 413, 415, 430, 431, 432, 434

Education, 350, 358, 416
Egg deposition, 365, 368, 369, 370
El Niño, 220, 232, 237, 238, 240, 294, 303
Electric fishing, 5, 6, 14, 15, 18, 20, 21, 40–44, 50, 51, 57, 90, 118, 128, 129, 145, 146, 161, 235, 284, 300, 352, 354, 365, 402, 405, 407
Emigration, 18, 126, 127, 139
Endangered species, 291, 428, 430, 432, 435, 436
Endocrine disrupters, 67, 352
Enforcement, 414, 415, 416
Entrainment, 183, 196, 222, 225, 229
Environmental degradation, 337, 340
Erosion, 293, 295
Escapement, 26, 27, 37, 364–371, 400
Eutrophication, 267, 352, 430, 432, 435
Exotic fishes, 420, 428, 432, 434
Exploitation, 207, 215, 251, 368, 370, 373–386

Fecundity, 205, 209, 326, 362, 403, 404
Fish behaviour, 27, 88, 91, 140
Fish pass, 62, 67, 98, 197, 269, 273, 274, 436
Fish poison, 408, 409, 414
Fish rescues, 224, 228
Fishing
    accords, 391–396
    effort, 63, 201, 203, 207, 336, 370, 377, 382, 391, 408
    mortality, 139, 215, 270
Flood defence, 56, 66, 144, 346

Flood pulse, 72, 76, 84, 201, 215, 255, 332, 333, 389, 390
Floodplain fisheries, 202, 388
Flow regime, 81, 183, 227, 247, 251
Flow regulation, 3, 65, 71, 82, 83, 261, 318, 319, 386, 432
Fly River, 232, 233, 236, 240
Frame survey, 203
Fraser River, 32, 34, 36, 37, 265, 267, 273

Gear efficiency, 4, 10
Gillnets, 41, 73, 102, 235, 338, 390, 391, 408
Gravel extraction, 428
Groundwater, 56, 218, 284, 293, 295, 296, 298, 302, 389

Habitat
    classification, 352, 357
    degradation, 26, 133, 326, 352, 388, 397, 400, 401, 418, 420, 423, 424
    improvement, 307, 349
    preferences, 143, 150, 158
    restoration, 68, 170, 280, 306, 309, 352
Halflog covers, 307, 309, 312
Hoop nets, 6, 40, 41, 43, 50, 51, 243
Hydraulic engineering, 201, 339
Hydroacoustics, 14, 26, 38, 41, 55, 57, 61, 354
Hydrology, 166, 269, 279, 290, 292, 295
Hydropower generation, 431

IFIM, 127, 129, 159, 172
Immigration, 126, 131
Impingement, 183, 196, 222, 225
Impoundment, 55, 87, 261, 269, 352, 386, 434
Integrated catchment management, 274, 353, 357
Inter-basin transfers, 180, 431
Introductions, 232, 331, 336, 341, 346, 355, 423
Irrigation, 334, 339

Jamuna River, 207
Joe Farrell's Brook, 307, 308, 312

Kenai River, 30, 31, 35

Land drainage, 144, 352
Large woody debris, 42, 143–155, 242, 252, 280–286
Lateral connectivity, 271, 333, 340, 341
Legislation, 339, 342, 346, 393, 425, 435

Levées, 201, 202, 333, 339, 389, 391
Log structures, 302
Logbook, 382, 384, 386
Longitudinal connectivity, 271, 333, 340, 341, 432
Lower Shire River, 400
Luxapallila Creek, 3, 4

Mark–recapture, 41, 88, 205, 211, 373, 376
Mesh size limitations, 343
Micromesh seining, 58
Migration, 35, 37, 55, 63, 64, 76–98, 177, 183–185, 193–198, 202, 211, 212, 261, 265, 267, 271, 298, 333, 339, 349, 409
Minimum acceptable flows, 346
Mississippi River, 3, 4, 10, 40, 41, 281, 283
Mortality, 18, 126, 127, 130, 133, 139, 140, 166, 171, 209, 211, 224, 229, 270, 318, 362, 365
Murray-Darling River, 270

Natural mortality, 137, 138
Navigation, 42, 55, 56, 58, 66–68, 89, 269, 282, 334
Neste d'Oueil, 127, 137

Ok Tedi, 232–240
Overfishing, 26, 232, 261, 264, 265, 332, 335, 337, 343, 352, 400, 413, 414, 432

Padma River, 207
Parana River, 71, 72, 81
PHABSIM, 126, 159, 161, 162, 169, 170
Pollution, 3, 23, 24, 65, 82, 88, 102, 228, 261, 267, 340, 349, 352, 396, 423, 428, 430, 432, 435
Population dynamics, 126, 138, 139, 158, 171, 201
Potamodromous fish, 74, 76, 81, 83, 84
Power generation, 334
Predation, 101,102, 109, 139, 152, 224, 226, 261, 270, 291
Proportional stock density, 245–247, 253
Public health, 219, 221

Radio tracking, 87–91, 96, 185, 194, 197, 177, 377, 384
Recreational fisheries, 57, 58, 63, 87, 88, 98, 252, 264, 274, 334, 339, 343, 350, 434

Recruitment, 24, 62, 64, 126, 127, 153, 158, 159, 164, 166, 171, 172, 188, 194, 201–207, 214–216, 224, 225, 240, 242, 247, 252, 283, 284, 315, 318, 326, 335, 340, 349, 352, 361–365, 369, 370, 405, 435, 436

Redd density, 158–172

Restoration, 67, 279, 285, 286, 291, 296, 297, 298, 299, 302, 303, 307, 342

Rio de la Plata, 72, 76

Riparian vegetation, 242, 251, 302

River
  basin management, 67, 349
  Bush, 365, 367
  channelisation, 242
  continuum concept, 332
  Danube, 283, 331
  Dee, 318, 319, 325, 373, 374
  Elbe, 14–19
  engineering, 269, 352
  Exe, 177–185
  Garonne, 269
  Great Ouse, 283
  Guadiana, 428–436
  Haddeo, 183, 190
  Kennet, 62
  Meuse, 98
  Nidd, 87–93
  Ouse, 61, 87–93, 98
  regulation, *see* flow regulation
  restoration/rehabilitation, 144, 261, 264, 267–275, 279–286, 298, 312, 314, 315, 318, 327, 331, 340–342, 350
  Rhine, 264, 331
  Thames, 55, 56, 57, 59, 65, 67
  Usk, 373, 374
  Wensum, 143–153
  Wharfe, 219, 223
  Whitewater, 143–155
  Wye, 373, 374

Sand/silt extraction, 432

Scoopnets, 408

Sediment transport, 170, 285

Sedimentation, 183, 232, 240

Seine netting, 102, 184, 235, 299, 325, 338, 391, 408

Selectivity, 40, 41, 43, 44, 45, 47, 50, 51

Siuslaw
  River, 290, 294, 297, 302

watershed, 292, 296

Snorkel surveys, 118, 300

Spawning, 71, 158, 167, 352
  targets, 358–369

Species richness, 236

Stocking, 58, 270, 331, 335, 336, 340–346, 349, 357, 436

Stock-recruitment, 362, 367

Subsistence fisheries, 252, 262, 233, 338, 339

Substrate, 42, 150, 161, 162, 170, 183

Surplus yield models, 337, 364

Survival, 61, 126, 133, 211, 214–216, 239, 362, 365, 366, 419

Suspended sediments, 172, 233

Tagging, 72, 78, 90

Telemetry, 87, 96, 98

Threatened species, 291

Total allowable catch, 339

Tourism, 415, 431, 435, 437

Trammel netting, 41, 377

Trawling, 41, 57

Turbidity, 195, 270

V-dams, 307, 309, 312

Water
  abstraction, 64, 423, 434
  depth, 57, 146, 161, 162, 190, 192
  quality, 15, 23, 56, 61–67, 116, 138, 143, 145, 158, 166, 171, 172, 180, 183, 192–195, 222, 224, 225, 228, 229, 230, 232, 233, 238, 267, 271, 298, 343, 346, 353, 413, 423, 430, 435
  resource development, 67, 428, 435, 436
  resource management, 177, 178
  temperature, 65, 71,74, 76, 81–83, 91, 97, 102–105, 109, 115–121, 133, 193, 224, 294, 296
  velocity, 23, 57, 90, 146, 150, 161, 162, 190, 192, 197

Watershed analysis, 291, 292

Weed-cutting, 144, 340

Western Cape Province, 418, 419, 420

Wimbleball Reservoir, 177

Yazoo River, 242, 243, 245, 248

Yield, 202, 216, 337, 339, 391

Yukon River, 30–37

Species index

Abramis brama, 20, 88, 184, 188, 268, 283, 319
Ailia coila, 212
Alburnus alburnus, 20, 90, 95
Alosa alosa, 59, 267, 269, 431
Anabas testudineus, 23, 209, 211, 214, 240
Anaecypris hispanica, 428, 431, 434
Anguilla anguilla, 55, 56, 62, 68, 147, 188, 196, 431
Astatotilapia calliptera, 404
Atlantic salmon, see *Salmo salar*
Austroglanis spp., 425

Barbatulus barbatula, 143–153, 188, 434
Barbus
  andrewi, 425
  arcislongae, 404
  atkinsoni, 404
  capensis, 423, 425
  cf. *lineomaculatus*, 404
  comiza, 432
  eurbescens, 424, 425
  johnstonii, 404
  microcephalus, 432
  sclateri, 432
  serra, 423, 425
  steindachneri, 431
  trimaculatus, 404
Bleak, see *Alburnus alburnus*
Blicca bjoerkna, 20, 58
Bream, see *Abramis brama*
Brook trout, see *Salvelinus fontinalis*
Brown trout, see *Salmo trutta*
Bullheads, see *Cottus gobio*

Carassius
  auratus, 432
  carassius, 58
Catla
  catla, 209, 214
  reba, 214
Chanda
  baculis, 212
  nama, 212
Channa
  marulius, 214
  striatus, 25, 209, 211

Chinook salmon, see *Oncorhynchus tshawytscha*
Chondrostoma
  lemmingii, 431, 432
  willkommii, 431
Chub, see *Leuciscus cephalus*
Chum salmon, see *Oncorhynchus keta*
Cichlasoma facetum, 432
Cirrhinus
  mrigala, 212
  reba, 212
Clarias
  batrachus, 240
  gariepinus, 404, 423
Clupisoma garua, 212
Cobitus paludica, 432
Coho salmon, see *Oncorhynchus kisutch*
Corica soborna, 212
Cottus gobio, 58, 143, 146, 147, 150, 152, 159, 188
Cyprinus carpio, 423, 270, 418, 432

Dace, see *Leuciscus leuciscus*

Eel, see *Anguilla anguilla*
Esox lucius, 88, 319, 432

Fundulus heteroclitus, 432

Gagata
  giuris, 203, 209, 211
  youssoufi, 212
Gambusia holbrookii, 432
Gasterosteus aculeatus, 143–150
Gobio gobio, 90, 95, 104, 103, 105, 107, 109, 159, 188
Grayling, see *Thymallus thymallus*
Gudgeon, see *Gobio gobio*
Gudusia chapra, 212

Hemigrammopetersius barnardi, 404
Heteropneustes fossilis, 212
Hilsa ilisha, 212

Ictalurus
  furcatus, 242, 244, 247, 253
  punctatus, 242, 244, 247, 252, 253
Ictiobus
  bubalus, 242, 245, 247, 252, 253
  cyprinellus, 242, 245, 247, 252, 253

Labeo
  bata, 212
  calbasu, 212
  cylindricus, 404
  mesops, 404
  rohita, 212, 214
  seeberi, 420, 423, 425
Lampetra planeri, 58, 143, 291
Lates calcarifer, 233, 238, 240
Lepomis
  gibbosus, 432
  macrochirus, 418, 423
Leporinus obtusidens, 73–78
Leuciscus
  cephalus, 20, 87–92, 96, 147, 284
  leuciscus, 87, 88, 90, 94, 95, 96, 97, 147, 184, 188, 301, 319, 325
  pyrenaicus, 432
Limnothrissa miodon, 346
Luciopimelodus pati, 73–76

Mastacembelus pancalus, 212
Micropterus
  dolomieu, 417, 418, 423
  salmoides, 423
Minnow, see *Phoxinus phoxinus*
Mystus
  bleekeri, 212
  vittatus, 212

Nemachilus botia, 212

Ompok pabda, 212
Oncorhynchus
  clarki, 290, 291
  gorbuscha, 30, 34, 36
  keta, 29, 30, 33, 34, 35, 36, 290, 291, 296
  kisutch, 170, 283, 290, 291, 296
  mykiss, 58, 115, 118, 122, 190, 270, 290, 291, 418, 423
  nerka, 30, 34, 36,265, 267, 268, 273
  tshawytscha, 30, 33, 35, 36, 290, 290
Opsaridium
  microcephalus, 400, 401, 404
  microlepis, 400–416
  tweddleorum, 400
  zambezensis, 400
Oreochromis
  mosambicus, 423
  shiranus, 404

Osmerus eperlanus, 58, 59, 63

Perca fluviatilis, 88, 319
Phoxinus phoxinus, 95, 143, 146, 147, 150–153, 188, 283
Prochilodus
  lineatus, 73, 75, 83
  reticulatus, 74–82
Pseudeutropius atherinoides, 212
Psuedobarbus
  burchelli, 423, 424
  burgi, 423–425
Pseudoplatystoma coruscans, 73, 74, 82
Pterodoras granulosus, 73, 74, 76
Puntius sophore, 205, 209, 211, 214
Pylodictis olivaris, 242, 244, 247, 252

Rainbow trout, see *Oncorhynchus mykiss*
Rhinomugil corsula, 212

Salaria fluviatilis, 432
Salminus maxillosus, 73–76, 81
Salmo
  salar, 55–62, 64, 141, 171, 180–197, 264, 267, 271, 291, 306, 309, 312, 315, 319, 336, 358, 361, 365, 367, 369, 373–377, 383, 400
  trutta, 88, 115, 118, 122, 126–129, 139, 143, 147, 158, 159, 167, 169–172, 180, 184, 188, 270, 370, 374, 423
Salmostoma phulo, 212
Salvelinus fontinalis, 306, 307, 309, 315
Sandelia capensis, 423
Scardinius erythrophthalmus, 58
Silonia silondia, 212
Smallmouth buffalo, see *Ictiobus bubalus*
Smelt, see *Osmerus eperlanus*
Sockeye salmon, see *Oncorhynchus nerka*
Stizostedion lucioperca, 58
Stolothrissa tanganicae, 346
Sturgeon, 264, 265, 291, 336, 431

Thymallus thymallus, 88, 184, 188
Tilapia sparrmani, 423
Tinca tinca, 88

Wallago attu, 205, 207, 209, 211, 212, 214

Xenentodon cancila, 212